Interferon

This innovative study charts the beginnings, history and fate of interferon – one of modern medicine's most famous and infamous drugs. Interferon is part of the medical profession's armoury against viral infection, cancer and MS. The story of its development and use is one of survival in the face of remarkable cycles of promise and disappointment as a miracle drug. By telling this story, Toine Pieters' book provides insight into the research, manufacture, and marketing of new bio-molecules that mark modern medical science.

Pieters' closely argued book adopts a multi-disciplinary approach in seeking to trace the extraordinary voyage of interferon. Through the lens of interferon's voyage, the book explores the interaction of the broad range of actors driving medical science:

- biological and clinical researchers
- the pharmaceutical industry
- high-powered government agencies
- doctors and patients
- the media.

The book demonstrates how research on interferon led to new clinical definitions of cancer and a new rationale for therapeutic use of the drug. 'Interferon' provides a marvellous insight into the development of one of the most controversial drugs of our time. It enhances our understanding of how medicine manufacture and marketing all played a part in pushing back the boundaries of research, from the post-penicillin era to the genetics revolution in medicine.

This study is of particular interest to undergraduates in the fields of History of Medicine, Pharmacology, Medical Genetics and History of Science.

Toine Pieters is Professor of the History of Pharmacy at Groningen University and Senior Lecturer in the History of Medicine at VU Amsterdam Medical Centre, The Netherlands.

Routledge Studies in the History of Science, Technology and Medicine
Edited by John Krige
Georgia Institute of Technology, Atlanta, USA

Routledge Studies in the History of Science, Technology and Medicine aims to stimulate research in the field, concentrating on the twentieth century. It seeks to contribute to our understanding of science, technology, and medicine as they are embedded in society, exploring the links between the subjects on the one hand and the cultural, economic, political, and institutional contexts of their genesis and development on the other. Within this framework, and while not favouring any particular methodological approach, the series welcomes studies which examine relations between science, technology, medicine and society in new ways, e.g. the social construction of technologies, large technical systems etc.

1 **Technological Change**
   Methods and themes in the history of technology
   Edited by Robert Fox

2 **Technology Transfer our of Germany after 1945**
   Edited by Matthia Judt and Burghard Ciesla

3 **Entomology, Ecology and Agriculture**
   The making of scientific careers in North America, 1885–1985
   Paolo Palladino

4 **The Historiography of Contemporary Science and Technology**
   Edited by Thomas Söderquist

5 **Science and Spectacle**
   The work of Jodrell bank in post-war British culture
   Jon Agar

6 **Molecularizing Biology and Medicine**
   New practices and alliances, 1910s–1970s
   Edited by Soraya de Chadarevian and Harmke Kamminga

7 **Cold War, Hot Science**
Applied research in Britain's defence laboratories 1945–1990
Edited by Robert Bud and Philip Gummett

8 **Planning Armageddon**
Britain, the United State and the command of Western Nuclear Forces 1945–1964
Stephen Twigge and Len Scott

9 **Cultures of Control**
Edited by Miriam R. Levin

10 **Science, Cold War and the American State**
Lloyd V. Berkner and the balance of professional ideals
Alan A. Needell

11 **Reconsidering Sputnik**
Forty years since the Soviet satellite
Edited by Toger D. Launius

12 **Crossing Boundaries, Building Bridges**
Comparing the history of women engineers, 1870s–1990s
Edited by Annie Canel, Ruth Oldenziel and Karin Zachmann

13 **Changing Images in Mathematics**
From the French revolution to the new millennium
Edited by Umberto Bottazzini and Amy Dahan Dalmedico

14 **Heredity and Infection**
The history of disease transmission
Edited by Jean-Paul Gaudillière and Llana Löwy

15 **The Analogue Alternative**
The electric analogue computer in Britain and the USA, 1930–1975
James S. Small

16 **Instruments, Travel and Science**
Itineraries of precision from the seventeenth to the twentieth century
Edited by Marie-Noëlle Bourguet, Christian Licoppe and H. Otto Sibum

17 **The Fight Against Cancer**
France, 1890–1940
Patrice Pinell

18 **Collaboration in the Pharmaceutical Industry**
   Changing relationships in Britain and France, 1935–1965
   Viviane Quirke

19 **Classical Genetic Research and its Legacy**
   The mapping cultures of twentieth-century genetics
   Edited by Hans-Jörg Rheinberger and Jean-Paul Gaudillière

20 **From Molecular Genetics to Genomics**
   The mapping cultures of twentieth-century genetics
   Edited by Jean-Paul Gaudillière and Hans-Jörg Rheinberger

21 **Interferon**
   The science and selling of a miracle drug
   Toine Pieters

22 **Measurement and Statistics in Science and Technology**
   1930 to present
   Benoît Godin

23 **The Historiography of Science, Technology and Medicine**
   Writing recent science
   Edited by Ron Doel and Thomas Söderqvist

24 **International Science Between the World Wars**
   The case of genetics
   Nikolai Krementsov

Also published by Routledge in hardback and paperback:

Science and Ideology
A comparative history
Mark Walker

Interferon
The science and selling of a miracle drug

Toine Pieters

Routledge
Taylor & Francis Group
LONDON AND NEW YORK

First published 2005
by Routledge
2 Park Square, Milton Park, Abingdon, Oxon OX14 4RN

Simultaneously published in the USA and Canada
by Routledge
270 Madison Ave, New York NY 10016

Routledge is an imprint of the Taylor & Francis Group

© 2005 Toine Pieters

Typeset in Bembo by BC Typesetting Ltd, Bristol
Printed and bound in Great Britain by
MPG Books Ltd, Bodmin, Cornwall

All rights reserved. No part of this book may be reprinted or reproduced or utilised in any form or by any electronic, mechanical, or other means, now known or hereafter invented, including photocopying and recording, or in any information storage or retrieval system, without permission in writing from the publishers.

British Library Cataloguing in Publication Data
A catalogue record for this book is available from the British Library

Library of Congress Cataloging in Publication Data
A catalog record for this book has been requested

ISBN 0–415–34246–5 (hbk)

To my sons, Joes and Kalle

Interferon★

Always just one demon in the attic
Always just one death in the village. And dogs
howl in that direction, while from the other way
the new born child comes, just one,
to fill the empty space in the big air.

Likewise, cells infected by a virus
send signals out, defenses
are mobilized, and no other virus
gets a chance to settle down and change the destiny. This phenomenon
is called interference . . .

Contents

List of figures and tables	xi
Preface	xiii
Acknowledgements	xv
Introduction	1
1 Interferon's birth	9

Starting a British/Swiss Collaboration in a London laboratory 12
Intimate interplay between thought and action in Room 215 18
Writing 'interferon' 26
Isaacs's interferon kitchen 29

2 Shaping a new field of research: investigating interferon(s) 33

Interferon going public 33
The antiviral penicillin 36
Misinterpreton 39
From the brink to revival: meeting the others 41
Debating interferon's mode of operation: the specificity effect 45
What's in a name? 50
The blossoming of a new subfield 53

3 Interferon on trial 58

A partnership originating in national interests 59
Establishing a collaborative programme for research on interferon 64
Preparations for an early trial in volunteers 68
Handling production and safety problems 70
The Lancet Report 76
Interferon losing its momentum 80
No 'magic bullet' 85

4 Managing differences   91

Portrait of a 'gift culture' 92
Creating a breeding-ground for standardization 94
An experiment in scientific communication 96
Safeguarding British interests 98
A Finnish detour 101
Wrestling standards 104
'Having to haul down one's colours' 106

5 About mice, malignancies and experimental therapies   110

Interfering with cancer 111
Hoping for patients to behave like laboratory mice 116
Mathilde Krim's interferon lobby 120
Staging a workshop on interferon in the treatment of cancer 124
Attempts to put interferon on the cancer map 128

6 Interferon, audiences and cancer   134

Capitalizing on a growing demand for unorthodox cancer remedies 134
Interferon, scientists and the media 140
Hailing a miracle drug 147
Gene dreams and the inflation of expectations 150
Coping with an imminent black market 154

7 Yet another twist: marketing interferon as a helpful neighbour   158

Beyond interferon 158
A drug looking for a disease 164
Naturalizing interferons as biological response modifiers and cytokines 168

8 Interferons in retrospect and prospect   176

Patients at work 176
Crafting interferon(s) 178
Seeking 'magic bullets' 181
'Doctoring' the media 183
Toward genetic medicine 186
Patients at risk? 189

Appendix   192
Notes   194
Bibliography   241
Index   256

Figures and tables

Figures

1	Front view of the NIMR	10
2	Jean Lindenmann (1957)	12
3	Alick Isaacs at work in his laboratory (1957)	14
4	Hen's egg incubator cabin (1950s)	15
5	Pieces of chicken embryo membrane in a Petri dish	19
6	Piece of chicken embryo membrane in test-tube	20
7	Roller-tube assembly (1950s)	20
8	Electron microscopist at work (1950s)	21
9	Electron-microscopic images of counts of influenza virus particles	22
10	Agglutination is characterized by the formation of a pattern on the bottom of the tube	23
11	Animal technician holding a laboratory rabbit in her arms at the experimental unit	31
12	Isaacs lecturing to the camera directly from his laboratory to explain his work on interferon to a mass audience in the 1958 BBC programme 'Eye on Research'	38
13	The first human trial of interferon was 'performed' in this Flash Gordon comic strip (dated early 1960)	40
14	The Common Cold Unit was housed in a former American army field hospital	46
15	Tyrrell's laboratory was located in one of the central barracks of the Common Cold Unit	47
16	Cover of the *Scientific American* May, 1961	54
17	Jan Vilcek, Jacqueline De Maeyer Guignard and Edward De Maeyer in the corridors of the Smolenice meeting	56
18	A round unvarnish'd tale	72
19	Dr R. F. Sellers supervises the production of monkey interferon at Wellcome Laboratories (1961)	73
20	The arm of a human volunteer before vaccination	75
21	Principle of the vaccination experiment with interferon	75
22	The arm of a human volunteer after vaccination	76

xii   *Figures and tables*

23 After having inhaled an interferon preparation the woman volunteer at the Common Cold Unit at Salisbury is infected with common cold virus   81
24 The production of leucocyte interferon in Helsinki (late 1960s/early 1970s)   104
25 The Expert Committee on Biological Standardization of WHO established 69/19 (MRC Research Standard B) as the International Reference Preparation of Interferon, Human Leukocyte   108
26 Gresser's conference slide with interferon treated and untreated female (Balb/c) mice inoculated with (2,000 RC19) tumour cells and sacrificed on the 19th day   115
27 Interferon drug development programme as part of the NIAID Antiviral Substances Program (October 1969)   122
28 Lowel Glasgow used this picture in 1970 to support his controversial claim that host resistance against viruses was multifactorial with multiple interrelationships of the interferon system and the immune system   127
29 A caricature of several interferon workers who all argue in favour of their own favourite hypothesis   128
30 The drawings in Figures 28 and 29 were based on this poem by John Godfrey Saxe   129
31 The large-scale production of interferon at the Finnish Red Cross Transfusion Service in the late 1970s   137
32 Mathilde Krim and co-worker in their laboratory at New York's Memorial Sloan Kettering Cancer Center   141
33 'The alert'   142
34 'All systems go: Interferon triggers a cellular anti-viral mechanism'   143
35 Charles Weissman delivers the news of the successful cloning of the human leucocyte interferon gene and its expression in biologically active form   151
36 *Sunday Mirror* front page on 1 June 1980   153
37 Photomicrograph of interferon $\alpha 2$ crystals   159
38 Record from clinical trial with Hoffmann-La Roche's recombinant alpha-A interferon, showing a patient's blood levels of interferon   165
39 Picture in the *Scientific American*, May 1994   171

Tables

1 The number of interferon-related papers published worldwide over a ten-year period (1957–1967)   55
2 Results of individual vaccinations in the vaccination trial of the Scientific Committee on Interferon   78/9

Preface

The history of interferon is a cascade of stories and anecdotes that time and again enchant laymen and professionals alike. I start with two exemplary personal experiences.

Shortly after a public presentation of my study on the history of interferon research in November 1999 I received a phone call from a Dutch farmer's wife. She had come across my name in a rather negative newspaper article on interferon. She was surprised to find that interferon was called 'the biggest flop in medical history'. As a supporter of orthomolecular medicine she firmly believed that interferon was in the same high potency category of natural substances as selenium. Her family and the flock of sheep had thrived on selenium for years and she was convinced that interferon would have similar beneficial health effects. She was not prepared to accept any shortsighted criticisms from the quarters of orthodox medicine on a substance that seemed to be a constituent part of the orthomolecular therapeutic universe. The issue raised baffled me. The only thing I was capable of was blaming the newspaper for selecting a misleading heading and telling her that, yes, I had been involved for years with interferon though not as part of regular medicine but as a medical historian. If it had not been for another incident that following day I would have put the matter to rest.

With the same almost 'religious' zeal as the proponent of alternative medical treatments had done before, the Dutch microbiologist with an impressive track record of 30 years of interferon research, Huub Schellekens, defended interferon in the Dutch newspaper '*Het Parool*' as an evidence-based asset to the list of essential drugs in modern medicine.[1] Most striking, however, was the editor's layout with centre-piece a photograph of two gloved hands holding out a nitrogen-frozen, smoky, crude interferon preparation. The image provoked an eureka experience. There was a magic quality to the picture that made me realize that, ever since Alick Isaacs and Jean Lindenmann publicly announced their discovery of a new virus interfering factor named 'interferon', both supporters and adversaries of orthodox medicine have attached to interferon 'talismanic' properties. Here was an answer to the intriguing question why interferon, despite its medical use being largely limited to specialized hospital care, enjoys a penicillin-like public name; in terms of its presence in almost every dictionary around the globe and its worldwide celebration as a milestone of twentieth century science and medicine. It was the encouragement I needed to produce a book that shows how interferon succeeded in being transformed from a nebulous problem at the laboratory bench into an imaginative biomedical entity that has entered and colonized the public consciousness.

Both events also made me acutely aware of what it meant to be dealing with history in the making. With most of the *dramatis personae* of the study, their allies and opponents included, still alive and in some cases professionally active, the historian may get caught up in on-going struggles over scientific and technical knowledge claims. In principle, all participants have something at stake in the historian's interpretation and presentation of the subject matter, but for some the stakes are higher than for others. However determined to keep a safe distance in writing about interferon in terms of retrospect and prospect I cannot avoid being contaminated by the 'battle scene'. It was my surprising appearance as a member of the core set of interferon researchers in the acknowledgment section of Stephen Hall's 1997 narrative *A commotion in the blood* that for the first time brought home the message to me that I had become part – albeit a tiny part – of the history I was writing about.[2]

Acknowledgements

The generosity and support of a collection of inspirational scholars, friends, researchers and archivists made this book project possible. I have to emphasize that the project would never have been undertaken without the initial generous support of the faculty of Arts & Culture at Maastricht University, so I would like to thank Wiebe Bijker first and foremost for enthusiastically backing up and restyling my crude research ideas. Wiebe provided me with invaluable training and criticism throughout the project. I am greatly indebted to the following people, not only for their high standards but also for their warm support: Huub Schellekens, for giving me an early introduction to the world of interferon research; Catherine Gardiner for transforming my approximate English into the standard idiom; Brian Burgoon for editing the extravagantly long and 'meandrous' early chapters far beyond the usual limits; Annemarie Mol and Rein Vos for reading and advising me on several drafts of the chapters; Eddy Houwaart for providing me with the opportunity to finish the book manuscript at the department of Metamedica (VU University medical center); the Pieter van Foreest Foundation for supporting my chair at Groningen University; and Lolkje de Jongh – van den Bergh for giving me a warm welcome at the research section of Social Pharmacy and Pharmaco-Epidemiology (Groningen University).

Special appreciation goes to a small circle of friends, helpers and supporters who bolstered my spirits (in times of low spirit), reacted to my ideas and pieces of writing (however wild and incomprehensible in nature) and encouraged me to carry on (whenever they heard some muttering about having a tough time in my 'isolating cell'): Roel Otten, Patricia Faasse, Harmke Kamminga, Jo Wachelder, Frans J. Meijman, Stephen Snelders and most precious of all my wife and dearest friend Karin Knippers.

I am also grateful to all those who aided my research and the preparation of this book. Staffs of archives and government offices in Britain (MRC and NIMR) and the United States (NIH, NIAID and FDA) provided friendly assistance with source materials. However, the most important input into the research came from scientists, doctors, drug company executives and policy makers involved in the development of interferon. The majority of whom were prepared to cooperate and share both their recollections and private correspondence far beyond my expectations. In particular I have to thank, Derek Burke, Kari Cantell, Norman Finter, Leon Gauci, Ion Gresser, Sussana Isaacs Elmhirst, Jean Lindenmann and David Tyrrell for their help. The full list of those interviewed can be found in Appendix A. Thanks also to the editors: John Krige, Joe Whiting and Tracy Morgan.

My long-lasting obsession with this biography of a wondrous family of biomolecules doubtless puzzled many friends and relatives. By asking repeatedly what it was about and why it mattered and to whom, they helped me more than might have been obvious. I doubt whether a manuscript would have emerged without their critical and sometimes sceptical questions and remarks. Special thanks goes to my mother and sisters.

Every attempt has been made to ensure permission to reproduce material within this book. The publishers would be interested to hear from any copyright holder who is not here acknowledged.

Introduction

This book is about the nine lives of interferon, one of the most newsworthy biomolecules that can be tapped from the medicine chest of substances within our own bodies. As 'godfather' of a family of therapeutic proteins interferon survived successive cycles of hope and disappointment and is currently considered a respected part of the doctor's 'bag'. In portraying the wondrous expeditions between bench, bedside and the public sphere I show how work on interferon fed into a new therapeutic rational as part of a new kind of molecular medicine: genetic medicine. As part of this new medical horizon we are faced with the challenge to correlate the evidence-based query language of 'can it work' in terms of laboratory outcomes and statistics to the everyday 'does it help' concerns of doctors and patients in clinical practice.

Interferon first entered into the public eye in the spring of 1958. After the British press reported the discovery of a substance which might one day play an important part in the fight against viruses, interferon became widely known as the potential new 'antiviral penicillin'. However, the British euphoria for interferon lapsed and by the late 1960s the hoped for 'magic bullet', that would do for viral disease what antibiotics had done for bacterial infections, was relegated to the category of obscure laboratory substances. In the late 1970s, interferon made a dazzling comeback. It was widely hailed as a 'wonder' medicine which promised to cure a universally feared disease, cancer. The kind of hyped atmosphere surrounding interferon is illustrated by the following words of a British reporter of the *The Weekly News* in December 1979: "At present it costs £720,000 for two teaspoonfuls. But to cancer patients, its value is beyond price. For this liquid, known medically as Interferon, may be a 'wonder' medicine".[1] As a result of the media frenzy interferon became a public football. Worldwide doctors and authorities struggled to cope with the public agony of expectations they could not meet.

In the early 1980s the euphoria surrounding interferon as a 'miracle' cure for cancer faded following reports about interferon's lack-lustre performance in large scale trials. While interferon rapidly lost its aura of fame and magic, the science writer Mike Edelhart in his bestseller *Interferon: The New Hope for Cancer* helped to give interferon's career yet another turn. In his view the parallels between the difficult early days of penicillin and those of interferon were remarkable: "they grow up slowly, surrounded by scorn from the short-sighted, skepticism from the scholarly, overinflation by the popular press, and panting expectation from potential patients".[2] According to Edelhart doubt and trouble are part and parcel of the twisting path to therapeutic prominence. "In their older years, today's young adults will almost certainly find interferon among the arsenal of common treatments their physician will have handy to help them." He added that in the same way as penicillin, interferon will lead the way for the creation of a whole new family of medicines. Reiterating the popular public comparison between interferon's capricious fortunes and the rocky beginnings of another wonder drug of its age, penicillin, lend additional credibility to future research on interferon and in retrospect helped to turn Edelhart's projection into a self-fulfilling prophecy.

In the 1990s interferon became 'the interferons', billion-dollar therapeutic biomolecules which are currently used, both alone and in combination with other agents, to treat a bizarre assortment of disorders and malignancies from chronic viral hepatitis, bladder cancer up to multiple sclerosis. And in 2004 yet another dreaded disease has been added to this list: Canadian, Dutch and German scientists say interferon could be the 'drug of choice' in the fight against the viral disease SARS (severe acute respiratory syndrome).[3]

Whereas interferon started as part of the conventional therapeutic framework, as an antiviral agent – with the theory of specific etiology and the notion of specific therapy as cornerstones – it has become a symbol of a different therapeutic paradigm; immunotherapy as part of a multimodality therapeutic framework. According to the American expert in biological therapy Jordan Gutterman, this has opened the door to a new way of thinking about cancer – as a complex and chronic molecular disease.[4]

The interferon story has all the ingredients that can be found in successful Hollywood movie scripts: the heroic struggle of individual scientists to get recognition for their work; the dramatic illness of Alick Isaacs, the co-discoverer of interferon, that brings him to an asylum and leads to an untimely death; the 'miraculous' effects of a seemingly forgotten biological compound; consecutive cycles of hope and disappointment and ultimately the commercial success of a drug 'looking for a disease'. In this portrait of a wondrous molecule, however, I am less interested in exploiting the dramatic qualities of the story (though without ignoring them) than in exploring how interferon played many different roles as an evolving family of biological agents.

In retracing the drug's history it is most common to view interferon's therapeutic success story as a result of the endurance and persistence of scientists, doctors and drug companies in tackling research and development problems. On this view, the problems posed by the promising but unruly interferon molecule are defined as essentially technical and economic in nature, and the history of policy making regarding interferon is portrayed as the history of how those problems were resolved. The ultimate success of interferon is presented as the self-evident evocation of modern research. This perspective can be found in Edelhart's book and more recently also in the retrospective witness account of Kary Cantell, the Finnish interferon researcher who was first in developing a production methodology for human interferon.[5]

By framing the question in terms of interferon's long obstacle-filled road from the laboratory to the pharmacy both Edelhart and Cantell were encouraged to seek out factors such as errors, mistaken beliefs, technical impediments and conflicts of interest that caused the delay in the molecule's development. The decisive factors in both interferon stories are 'science and its practitioners' assumed to be assigned to the difficult but heroic task of producing the 'greatest benefit to mankind'; making major contributions to the advancement of modern medicine. In addition, 'money and politics' are assumed to harm as well as help a just cause. In particular Cantell's saga can be cast as a marathon struggle to find the truth between bench and bedside.[6]

This book starts from the premise that broadening our understanding of interferon's multi-dimensional historical trajectory requires an account that goes beyond the scientific or medical career of a drug.[7] Therapeutic drugs show in their conception, making, marketing and their uses that they are far more than just medical commodities. They also reflect developments and transformations in the science and art of healing as a cultural process.[8] However scientific and modern in nature, medicine and its therapies are bound up with dynamic mixtures of opinions, practices and rituals that shape the gardening of our health care landscape. Produced by the state-of-the-art knowledge of various sciences and increasingly visible by the resources of big science and industry, and consumed in increasingly media-permeated contexts of use, medicines have become an important subject of public and private mediation and imagination. They have taken on a special symbolic significance as icons of the increasing healing powers of modern medicine.

Exemplary in this respect is one of the most potent cultural metaphors: the representation of highly effective drugs as 'magic bullets'.[9] The 'magic bullet' imagery, which can be traced back to Paul Ehrlich's turn of the twentieth century concept of so-called *Zauberkugeln* ('magic bullets') drawn to specific disease targets, suggests a medical horizon of controlling and vanquishing disease.[10] This kind of promising cultural conceptions of drug development and drug use has stimulated medical innovation and helped to enhance expectations of doctors and patients alike, as well as to propagate belief in medicine as an enterprise for public and private good. But popular representations of pharmaceuticals do not only reflect unqualified success stories of breakthroughs in scientific medicine. Pill making and pill taking is a continuous site of public contestation; from hailing miracle drugs and applauding medical innovation up to warning against disease mongers and the hidden dangers in your medicine cabinet.[11] These glaring contrasts are also illustrative of the way drug development and use are part of the political economy of health care, which is essentially cyclical in nature and subject to supply and demand interactions.[12]

Since the emergence of the pharmaceutical industry in the late nineteenth century to the appearance of most new therapeutic drugs a common cycle of events can be observed.[13] The medical market is made receptive by a growing number of favourable scientific reports, by discussions of the new drugs by influential members of the medical profession and by coverage in the media. A wave of public enthusiasm makes the drug fashionable. The growing use of the drug and high expectations of its therapeutic effects are usually followed by a downturn. In most cases this is a gradual process that involves the accumulation of less favourable experiences with and reports of the drug's effectiveness and adverse reactions in everyday practice. Moreover, over time newer and presumably better alternatives claim attention. Mostly the therapeutic life-cycle of a drug ends once the patent has expired and there is little or no business reason to keep it on the market owing to a steady erosion of its use. On average this amounts to a 20–25 year therapeutic life-cycle.[14] But there are exceptions to the rule. A number of drugs have come to an untimely end due to the reporting of unexpected and severe side-effects. Most notoriously in this respect is the global tragedy of thalidomide (Distaval(R), Softenon(R)) in the early 1960s.[15] On the other end of the scale there are drugs that reach an exceptional old age.[16] Aspirin(R) is a case in point and has survived more than a hundred years of rampant competition on the medical market.[17]

Occasionally, in the process of circulating between bench, bedside and the public sphere new therapeutic substances have aroused expectations which are amplified beyond measure, thereby yielding the label 'miracle drug'. Achieving this 'devine' quality requires a successful alliance between a molecule's scientific potential as a golden asset to the doctor's bag and its potential cultural qualities as a carrier of the message of promise and hope.

In concerning itself with interferon's career paths this book enhances understanding of the dynamics of hailing a modern miracle drug and more in general of the cyclical dynamics of drug development and use. Given the repetitive nature of the public boom-bust process in the case of interferon this study offers an opportunity to do a comparative analysis of successive cycles of promise and disappointment in the history of a particular modern drug. As such it sets the scene and provides a framework for the exciting future endeavour to compare twentieth century drug careers as indicators of the changing scientific, political and social economies of healing in modern times.

I shall follow interferon on its twisting roads through science, medicine, industry and the media; reconstructing the subtle webs of technologies, people and policies which have gone to make up the dynamic weave of the science and selling of a modern-day family of therapeutic drugs. Interferon's 'biography' is an exemplary historical journey through the changing worlds of developing, testing, marketing and using therapeutic drugs in the second half of the twentieth century. In cutting through these various drug-shaping worlds this book's narrative sketches a detailed picture of the changing alliances that built up around the interferons as part of a specific category of contemporary drugs that originate in the preparation and purification of biological material, so-called 'biologicals'. In closely following around the interferon researchers and their objects of study, between bench, bedside and the public sphere, this ethnographic approach enables me to account for the full array of interferon's multiple identities as a research object, medical tool and commodity but also as an icon of twentieth century miracle medicine.[18]

Granting 'life' to interferon as a dynamic protagonist is meant to underscore the deep sense in which medicines have invaded practical consciousness in modern medicine. In attaining ever more active roles as flying colours of progress in science and medicine therapeutic drugs like interferon help to shape healing practices and cultures, and to drive biomedical innovation. As historic markers the evolving menagerie of interferons serve to enhance our understanding about the blending into one another of science, medicine, industry and the media; from the post-penicillin era to the genetics revolution in medicine.

This study of a present-day remedy in the making explains the various ways in which interferon's scientific, medical and cultural profile is established as part of a dynamic health care landscape. The book makes manifest the dynamics of energizing, halting or transforming work on interferon as part of a process of 'molecularizing' medicine; opposing as well as combining reductive and more complex molecular approaches. This also implies accounting for the possible role of interferon in redrawing the map of diagnostic and therapeutic regimes and dependencies in medicine – as a prototype of a specific category of modern-day drugs which originate in the engineering, preparation and purification of biomolecules, so-called 'recombinant protein drugs'. As advertised products of molecular biology the interferons fitted in well with the 'molecularization' of medicine and this helped to make them 'work' at the bedside.[19] I point to the solidarity implicit in the emphasis of both laboratory researchers and clinicians on the importance of linking through interferon bench and bedside as part of a new therapeutic modality: immunotherapy.

At the beginning of her 1984 book *The Interferon Crusade* the policy analyst Sandra Panem quotes a science administrator saying that "interferon is like the Lindbergh flight to Paris. It demonstrates a principle". Given the book's last phrase about the 'interferon soldiers' who have 'regrouped' to 'lead the lymphokine crusade' Panem must have been rather confident about interferon's chances of opening up yet other horizons for research and development as a robust and at the same time elastic principle. But how did interferon take on new meaning and agency as the showpiece of this new crusade? And to what extent do Panem's final words relate to her seminal observation that several stories are interwoven in the 'interferon crusade'?

I argue that addressing these kind of questions requires mapping the complex and dynamic relations between all those involved in shaping and reshaping scientific, medical and public features, meanings and images of interferon. Worldwide scientists, doctors, entrepreneurs, politicians, journalists and patients have been in the process of producing, evaluating or disseminating new interventions and knowledge associated with interferon. Insofar as those involved have attributed new features, meanings and identities to interferon it is as a consequence of the mangle of practice and culture.

Following the sociologist of science Andrew Pickering, I shall use the word 'practice' in a generic sense, around which all that follows is organized in science and medicine. In this sense it refers to the realm of realtime efforts of scientists and doctors to engineer workable and stable configurations of theories, phenomena, instruments and therapies through a dialectical process of "resistance and accommodation".[20] The term 'culture' in turn refers to a zone for discourse in which productive relations are established between the social world of researchers, physicians, patients, policy-makers and journalists and the material world of phenomena, instruments, tumours, laboratory animals and therapeutic drugs.[21] The mangle of practice and culture refers to the dynamics of communicating vessels that involves flows of information, ideas, procedures or interventions in words and images by means of speech, writing, print, radio, television and most recently by the Internet. As a result medical features, understandings and images may emerge that become part of the public sphere.

In forming an active part of the mangle of medical practice and culture interferon has contributed to the production of new differences and similarities that affect the ways we give meaning to and handle disease, illness and health.[22] Sorting out, however, how, through complex lines of interaction both interferon and its 'environments' have been modified was far from easy. The moment I started following the things called interferon 'I lost them'; interferon's scientific, medical and cultural identities had changed and they would continue to change during my research journey.

The interferon story starts with unravelling the process, whereby interferon as a current matter-of-fact was shaped in one of the laboratories at the National Institute of Medical Research (NIMR) in London between 1956 and 1957. On the basis of the original laboratory notebooks I follow the investigative pathways through which the British virologist Alick Isaacs and Jean Lindenmann, a visiting research worker from the Institute of Microbiology of the University of Zurich, arrived at the claim of the discovery of a novel biological factor that was involved in the well-known inhibitory phenomenon called 'viral interference'. My historical reconstruction of the events at the laboratory bench reveals how the state-of-the-art research environment of the NIMR was strongly conducive to innovative collaboration between the two biomedical researchers, the 'expert' Isaacs and 'novice' Lindenmann.

Chapter 2 discusses the fierce fight involved in transforming interferon from an obscure biological factor at the margins of viral research into a laboratory phenomenon that legitimized the 'blossoming' of a new field of research. Despite the established reputation of both Isaacs and the NIMR, interferon initially received a cool reception within the international scientific community. Quite some coordinating and coaching on the part of Isaacs was required to transform interferon from a local into a trans-national issue as part of a new international line of investigation within the field of animal virology.

Subsequently, chapter 3 analyses how interferon on its trails through the ever more closely related worlds of science, medicine, industry and the public sphere 'took on' new meaning and agency; and in a metaphorical sense 'the word became flesh'. The main focus is on how the various parties tried to cooperate across the domains which those involved routinely demarcated as 'science', 'policy' and 'industry', in the early 1960s. I point out that penicillin served as a powerful cultural symbol that set the stage for subsequent collaborative efforts of the British Government and the drug industry to develop interferon as an antiviral drug in Britain.

Chapter 4 is concerned with how standardization played a part in mastering and establishing differences in interferon research. Apart from production and purification problems, interferon researchers had a difficult time comparing experimental results. As we follow the subsequent attempts to manage differences, we find that the researchers involved devised a myriad of accountability strategies to facilitate the circulation of research results and materials. I show how the intrusion of powerful actors, such as the pharmaceutical industry, into the relatively private domain of laboratory life was instrumental in breaking down informal practices of measurement and in creating a demand for formal forms of standardization.

Chapter 5 describes the emergence of interferon as an anti-cancer agent within the American context of the early 1970s. Focus will be on how in working between bench, bedside and the public sphere French and Scandinavian laboratory researchers and clinicians with the help of the self-appointed American lobbyist Mathilde Krim succeeded in reshaping interferon's profile; from a dime a dozen biological research object into a promising lead in the search for anti-cancer therapies. I show that in comparison with interferon's short-lived British career as the 'new anti-viral penicillin' the assessment and evaluation of the therapeutic potential of interferon as an anti-tumour agent in America in the 1970s was for the greater part a different story. The course of events reflected the preoccupations of a different society in a different epoch.

Chapter 6 discusses the role of the mass media in establishing and amplifying associations between interferon, cancer, penicillin, 'miracle' healing, genetic medicine and natural remedies. In this chapter the focus is on reconstructing and analysing the cultural process through which the status of wonder drug was achieved in the case of interferon. Special attention will be paid to the channels of communication and how in the process the breakthrough imagery intensified and weakened.

Chapter 7 examines the question of what ultimately constituted therapeutic success in the case of interferon. Despite having failed to live up to its public promise as a therapeutic breakthrough, interferon succeeded in finding a niche in clinical practice in the 1990s. The primary focus of this chapter will be the interaction between interferon and the existing biomedical culture – how this interaction both modifies the potential medical innovation and its environment. Special attention will be paid to the role clinical trials have played in naturalizing the interferons as part of medical practice.

Finally, chapter 8 is meant as an extended commentary on this 'portrait of a wondrous molecule'. It will analyse how through the mangle of practice and culture, interferon continued to change shape in its twisted existence as an evolving family of biological molecules. In an effort to capture the nuances of research between bench and bedside, I relate work on interferon to the aims, choices and interests of the investigators involved as well as to the constraints and resources within which they operated and on which they thrived.

Subsequently, I analyse how the laboratory scientists and clinicians involved tried to align their research activities with broader elements of society, such as scientific institutions, governments, the public, pharmaceutical companies and regulatory bodies. Special attention is paid to the cultural process through which the status of miracle drug was achieved. In addition, the therapeutic evaluation of interferon within the context of the molecularization of medicine is stressed.

I shall end the book by drawing out some 'real world' implications in a final section called 'patients at risk'. In combining scientific and medical, social and commercial aspects I draw attention to changing powers and dependencies in modern healing cultures as part of a new kind of medicine: genetic medicine.

1 Interferon's birth[1]

On an early Monday morning in June 1956, Jean Lindenmann arrived at the gates of the National Institute for Medical Research (NIMR) at Mill Hill, London, for the first time. The Institute's enormous main entrance and seven-storeyed building, which overlooked both London's northern suburbs and the adjacent countryside, made a profound impact on the young medical researcher from Switzerland (see Figure 1). On his way to the Division of Bacteriology and Virus Research on the second floor, Lindenmann became even more impressed when he glimpsed at the interior of some of the laboratories. All imaginable equipment seemed to be available. This was confirmed later on when Lindenmann was shown around the Institute by the head of the Division of Bacteriology and Virus Research, and Deputy Director of the NIMR, Christopher Andrewes.

Andrewes belonged to the group of 'big names' in biomedical research, which was accommodated at the NIMR, and formed part of a staff of approximately one hundred doctoral and post-doctoral workers, and an auxiliary workforce of about three hundred technicians, secretaries and others. The Institute ranked among the world's best centres for biomedical research. It provided a state-of-the-art research environment aimed at stimulating internationally competitive research projects at the frontier of fundamental biomedical sciences. As a result there was great interest in doing research work at the NIMR.

The regular influx of visiting scientists was officially encouraged by the Director of the Institute, Sir Charles Harington. Harington believed that the resulting exchange of ideas would avoid scientific inbreeding, maintain a steady influx of new methods and techniques and keep researchers flexible in respect to incorporating new projects into their division's research repertoire. Like most other Anglo-Saxon biomedical research centres, NIMR's research agenda was primarily driven by motivations of scientific significance and impact, thereby favouring fundamental rather than applied research. At the same time the research agenda was often justified towards funding agencies for its potential medical therapeutic spin-off.[2]

Lindenmann had never seen such a collection of sophisticated equipment like ultracentrifuges, freeze dryers, electric incubators and deep freezes, phase contrast microscopes and an electron microscope. The NIMR also had a special

10   *Interferon: The science and selling of a miracle drug*

Figure 1 Front view of the NIMR. Courtesy of NIMR.

wing with large animal breeding facilities. The laboratory animals division assured a standardized supply of healthy laboratory animals such as dogs, ferrets, rabbits, rats, mice and poultry. Furthermore there was a well organized kitchen at the heart of the Institute on the lower ground floor, which was primarily responsible for cleaning and sterilizing glassware and preparing various standard solutions and culture media for the various research departments. The kitchen also produced special batches of media and solutions on request. This gave researchers a certain freedom to tinker with media and fluids during their experimental work. The kitchen was thus as much a central organ of the NIMR as the library on the fourth floor. This central storage for scientific paperwork was the last stop on their tour through the Institute.

In comparison with the rest of the Institute, the Division of Bacteriology and Virus Research had relatively few highly sophisticated laboratory instruments and facilities at their immediate disposal. However, Andrewes – or 'Cha', as he was informally known in his department – was not at all bothered by this state of affairs. On the contrary, since his research group was mainly interested in the qualitative, phenomenological aspects of virus research, there simply was not that much need for high-tech equipment. With a regular supply of fertile eggs, laboratory animals, nutritive media and glassware, and facilities for cold and warm storage (hot and cold rooms, an electric deep freeze for biological specimens, incubators for eggs and tubes) and for sterile work (a hood for handling tissue cultures and viruses), the division was thought to provide the standard equipment and materials for work on animal viruses and bacteria. If on occasion one needed to do some work which required facilities and equipment for complicated procedures, such as ultracentrifuges, an electron microscope, or an arrangement for electrophoresis, one could always ask the divisions of Chemistry, and Biophysics and Optics, for help. Quite often researchers in Andrewes's division would joke about their servants over there, who were so skillful in carrying out the quantitative jobs. Andrewes made no secret about the fact that he did not think a training in science was of any help for members of his division, who were predominantly graduates in medicine or veterinary medicine.[3]

Lindenmann had been working under quite different conditions in Hermann Mooser's laboratory that was part of the small institute of microbiology of the University of Zürich. The 'Hygiene Institut', as it was called, lagged years behind its counterparts in England with its rather old-fashioned laboratories and kitchen and no access to electron microscopes, ultracentrifuges and the like. In a way this is remarkable, since both Swiss and English science had come through the war relatively unharmed – without being cut off from a regular supply of American journals and without a dramatic drop in research activities. According to Lindenmann, Medical Microbiology was one of the sciences which remained ossified at a pre-war level until the 1950s, mainly due to the intellectual isolation of the leading scientists involved.[4]

Unlike the Division of Bacteriology and Virus Research at the NIMR, the 'Hygiene Institut' did not have a virology section, as it was dominated by old-fashioned bacteriologists. They still regarded viruses as ultramicrobes: living, autonomous infectious entities, which, like bacteria, multiplied autonomously by a process of binary fission.[5] As such, viruses were thought to belong to the domain of bacteriology, thereby ignoring the international trend which recognized virology as an independent field of research. For the most part, the scientific staff at the 'Hygiene Institut' seemed to have missed the emerging consensus concerning the nature of viruses among American, British and French microbiologists. Viruses no longer were regarded as ultramicrobes but as infectious, potentially pathogenic, nucleic acid-containing entities of protein nature. Unlike bacteria, viruses were considered to be dependent upon the host cell for their reproduction.[6]

These were anything but favourable circumstances for a young biomedical researcher with a vivid interest in the relatively new and dynamic field of animal virology – the study of viruses that prey on animals and human beings. So, after some unsatisfactory virus experiments in Zürich, Lindenmann asked permission to pursue his virus studies somewhere abroad. The head of the laboratory, the pre-eminent bacteriologist Mooser, was well aware of the limited possibilities for advanced virus study in the 'Hygiene Institut' and agreed to contact some British virologists whom he had met at an international microbiology meeting. One of the pioneers in the field of animal virology, Christopher Andrewes, who had acquired a worldwide reputation for innovative studies of animal viruses, agreed to let Lindenmann join his research group as a visiting worker for a period of one year. Subsequently, with Andrewes's letter of intent and Mooser's references, he was able to obtain a fellowship from the Swiss Academy of Medical Sciences and could start with preparations for his passage to England.[7]

Starting a British/Swiss collaboration in a London laboratory

With the intent to learn as much about animal viruses as possible, Lindenmann started working in Andrewes's division at the NIMR (see Figure 2).[8] As a relative novice to the field of virology, he was assigned to a rather low-key research job which would give him the opportunity to learn a variety of viral techniques and at the same time make a contribution to one of the division's research projects. Andrewes wanted him to work in his laboratory and do a series of experiments on the growth of polio virus in cultures of mouse and rabbit tissue.[9]

In quite a number of laboratories around the world polio virus was produced on a daily basis in human and monkey tissue cultures for the large scale production of vaccines and for research purposes.[10] Since polio virus had been shown to be host-specific and not transmittable to mice or rabbits it had long been taken for granted that it would not grow in rabbit or mouse tissue culture either. However, with Donna Chaproniere, Andrewes had been able to demonstrate that myxomatosis virus did grow in guinea pig tissue cultures, in spite of the fact that the virus was known to be not transmittable to guinea pigs.[11] Andrewes had picked information up at a recent meeting of the Society for

Figure 2 Jean Lindenmann (1957).
Courtesy of NIMR.

Interferon's birth   13

General Microbiology on successful experiments concerning the growth of polio virus in cultures of rabbit tissue, despite the fact that rabbits were not prone to the disease. This suggested the possibility that the specificity of viruses for particular hosts might be lost in tissue culture. Andrewes became interested in whether polio could indeed be grown in animals not prone to the disease. As a novice scientist at the frontier of virus research, Lindenmann was asked to show that polio virus would multiply in cultures of rabbit and mouse tissue. Lindenmann's education as an M.D. and his training as a postgraduate student in diagnostic bacteriology was believed to provide him with enough resources and previous experience to tackle this new research problem.[12]

Lindenmann's first two months in Mill Hill were taken up by learning through hands-on apprenticeship the craft of working with viruses as it was being practised by workers in Andrewes's division. At the same time, he initiated efforts to grow polio virus in non-specific tissue cultures. Despite frequent changes of experimental conditions and procedures, Lindenmann was unable to find any evidence showing that polio virus multiplied in these tissues. In a first report to the Swiss Academy of Medical Sciences he stated rather optimistically: 'Negative results in those experiments don't mean a lot, as only small changes in the experimental conditions can make the difference between success and failure'.[13]

However, to the best of Lindenmann's memory the research project became increasingly frustrating. Despite numerous follow-up experiments, he still did not manage to produce any positive results: he could only show that polio virus did not multiply in his experimental arrangement. Andrewes strongly believed that Lindenmann's unsuccessful attempts to replicate the claimed results regarding polio virus losing its host specificity in tissue cultures were due to an as-yet undetected artefact or some uncontrolled aspect of Lindenmann's experimental set-up.[14] Andrewes therefore did not think much of the idea of abandoning the project in the face of the repeated negative results. As a visitor to the Institute and novice to the field of virology, Lindenmann found it difficult to oppose Andrewes and to ask him to end the project.

Sometime during this period he was introduced to Alick Isaacs, a neighbouring researcher who had just returned from holidays. Only slightly older than Lindenmann, Isaacs at the age of 35 was already a distinguished virologist and an expert in influenza viruses who, like Lindenmann, had started his research career in a bacteriology department after being trained as a physician. Isaacs was in charge of the World Influenza Centre in room 215 which consisted of a large laboratory workspace with a rather small office corner for the necessary paperwork (see Figure 3). The World Influenza Centre had been set up in 1947 by Andrewes in collaboration with the United Nations' World Health Organization (WHO) as the centre of a global 'early warning' network of collaborating laboratories to study and monitor outbreaks of influenza ('flu') epidemics throughout the world. The main task of the World Influenza Centre was to gather influenza virus specimens from all over the world and collect all sorts of data concerning the outbreak and spread of flu epidemics

Figure 3 Alick Isaacs at work in his laboratory (1957). Courtesy of Dr S. Isaacs-Elmhirst.

with the aim to study the nature and epidemiology of influenza. Ultimately, the research efforts of Isaacs and his collaborators were aimed at forecasting and controlling the outbreak of flu epidemics, thereby preventing a recurrence of the influenza pandemic of 1918 in which over twenty million people had died worldwide.[15]

During their first encounter, Isaacs asked Lindenmann about his research work in Zürich. Isaacs showed real interest the moment Lindenmann mentioned his yet unpublished study of the virus interference phenomenon.[16] This biological phenomenon had first been named by the British scientists Gerald Findlay and Frank MacCallum in 1937.[17] Findlay and MacCallum had observed that infection with Rift Valley fever virus protected monkeys from infection with the unrelated yellow fever virus. Their phenomenon was defined as the interference of one virus with the pathogenic action of another, hence the name 'interference phenomenon'.[18]

Viral interference had been studied most extensively in the developing chicken egg since the 1940s when the German war refugees Werner and Gertrude Henle, who worked at the Children's Hospital of Philadelphia, had shown that the interference phenomenon could be studied most conveniently in that test system. Laboratory animals often became cross-infected, carried undetected virus infections and were argued to have more complex immune reactions. Moreover, studying animal viruses in fertile hen's eggs – or what were referred to as 'embryonated eggs' or simply as 'eggs' – was more eco-

Figure 4 Hen's egg incubator cabin (1950s). Courtesy of NIMR.

nomical and allowed for more detailed study of viruses in their specific cellular environment. Christopher Andrewes called the fertile hen's egg 'the new experimental animal' which yielded by far the best dividends'.[19] By the term 'dividends' Andrewes referred to the successful exploitation of chick embryo techniques both in viral research and in the large-scale production of viral vaccines – in particular, with reference to work on influenza virus and the development of an influenza vaccine (see Figure 4).

Isaacs had published a series of papers on the subject of viral interference. So he was eager to hear more about Lindenmann's interference experiments. Isaacs probably then asked Lindenmann to join him for lunch. Since most researchers worked mainly within the confines of their laboratory and division, the lunch hall on the top floor was a prime marketplace for the exchange of information. Here researchers from all laboratories and research divisions mingled quite freely while communicating the latest shop talk and the Institute's news.[20]

Lindenmann told Isaacs that in May or June 1955 his boss in Zürich, Hermann Mooser, had seen a paper on the interference between strains of rickettsia, indicating that interference of one strain with the propagation or reproduction of the other was not brought about by competition for or blockade of a receptor.[21] Mooser thought it conceivable that the interference phenomenon between rickettsia might be rather similar to the interference between influenza viruses. However, to the best of his knowledge, the viral interference phenomenon was still explained in the literature as competition for or blockade of a cellular receptor. In the context of this view of the phenomenon of interference it seemed interesting to consider an interference experiment with influenza virus in fertile hens' eggs, which would put this 'receptor hypothesis' to a test. If Mooser had been more familiar with the viral interference literature, he would have known that by 1955 most scientists in the field had already abandoned the 'receptor hypothesis' in favour of the so-called 'key-enzyme hypothesis' – viral interference was claimed to be due to some kind of competition for or blockade of a key element in the viral multiplication process.

In discussing his idea with Lindenmann, Mooser had proposed the following experiments: starting with the standard procedure of growing influenza virus in embryonated eggs, Mooser and Lindenmann would first try to reproduce the interference experiment between heat-inactivated, the interfering virus, and live influenza virus, the challenging virus, as described in the materials and methods section of an article in the *Australian Journal of Experimental Biology* that Mooser had come across.[22] Mooser emphasized the fact that the chick embryo–influenza system – involving inactivated (non-infectious) and active (infectious) influenza virus particles in fertilized hen's eggs – was widely considered a standard laboratory tool for studying interference among animal viruses.[23] Once they had succeeded in inducing interference in their own chick embryo–influenza system, Mooser and Lindenmann would use heat-inactivated influenza virus stuck to red blood cells in order to see whether these virus-coated red blood cells would also be able to induce interference. If they were able to show that despite the physical impediment, virus-coated red blood cells inhibited the growth of live influenza virus, they could make a case for refuting the 'receptor hypothesis'.

Mooser left the work at the laboratory bench to Lindenmann, while he travelled to Africa for his annual holidays. Upon Mooser's return, Lindenmann had managed to do a number of experiments showing that heat-inactivated influenza virus interfered with the growth of live influenza virus in the egg, even when stuck to the surface of red blood cells. Lindenmann thought that he had done a fairly good job and that the results were ready for publication. When talking over the series of experiments with the rigorous and imperious Mooser, he gradually lost his confidence. Basically, Mooser told him that the experiments had too many flaws. The most important problem was that Lindenmann could not make sure that no virus became disentangled from the red blood cells during his experiments. If virus had become unstuck during the experiments it would imply that the interfering virus had not been fully hindered from interacting with the challenging virus and that they were most likely dealing with an artefact, in other words in that case the experiments had gone wrong. In principle it would be possible to do additional experiments but this would require state-of-the-art instruments which were not available in the Institute. Mooser therefore decided to abandon the project and not to publish their existing results.

Lindenmann must have presented a somewhat similar account of his work at the 'Hygiene Institut' to Isaacs, who according to Lindenmann, remarked that he knew the authors of that Australian article quite well. In fact, he had published it himself in collaboration with Margaret Edney during his two-year fellowship as a visiting researcher in MacFarlane Burnet's laboratory at the Walter and Eliza Hall Institute in Melbourne.[24] Isaacs had spent a great deal of his time over there extensively studying the interference phenomenon with heat-inactivated influenza virus as the interfering virus and live influenza virus as the challenging virus. In these studies, he and Edney had ascribed the interference phenomenon to the competition between the interfering and the challenging virus for some key constituent within the cell, while rejecting a possible competition for or blockade of cellular receptors. However, he had neither been able to produce convincing experimental data in support of this hypothesis in Burnet's laboratory, nor could he on his return to the World Influenza Centre at the NIMR.[25]

In response to Lindenmann's reconstruction of the interference experiments performed in Mooser's laboratory, Isaacs suggested two possible explanations for Lindenmann's observation that inactivated influenza virus attached to red blood cells was a good interfering agent. First, in accordance with Mooser's ideas, the virus might have become detached from the red blood cells during the experiment so that it could induce interference by entering the cell as a whole. The second and rather novel assumption was that the virus coat, the 'virus haemagglutinin', might have remained firmly attached to the red cells, while only the virus content, the 'virus nucleic acid', had entered the cell and induced interference.[26] Isaacs told Lindenmann that this idea was based on a rather well-known American study, which was presented at the prestigious Cold Spring Harbor Symposium, in 1952, the year before Francis Crick and James Watson published their famous *Nature* papers on the structure and function of DNA.[27] Upon conducting a study with radioactively labelled bacterial viruses the Americans Alfred Hershey and Martha Chase claimed that during bacterial virus infection only the virus nucleic acid entered the bacterial cell, while the protein coat, the 'virus envelope', remained at the cell surface. This study provided a most compelling argument that virus nucleic acid, which could be either DNA or RNA, directed viral reproduction within the host cell.[28]

Within the field of virology there was cautious consensus about bacterial virus infections as a useful and simple experimental model for studying viral reproduction as a possible key to the elucidation of other virus–host systems. By suggesting with his second explanation that animal virus infections proceeded in a similar way to bacterial virus infections – the virus unit attached to the cell and injected its genetic information into the cell, leaving an empty protein envelope attached to the cell wall – Isaacs carried the analogy between animal viruses and bacterial viruses quite a bit further. This idea was rather controversial and hotly debated within the field of virology.[29]

Lindenmann's and Mooser's experimental set-up somehow seemed to provide a starting point to test Isaacs's assumption that during interference with inactivated influenza virus only the virus nucleic acid entered the host cell, while the virus coat remained outside. Most likely, this played a major role in Isaacs's decision to ask Lindenmann whether he would like to collaborate in work on viral interference, in particular, to help him figure out whether this idea, what I will label the 'nucleic acid' hypothesis, made any sense. As his poliomyelitis work was as yet far from promising, Lindenmann was pleased to hear Isaacs's research proposal. After some more talking and with Andrewes's permission, they began working together at the beginning of September 1956.[30]

In the meantime, Lindenmann would continue with his poliomyelitis work, and Isaacs had to make sure that the World Influenza Centre was properly run and fulfilled its research obligations towards the WHO. In addition, Isaacs had to pay regular visits to the Division of Biophysics and Optics, where he collaborated with the electron microscopist, Robin Valentine, in studies on the structure of influenza viruses.[31]

Isaacs's decision to collaborate with Lindenmann did not seem to be spurred by interests of political nature. Collaboration with the latter would bring Isaacs neither greater credibility nor would he be eligible for resources that would otherwise be missed. Furthermore, compared with Isaacs, the established influenza virus expert, Lindenmann was a relative novice at the frontier of virus research with little expertise to offer. Apparently, recruiting Lindenmann to work as a 'co-labourer' toward a common purpose was motivated by a common research interest in the phenomenon of virus interference. Moreover, Lindenmann had worked with an experimental set-up that seemed to offer a lead towards testing a then hotly debated idea: namely that infection by influenza virus was initiated by injection of the viral nucleic acid into the cell, leaving an empty virus envelope attached to the cell wall.[32]

Intimate interplay between thought and action in Room 215

> Twenty years after Findlay and MacCallum had described virus interference its mechanism was still a mystery. How did infection of a cell by one virus prevent infection by a second? . . . In 1957 Jean Lindenmann and I, working at the National Institute for Medical Research in England, were investigating the action of heat-killed influenza virus when we found an unexpected handle on the problem. We found that a few hours after we had treated a cell culture with killed virus the cell-free culture medium had acquired a surprising property. When the medium was mixed with fresh cells, it made them resistant to virus infection. This resistance had all the earmarks of virus interference, since the fresh cells proved resistant not just to one virus but to many different viruses. We were soon able to isolate the active substance responsible for conferring resistance, and we named it interferon.[33]

Reading this 1961 portrait of what the author, Alick Isaacs referred to as the 'discovery' of interferon, brings up the following questions: How did they arrive at an 'unexpected handle on their problem', and in what way was this related to the 'isolation of the active substance' named 'interferon'?[34]

Isaacs's notebook has an entry for 4 September 1956 where he and Lindenmann are said to have started with an experiment to see whether virus-coated red blood cells can induce interference in the test-tube.[35] Technically, this first collaborative experiment differed substantially from Lindenmann's experiments at the 'Hygiene Institut'. During a meeting to discuss both planning of the course of investigation and design of the first experiment, Isaacs proposed a number of changes to Lindenmann's original set-up. Instead of doing the experiment *in vivo* Isaacs preferred to employ a relatively novel *in vitro* technique.

This technique required the following manipulations.[36] Fertile hens' eggs containing embryos 10–11 days old (but all of the same age for a particular experiment) were opened and the embryos tipped out and discarded. In addition the part of the chicken embryo membrane adhering to the egg shell was removed, cleaned ('washed') and cut up with scissors into six or seven pieces. These pieces were then pooled in a dish containing an aqueous solution and added, one at a time, to a test-tube filled with a special nutrient fluid (see Figures 5 and 6). This could be done very fast by a laboratory technician. Subsequently, the virus or the controls were injected into the test-tubes, whereupon the tubes were stored and kept in controlled conditions, at 37°C for 24 hours ('incubated' or 'incubation') in slowly rotating tube-containers,

Figure 5 Pieces of chicken embryo membrane in a Petri dish. Courtesy of Dr D. Tyrrell.

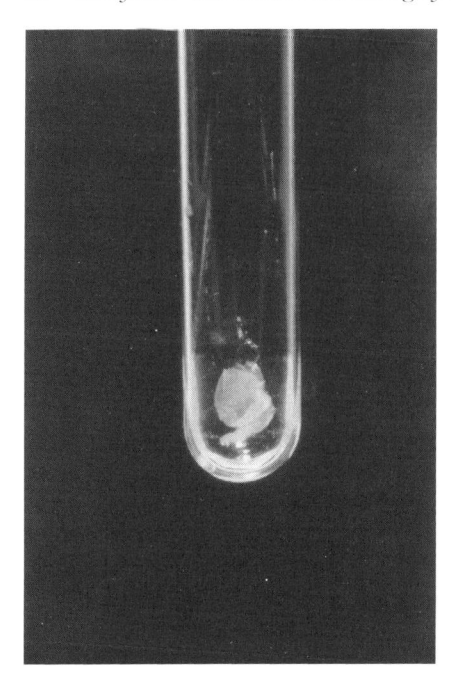

20   *Interferon: The science and selling of a miracle drug*

Figure 6 Piece of chicken embryo membrane in test-tube. Courtesy of Dr D. Tyrrell.

so-called 'roller drums'. Previous experimentation had shown that through rotation or rolling of the tubes ('roller-tube method') virus multiplication could be stimulated (see Figure 7). The next day the fluid could be taken out of the tube and the amount of virus quantified.

Isaacs had applied this experimental technique successfully in studies of virus interference, which he had done together with Forrest Fulton, a fellow

Figure 7 Roller-tube assembly (1950s). Courtesy of NIMR.

virologist working at the London School of Hygiene and Tropical Medicine.[37] According to Isaacs it had decisive advantages over the egg. The *in vitro* technique required roughly one-sixth of the usual amount of fertilized eggs, thus saving both money and space. In addition, experiments in eggs required far more skill and time than handling the pieces of membrane in the test tube. Further, he argued that this *in vitro* model could be more readily quantified and manipulated.[38]

A second change in the procedure of the 'original' experiment would be to stick the inactivated influenza virus to haemolysed red blood cells, so-called 'red cell ghosts', instead of normal blood cells.[39] By attaching influenza virus particles to red cell ghosts, Isaacs and a researcher from the Biophysics and Optics Group, Heather Donald, had managed some years before to visualize influenza virus through the use of the electron microscope (see Figure 8).[40] They had made a series of electron micrographs of influenza virus preparations (see Figure 9), which were said to show red cell ghosts with distinctive little round structures on them, representing virus particles. This interpretation is clearly an accomplishment of two experts familiar with interpreting electron microscopic images of virus samples. Without Donald's and Isaacs's guidance these pictures look rather more like exotic art photographs, showing intriguing patterns of lines and dots.[41] Isaacs thought that by sticking inactivated influenza virus to red cell ghosts it would also be possible to visualize influenza virus particles in his experiment with Lindenmann. If Isaacs's assumption that during interference only the virus nucleic acid entered the host cell while the virus coat remained firmly attached to the red cell ghosts was correct, then they

Figure 8 Electron microscopist at work (1950s). Courtesy of NIMR.

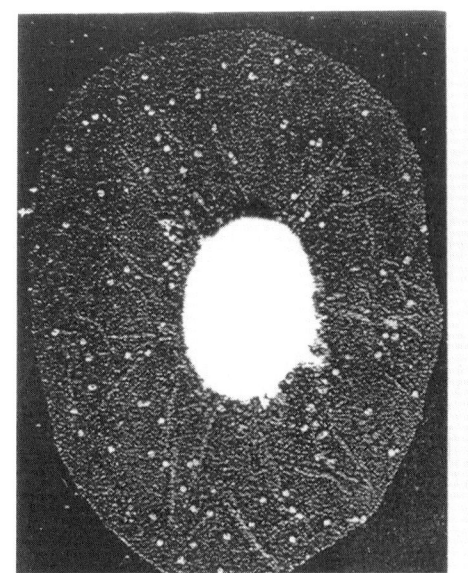

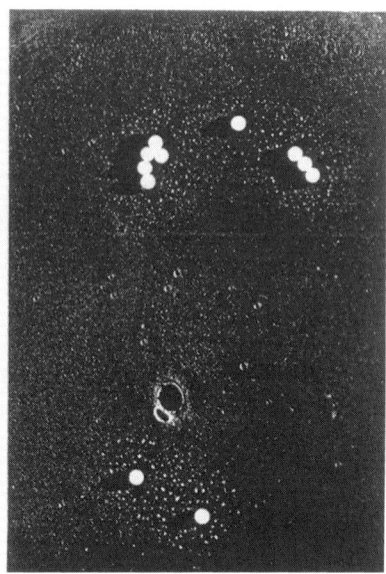

Figure 9 Electron-microscopic images of counts of influenza virus particles; H. B. Donald and A. Isaacs (1954). Courtesy of NIMR.

should be able to see full round virus structures before, and collapsed structures after, interference had been induced. Furthermore, Isaacs proposed to use the same strain of influenza virus and the same method of viral inactivation that he had used successfully in previous interference experiments. Isaacs's infectious enthusiasm made his proposals sound more than convincing and Lindenmann accepted the changes without much discussion.

Standing face-to-face at the laboratory bench in room 215, Lindenmann, Isaacs and a laboratory technician began by opening a batch of embryonated eggs, removing and washing the chicken embryo membranes, cutting them up into pieces, heat-inactivating batches of influenza virus, preparing red cell ghosts, and sticking part of the inactivated virus to these ghosts, which were now designated as 'virus-coated ghosts'. After this preparatory work, they did a first experiment to see whether they could induce viral interference with these virus coated ghosts *in vitro*. One by one, pieces of membrane were put in test-tubes together with nutrient fluid and either inactivated virus stuck to red cell ghosts, free inactivated virus or no virus. The last was referred to as the control preparation.[42] By including 'controls' with nutrient fluid and nutrient fluid plus red cell ghosts only, Lindenmann and Isaacs wanted to check for possible unanticipated effects on the virus growth due to the nutrient fluid or the ghosts.

The groups of test-tubes were then rotated in the roller apparatus. After 24 hours in the roller-drum the test-tube fluids were discarded, while the membranes were transferred to dishes containing an aqueous solution in order to wash off the 'old' fluid ('washing the membranes'). Subsequently, these membranes were put in test-tubes one by one together with live virus and fresh nutrient fluid, and incubated for another 24 hours in the roller-drum. Halfway through the first week, after collecting the tubes from the roller-drum, they subjected the tube fluids to a so-called 'haemagglutination titration', a specific bioassay for determining the relative rate of influenza virus multiplication. For agglutination to occur it took hours, leaving time for discussions or other research work (see Figure 10).

> The experiments took hours to titrate . . . and this left time to talk. Alick was the leader in conversation, and ideas for new experiments, political discussion, or identification of snatches of opera that he would sing made the time pass quickly. Alick, too, was adept at determining where the end point was, with the aid of a hand lens, long before the rest of us could do it, and he had often planned the next experiment before the red cells had really settled.[43]

The data output of the first experiment was encouraging. The experimental results were similar to the results Lindenmann had obtained in Zürich: both free inactivated virus, and inactivated virus stuck to red cell ghosts (virus-coated ghosts) were able to induce viral interference and inhibit the multiplication of live virus in their 'test-tube arrangement'. After another three weeks of

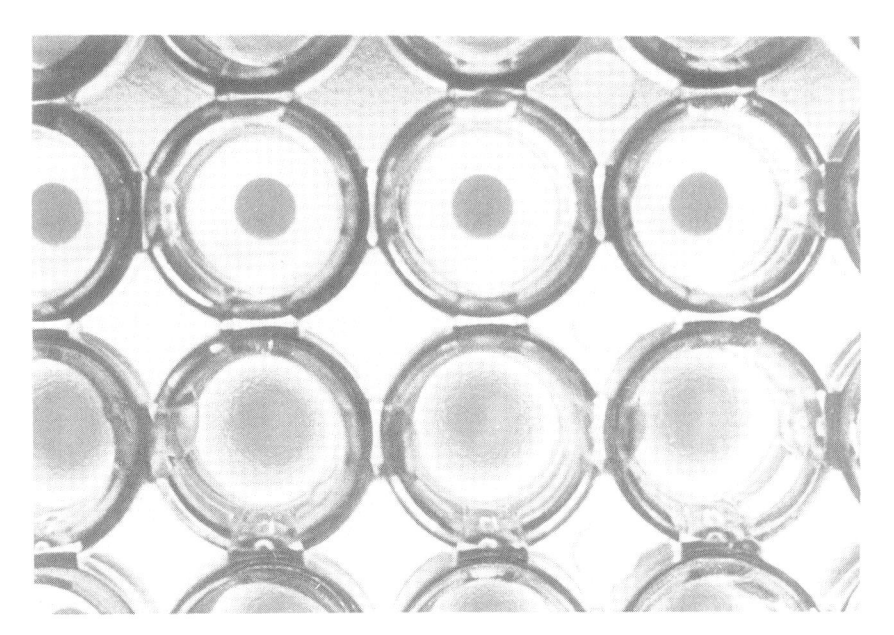

Figure 10 Agglutination (lower row) is characterized by the formation of a pattern on the bottom of the tube. Non-agglutinated blood cells settle down to a small dot (upper row). Courtesy of Dr D. Tyrrell.

repeated trials to fine tune their experimental arrangement – setting the optimal experimental conditions for the induction of interference with virus-coated ghosts – Isaacs and Lindenmann decided to start efforts to visualize the virus by electron microscopy. Before and after the induction of interference in their experimental set-up, samples would be taken and sent to the electron microscopist who would be asked to produce electron microscopic images of red cell ghosts and virus particles. Again, if Isaacs's assumption that during interference only the virus nucleic acid entered the host cell while the virus coat remained firmly attached to the red cell ghosts was correct, then they should be able to see full round virus structures before, and collapsed structures after, the induction of interference.

By special request, Valentine immediately processed electron microscopic images of the samples from Isaacs's laboratory. The pictures of the first samples, which were taken at the start of the experiment, looked promising. Clearly visible structures of red cell ghosts with virus particles attached to them could be distinguished, and Isaacs enthusiastically showed Lindenmann that it was even possible to count the number of particles per ghost. However, the electron micrographs of the other samples taken after the induction of interference were disappointing. The red cell ghosts looked rather smudgy with no more than a blurry outline of virus particles. This made the photographs hard to interpret. Neither Isaacs nor Valentine could tell the difference in structure between virus particles before and after induction of interference (half-way through the experiment). Furthermore, on the basis of Valentine's electron micrographs it was impossible to figure out whether in the 'control' groups without membranes, which contained either virus-coated ghosts or a mixture of virus-coated ghosts and empty ghosts, virus particles had become disentangled and were transferred to the empty ghosts. In Isaacs's notebook two question marks serve to emphasize the problems of interpreting the electron microscopic images.[44]

Despite subsequent attempts to refine the experimental procedures, the electron microscopic images remained difficult to interpret. Analysing the electron micrographs became a rather frustrating activity. However little, eventually they did get something out of these efforts. According to Valentine there were at least hints that in the mixture of virus-coated ghosts and empty ghosts a transfer of virus particles from one to the other had taken place. In discussing the recent disappointing experimental events Isaacs and Lindenmann agreed that if they took the hints seriously that some virus particles had become disentangled during the experiment, it would undermine their efforts to test Isaacs's idea – that only the virus nucleic acid entered the cell and induced interference, leaving the empty virus envelope outside. Without dropping this 'nucleic acid' hypothesis though, an additional research question was raised. Could the interference induced by virus-coated red cell ghosts be due to virus units which became detached from the ghosts during the experiment? It was the same kind of question that had bothered Lindenmann in Zürich. They subsequently developed the following argument: 'let us assume this to be the case and suppose that the few virus particles that become disentangled really suffice to induce interference, then it might be possible to induce interference with the same virus-coated ghosts more than once.'

According to Lindenmann, 'at a certain moment in November 1956, after about two months of experimentation, Isaacs felt that something unusual was happening while re-using the test-tube fluid with inactivated influenza virus in a second interference experiment'.[45] The residual interfering activity in the supernatant fluid that had been separated from the virus-coated ghosts was far higher than Isaacs would have expected, judging by his extensive research experience with inactivated influenza virus as an interfering agent. Without the large excess of interfering power that was believed to be associated with the ghosts, the supernatant should have lost its interfering capacity more rapidly with time. Even more surprising was the observation that a sample of the supernatant fluid, which was taken off immediately after the separation procedure and tested for its virus content – inactivated virus that had become disentangled – seemed to contain only minute amounts of virus. The following question must have racked their brains: how could the inhibition of virus growth have come about if the inactivated virus as the interfering agent had apparently been absent?

While discussing the readings in his notebook with Lindenmann, Isaacs apparently suggested the possibility that somehow new interfering activity had been generated. According to Lindenmann Isaacs was the first to admit that the observed effect could just as well be due to a change in pH or to the exhaustion of nutrients in the test-tube fluid, as suggested by Lindenmann. Isaacs also indicated that he knew by experience that the presence of virus in the supernatant fluid could easily have been masked by a certain membrane substance, which was known to interfere with the type of virus measurement they had used. Eventually, they had to concede that on the whole their experiments had produced more questions than they had resolved.

However, despite the fact that both researchers seemed to agree that the data present were not much to go on, each judged the experimental situation differently. Apparently, it was a perplexing experience.[46] Whereas Isaacs, primarily, became fascinated by the idea that, somehow, new interfering activity had been produced in their experimental arrangement, Lindenmann had a more cautious attitude and thought more in terms of possible artefacts that might have disturbed the expected course of events. Once again they would bring the matter up for discussion. As a joke, during one of these sessions, Lindenmann started calling Isaacs's mysterious interfering activity by the name of 'interferon'. 'Before studying medicine, I had studied physics for a while. Of course in physics terms ending in '-on' were very popular, like electron, positron, muon etc. Because of this old love for physics I believe I invented the term 'interferon' . . . It was not to be taken serious'.[47]

The sequence of events clearly shows that the work at the laboratory bench was far from an all or nothing effort to confirm or refute hypotheses. Isaacs and Lindenmann did not change course because one hypothesis seemed to make more sense than the other. The range and priority of hypotheses and research questions with which the experiments were associated was subsidiary to and dependent on the events at the laboratory bench. The unexpected handle on their initial research problem resulted recognizably from a repeated failure to meet prior assumptions and expectations due to resistances arising from the non-compliant nature of the instruments and materials at hand. The relative robustness of the new phenomenon was going to change the order of events in Isaacs's laboratory by sparking a new line of research.

Writing 'interferon'

Isaacs's notebook has an entry for 6 November 1956, starting with the sentence: 'In search of an interferon'.[48] Probably, these words marked the decision to put to the test Isaacs's vague idea about some sort of new interfering activity that was generated in their experimental set-up. Isaacs was strongly in favour of the idea that one way or the other new interfering activity had been generated. However, Lindenmann remained more sceptical and played the devil's advocate by arguing that the virus interfering activity could just as well be due to traces of inactivated virus.

It is difficult to make sense of the subsequent series of experiments. Basically, Isaacs and Lindenmann varied their experimental set-up in a number of ways in an attempt to eliminate all possible background effects they could think of. One of the things they did was to check for the possibility that the 'new' interfering activity was due to the presence of inactivated virus that might be resecreted into the tubes, as Lindenmann had suggested.

According to Lindenmann, in the process they became more and more convinced that there was something, some factor in the biological soup that was responsible for the phenomenon, but they did not know what it was. Most likely, the apparent consistency of the phenomenon made them believe that 'Isaacs's interferon' might indeed be an interesting new experimental feature, a signal instead of a noise, which was worth pursuing. In an attempt to characterize their elusive factor, they then initiated a series of 'properties of interferon' experiments. It may look as if it was a rather straightforward and unproblematic series of experimental trials but again and again experiments went wrong for no discernible reasons. Isaacs and Lindenmann had a difficult time in making the experimental arrangement work in a reproducible way, in other words, to produce workable amounts of the viral interference activity-containing fluids which they referred to as 'interferon preparations'.

Isaacs and Lindenmann tested the stability of their interferon preparations at different temperatures, by heating at 60°C for one hour (result: the interfering capacity decreased sharply) or at 37°C for 24 hours (result: the interfering capacity remained stable) and by storing at 2°C (result: the interfering activity remained stable for at least 14 days). They made dilutions and subsequently compared the interfering activity of these dilutions (result: the activity decreased with dilution), or subjected interferon preparations to haemagglutination tests to see whether their putative factor would be able to agglutinate red blood cells like inactivated influenza virus did (result: no agglutination could be observed).

In addition they tried to neutralize the interfering activity with specific viral antiserum, which was to dissolve the interfering capacity of inactivated influenza virus (result: the interferon fluids turned out to be resistant against this viral antiserum). The researchers also ran filtration and centrifugation trials (result: the interfering activity of an inactivated virus preparation completely settled down in the test-tube after centrifugation, whereas the activity of an interferon preparation did not; there was no clear-cut difference with regard to filtration behaviour).[49] Furthermore the viral interfering activity of interferon preparations was tested with a number of different viruses, which had also been shown to be susceptible to inactivated influenza virus as interfering agent (result: the interferon preparations showed an inhibitory activity against all viruses tested, such as the pox virus, 'vaccinia').

It is worth noting that in between these series of *in vitro* trials an isolated attempt was made to see if the viral inhibitory effect of interferon preparations on influenza virus could also be demonstrated in mice (*in vivo*). However, no inhibitory effect could be observed. Unfortunately Isaacs's notebook does not give any hint with regard to the motives behind this ad hoc *in vivo* trial.[50]

These kinds of experimental trials were performed over a two-month period. They involved the most practical kind of skilled laboratory craftsmanship, a sort of cookbook benchwork, using various recipes for identifying interfering agents. As such it was part of a slow and laborious process of gaining experience and learning to read the data concerning a new laboratory factor. Despite attempts to reproduce experiments in detail, a certain variability in results was common. At times when experiments had produced no workable data, they would sit down and discuss the dangers of 'chasing a red herring', an anomaly of artefactual nature. After all, it might very well not be the presence of something but the lack of something that kept them busy. Maybe a nutrient was being used up, thus preventing viruses from growing.

However, through the steady accumulation and superposition of data stemming from the physical and biochemical manipulations at the bench, the idea that the 'interferon' was a transitory factor, an artefact or noise, disappeared. Instead, interferon as a signal, a new kind of interfering agent, was articulated in accommodation to a series of non-trivial, data-generating experimental events. These were regarded as direct indicators of the factor under study. In particular, the consistency of the interferon phenomenon in the face of widely differing physical conditions was instrumental in solidifying their sense of having found a novel biological factor. The anomaly became increasingly transparent and self-evident and in the process a primary profile of Isaacs's and Lindenmann's elusive factor or activity was shaped. Apparently a kind of 'soluble interfering factor' was produced in their experimental set-up, which displayed properties different from other interfering agents such as the inactivated influenza virus.

At the time of Lindenmann's introduction of the name 'interferon' neither he nor Isaacs had the slightest idea about the nature of this phenomenon they called 'interferon'. Something was involved in the experimental anomaly, but what was it? Another viral interfering factor or factors, or something quite different? It felt like moving very fast and having no fixed markers to give an idea of position and speed.

Within this experimental situation, their status of novice and expert were relative to each other. I initially called Lindenmann a 'scientist novice at the frontier of viral research', and Isaacs, who was widely known as a distinguished virus researcher, a 'scientist expert'. However, we saw that when exploring and manipulating their experimental arrangement, both Isaacs and Lindenmann were, in a certain sense, novices. The two men had yet to discover how to go about experimenting and what experimental success meant in their case. In modelling the experimental arrangement on a permanent basis they accumulated manipulative skills that would bring the new and unexpected under control.[51] As such, both Isaacs and Lindenmann were constantly engaged in producing, mastering ('domesticating as it were') and communicating new procedures, ideas and phenomena.[52] In a way they had ordered chaos by acquiring skills in the manipulation, interpretation and representation of a phenomenon which in November had still been perplexing.

By the end of February 1957, in consultation with Andrewes, it was decided that the experimental data looked consistent enough to argue against the interpretation of their 'interferon' as an artefact and to warrant writing up their research work for publication.[53] Andrewes offered his support as a Fellow of the Royal Society in submitting the final papers to the highly prestigious *Proceedings of the Royal Society*.[54] In this series of two articles, which were published in October 1957, Isaacs and Lindenmann claimed to have found a new interference-inducing factor named 'interferon' during a study of the interference phenomenon.[55] They argued that because of observed analogies between interferon and virus production they would provisionally consider interferon as an abortive product of virus multiplication which blocked the synthesis of live virus and had many different properties from those of the inactivated influenza virus – the other interfering agent present in their arrangement.[56]

The knowledge claims in the resulting scientific articles bear relatively few marks of their generative production process. The twists and turns at the bench in Isaacs's laboratory and the flaws the researchers had experienced in manipulating their experimental arrangement were omitted from the final papers. Instead, the occurrence of what was called, alternately, the factor or substance 'interferon' had been caught in a web of mostly post-hoc reasonings. Thus the significance and identity of materials and methods had changed retrospectively in the presence of a solution: the discovery of the substance named 'interferon'. This is in line with a change in meaning of the novel object under study: from interferon as a transitory factor, a 'mysterious' kind of biological activity, to interferon as a new interference-inducing factor. As we will see the meanings attributed to interferon in terms of what it signified and what its biomedical significance was, not only altered but also multiplied.

Isaacs's interferon kitchen

Around the time when final drafts of the first two papers on interferon were being submitted to the *Proceedings of the Royal Society*, Isaacs was approached by Derek Burke a post-doctoral worker from the Chemistry Division, two floors down on the ground floor of the NIMR. Isaacs had known Burke for nearly two years as an assistant to James Walker, a senior organic chemist with whom Isaacs had been working on the possible effects of various synthetic organic compounds on influenza viruses. Burke, who had spent two years as a post-doctoral fellow at Yale University doing 'natural products chemistry', was responsible for studying the possible effects of the chemical agents, synthesized by Walker, on the nucleic acids of influenza viruses.[57] Once familiar with the basic techniques used in experimental virology, such as growing and harvesting influenza viruses in fertile hen's eggs and determining virus titres, and testing the first series of chemical agents, Burke had become seriously interested in studying the biochemistry of viruses. At the same time he gradually began to lose interest in Walker's and Isaacs's collaborative research project.

For some time, Burke had had an idea for a project involving the biochemistry of the influenza virus infection, and he wanted to talk it over with Isaacs. However, when asked for his opinion Isaacs did not seem much interested, and, instead of carefully listening to Burke's proposals, he started talking about his favourite research project. Together with Lindenmann he had been working for some time on a rather new and interesting viral interfering agent, which they were in the process of characterizing. However, he and Lindenmann were rather amateurs when it came to the task of identifying, purifying and isolating biological substances.

Both Isaacs and Lindenmann had been trained in medicine rather than in science with a rather basic knowledge of chemistry. Moreover, neither of them thought much of the chemical work, which they considered a necessary evil to characterize their putative biological factor: such work, they felt, kept them from working on the more interesting and rewarding fundamental biological research questions such as the elucidation of the mechanism of action. They would be far better off, they reasoned, if they handed over the chemical work to one of the postdocs in the NIMR's divisions of Chemistry or Biochemistry.[58]

This in mind Isaacs asked Burke to help with the biochemical work. Despite the poor hearing he had had from Isaacs for his own plans, the young scientist Burke considered it an honour to be asked for help by a highly regarded senior researcher like Isaacs.[59] Thus, by the end of March 1957 the biochemist Burke joined Isaacs's and Lindenmann's investigations into the 'interference inducing factor', named 'interferon'. Burke's laboratory records indicate that he mainly worked on the characterization of interferon, subjecting interferon preparations to various chemical manipulations. Most of these were routine in biochemistry to identify natural substances – for example, testing the effect of organic solvents like ether, testing the effect of protein degrading and nucleic acid degrading compounds or testing the pH stability, i.e. acid and base stability.[60] First, however, Burke enlisted the help of Isaacs's laboratory technician to familiarize himself with the production of interferon through replicating the material realization, or performance, of the interferon system.[61]

With the help of Burke's biochemical toolkit the list of biochemical and physical properties ascribed to interferon grew steadily. 'Interferon was stable in fairly strong acid solution . . . Shaking with an equal volume of ether caused complete loss of activity . . . Trypsin considerably reduced its activity.'[62] According to Burke the sensitivity to the protein-degrading enzyme, trypsin, suggested that interferon might contain protein, but the other experimental properties did not allow any further conclusions regarding interferon's chemical nature.[63] It seemed unlikely that they would learn much more about its chemical composition until interferon could be prepared in a purified form. Burke also helped in scaling up the experimental set-up in order to prepare interferon in bulk, as the production of interferon was lagging behind the increased research demand.

Isaacs and Lindenmann – who coined Burke the name 'Calamity Joe' because whenever he handled glassware, there was a fair chance that he would break it – in turn, shifted their attention to the phenomenological aspects of interferon research, such as the mode of production and the mode of action of interferon *in vitro*, in test-tubes each containing at least one piece of chick embryo membrane.[64] Through this research work unexpected differences emerged between interferon and virus. For example, membrane preparations could be induced repeatedly to produce the same amounts of interferon, whereas at the same time the preparations showed a greatly reduced ability to support virus multiplication. These experimental results were thought to suggest that only part of the 'complex machinery' that was required for viral synthesis was required for the production of interferon.[65] The results strengthened Isaacs's and Lindenmann's beliefs that interferon was most likely an abortive product of virus multiplication, which passed through only part of the viral reproductive process. As such, interferon was regarded as an abnormal viral entity that played a major role in mediating viral interference.[66]

Apart from studying the mode of action of interferon *in vitro*, Isaacs also initiated a series of experiments to test the effect of interferon preparations *in vivo*. According to Lindenmann, as soon as Isaacs learned that interferon preparations were capable of inducing viral interference against vaccinia virus *in vitro*, in the test-tube, he began suggesting plans to test interferon's action on vaccinia *in vivo*, on rabbits at the experimental unit of the laboratory animals division (see Figure 11). Such experiments were argued to be perfectly in line with trials Isaacs had performed in his laboratory about three years earlier. In the autumn of 1953, he succeeded in showing that inactivated influenza virus interfered with the development of vaccinia virus not only *in ovo*, in fertile

Figure 11 Animal technician holding a laboratory rabbit in her arms at the experimental unit. Courtesy of NIMR.

hen's eggs, but also in laboratory rabbits. He figured out that if he injected rabbits intradermally – into the skin – with inactivated influenza virus preparations, and one day later made injections with vaccinia virus preparations in the same areas, the development of skin lesions, normally expected after a single vaccinia virus inoculation, was inhibited significantly.[67] Since interferon was considered to be an interfering agent with a viral interfering activity similar to inactivated influenza virus, Isaacs expected to see a similar effect when substituting inactivated influenza virus for interferon in the aforementioned rabbit experiment.[68] However, substituting inactivated influenza virus for interferon did not initially meet with Isaacs's expectations. Unlike inactivated influenza virus preparations, interferon fluids did not seem to have any effect on the development of vaccinial lesions in the rabbit skin. Only after many unsuccessful trials, changing the experimental conditions a number of times, did Isaacs finally manage to observe a minor inhibition of lesions with one of his laboratory rabbits. Subsequently a number of other test series were required to substantiate the observed effect, though the overall inhibitory effect of interferon appeared to be far less significant compared with similar experiments using inactivated influenza virus. Isaacs was somewhat puzzled by the apparent difference in the protective effect between interferon and inactivated influenza virus.

After all, the viral interfering effect of the latter was thought to be mediated through interferon, which was thought to be released as a result of the virus–host interaction. Isaacs and Lindenmann discussed the rabbit experiments over and over again, but they could not think of an explanation other than that far more interferon was present in the skin of rabbits treated with inactivated influenza virus.[69]

The persistence with which the animal experiments were performed suggests that the idea of the practical medical application of interferon preparations to virus disease must have crossed Isaacs's and Lindenmann's medically trained minds. In principle the laboratory rabbit is as much a model of the real situation as the test-tube, but biomedical researchers have a 'natural' propensity for linking the experimental output of *in vivo* experiments, more easily than *in vitro* experiments, to practical medical applications.[70]

Most likely the *in vivo* experiments with laboratory rabbits were the first instances in which interferon's effects were translated into practical medical terms. As I will show in the next chapter medical concerns would continue to play a role in structuring interferon research, exploring the interferon phenomenon in terms of its possible value as a tool to be used in clinical practice. The fact that animal virology was institutionalized predominantly as part of the medical field of microbiology, whose major professional goal was the prevention, diagnosis and treatment of viral diseases, played an important role in this respect.[71]

While being 'hammered out' in the local context of Isaacs's laboratory, interferon seemed largely inseparable from its 'home ground'. However, by moving rather easily between the 'world' and his laboratory, Isaacs would play a major role in linking interferon to a growing number of events and people 'outdoors' and thereby in accomplishing a separation that mobilized interferon beyond the physical space of Room 215. A closer look at this mobilization process will give us a clear insight into how interferon developed as the organizing principle for a new line of investigation within the international field of animal virology and more in general into the early stages of the evolution of a scientific subfield. To this end I shall investigate the ways in which work on interferon was constituted and modified through actions inside *and* out of Isaacs's laboratory, exploring both the private spaces of laboratories and the public spaces in which interferon met its audiences.

2 Shaping a new field of research
Investigating interferon(s)

Interferon going public

Early in June 1957 Lindenmann's term as a visiting researcher came to an end and he returned to Switzerland. Before resuming work in Mooser's laboratory at the 'Hygiene Institut' in Zürich, Lindenmann presented a paper on his foreign research work at the annual meeting of the Swiss Society of Microbiology. It was interferon's first official and extramural appearance.

In his paper presentation entitled 'trials on virus-interference' Lindenmann claimed to have found a product of influenza viral interference, named 'interferon', which interfered with the growth of live influenza virus in the test-tube. He indicated that little was known about the mode of production and action of interferon and about its chemical composition. His lecture gave rise to a rather lively debate among the members of the audience. Prof. Hallauer, one of Switzerland's leading bacteriologists and virus experts, started the discussion by asking Lindenmann: 'What kind of theoretical interpretation should we give to this experimental feature?' and 'Do we have to question the fact that viral interference is induced by the virus itself?' Lindenmann replied that it was the virus that was at the centre of their experimentation too, and that he believed interferon to be an intermediary product of the virus interference reaction.[1] This answer did not satisfy Hallauer, who retorted that 'as this [interferon] was utterly new and contrary to all he had read, it must be rubbish'.[2] Hallauer's fierce critique met the approval of most of the scientists present. They all considered viral interference an inter-viral competition phenomenon, due to the competition between interfering and challenging virus for a key element of the viral reproductive process, which was present in limited amounts in the host cell. This was in general agreement with the concept of interference as presented in the latest international viral handbook.[3] The idea of an autonomous viral interfering factor, whether or not of viral origin, just did not seem to fit into the textbook concept of viral interference, and it was therefore rejected.

One of the few people who did not share the general negative verdict on Lindenmann's paper presentation was the virologist Dr Kradolfer from the drug company Ciba. In March 1957, Ciba observers in London had sent a promising report to Ciba's headquarters in Basel, which associated interferon's workings in the laboratory with potential therapeutic effects in the clinic. Upon assessment of the report Kradolfer was assigned to the job of exploring interferon's therapeutic potential. Kradolfer's interferon project was a product of Ciba's firm belief in their own strategical observation and screening system, which aimed to signal and establish links between fundamental biomedical research and clinical problems as early as possible, as a means to the end of bridging the gap between the laboratory and the clinic. In addition, the rather severe Asiatic flu epidemic that was spreading round the world, striking at least 10 million people by June 1957, encouraged drug companies in the search for antiviral drugs.[4]

In a corridor chat with Lindenmann during the Swiss meeting, Kradolfer indicated that he had been working on interferon for about three months and that he wanted to exchange information with Lindenmann on this possible lead toward antiviral therapy. Lindenmann expressed his willingness to do so, though he was far from impressed by Kradolfer's research efforts. According to Lindenmann, Kradolfer could confirm the appearance of new interfering activity by using a similar experimental system that he and Isaacs had used, but only as a transient phenomenon.[5] Lindenmann, in turn, suggested the possibility that the transient nature of the phenomenon might be due to a difference in the strain of virus which was employed in Kradolfer's experiment. In a letter to Isaacs, Lindenmann wrote about his contacts with the Ciba researcher and asked him whether he was against exchanging information with Kradolfer.[6] Isaacs's reply was quite explicit: 'I think it is a very bad idea to show our manuscript to anyone in a commercial firm. I would be very much against your doing so in future'.[7] On first thoughts this answer seems to reflect a general tendency amongst British academic workers in the 1940s and 1950s to view collaborative research with an industrial firm as an intrusion upon their scientific work. But there was more to it, as the following excerpt from Isaacs's next letter illustrates:[8]

> In this country we are all very sensitive about the fact that penicillin is a British discovery, but that we have to pay royalties to American commercial firms for every gram of penicillin we use. I am not suggesting that interferon is in the same class as penicillin but on general grounds I don't think it is a good thing to deal with commercial firms. It is much better that they should wait until our information is published and then they can compete on even terms.[9]

The way Isaacs associated interferon with penicillin in this second response to Lindenmann not only shows rather high hopes for interferon's possible medical applications, but also reveals a cultural condition that was instrumental in shaping Isaacs's policy toward interferon. There was a general feeling amongst government and academic workers in Britain that, because of an inability to control the patents, penicillin had been given free of charge to commercial drug houses and, in particular, American commercial firms. Even worse, royalties or licensing fees had to be paid to American drug companies for using penicillin in British medical practice, while Britain's dollar deficit was a constant source of political concern and irritation.[10] To avoid another penicillin-like 'patents affair', novel scientific findings were handled, at least within Britain's national research institutes, with exceptional caution until results had been published and patents had been filed. In particular within the field of biomedicine, developments were closely monitored by government officials and taken up actively at the very first sign of a potential application in practical medicine. Moreover, as an additional and legal safeguard to secure the development of inventions resulting from public research in Britain, an entirely new type of government agency, the National Research Development Corporation (NRDC), had been established under the British Industries Act of 1948. The NRDC's task was to help to obtain patents and to provide financial support for the development of inventions up to the stage where industry was prepared to support the project completely. In facilitating the commercial development of state-aided research, the new agency tried to meet everyone's wishes to make better use of patenting in the medical field.[11]

Almost simultaneously with interferon's public appearance in Switzerland, Isaacs's laboratory was visited by a group of senior officers from the Medical Research Council (MRC), who were on their yearly inspection visit to the NIMR.[12] On these occasions researchers were expected to give information about the current state of research affairs.[13] Isaacs had prepared a demonstration of interferon's effects to convince the official visitors that he and Lindenmann had recently discovered a biological substance with distinctive antiviral properties. Several experimental set-ups were on display to show that interferon preparations had an inhibitory effect on a number of viruses both *in vitro* and *in vivo*. While showing the MRC officers around, Isaacs enthusiastically speculated about the possibility that this interferon might offer a potential new approach to the medical treatment of virus infections.[14]

Most likely, Isaacs's speculation about interferon's clinical potential was more than a mere expression of his professional interest as a medical doctor in practical problem-solving. He must have been aware of the fact that within the MRC there was a tendency to favour research projects with practical medical promise, even though MRC's official policies were directed more to basic than applied research.[15] It seems that the Secretary of the MRC, Sir Harold Himsworth, as a former practising clinician, emphasized the need to link basic research to clinical problems. The penicillin story was Himsworth's favourite case to convince his staff of the necessity to be on alert for claims linking fundamental research with practical medicine.[16]

The subsequent news of Ciba's interest in interferon as a lead towards antiviral therapy undoubtedly must have strengthened Isaacs's feeling that, however premature, he was justified in making the claim to the MRC that interferon could have therapeutic potential. In one of his letters to Lindenmann Isaacs emphasized the fact that the MRC had been very impressed with the demonstration.[17]

Thus, while keeping the Swiss drugmaker at a distance, Isaacs actively tried to draw the MRC into the interferon project. Obviously Isaacs was concerned to safeguard British interests and at the same time win wider interest in and public support for further research on interferon.

In the meantime Lindenmann was defending his scientific existence. From the first moment he resumed work at the 'Hygiene Institut' his boss, Mooser began behaving in a peculiar way. On the one hand, Mooser emphasized the fact that he did not think much of Lindenmann's efforts to continue research on interferon by saddling him with so much other work that Lindenmann had to conduct his interferon research in the evening. Furthermore, he told Lindenmann right away that if he intended to use large numbers of laboratory eggs on a regular basis he had to pay the expenses out of his own pocket.[18] On the other hand, Mooser repeatedly entered the laboratory and asked Lindenmann to give yet another detailed account of his interferon experiments with Isaacs at the NIMR. The fact that Lindenmann had difficulty in repeating the interferon experiments he had worked on in London only complicated matters.

The situation took a turn for the worse after the reprints of the *Proceedings of the Royal Society* papers arrived. Mooser immediately sent for Lindenmann and asked him why he was not quoted in either of the two publications. Mooser had taken great offence as he firmly believed that the idea of an interference inducing factor had originated from his experiments with Lindenmann in 1955. He indicated that he deserved at least as much credit for his pioneering laboratory work as Lindenmann himself. In a way, he even believed he was a co-discoverer of this 'interferon', and despite Lindenmann's efforts to provide evidence to the contrary, Mooser persisted in this belief. Eventually the situation became unbearable and Lindenmann had to leave his position at the 'Hygiene Institut' and accept a job as a microbiologist at the Federal Health Department in Berne.[19]

The antiviral penicillin

While Lindenmann was fighting a losing battle to keep his university position, most of Isaacs's time was consumed by running the World Influenza Centre during the 1957 Asiatic flu epidemic.[20] Nevertheless, with Andrewes's support and with both Burke and a technician at his side, Isaacs managed to keep research on interferon going and to involve himself in operations to raise the factual status of interferon inside and outside the walls of the NIMR. He organized an Institute colloquium on interferon, was responsible for submitting another two articles dealing with studies on interferon, and he also sent out reprints of interferon publications to all influential virologists he could think of.[21] Furthermore Isaacs prepared a short paper on the discovery of the substance 'interferon' for the *MRC Annual Report*, in agreement with the demonstration for MRC officers.

In this report attention was primarily focused on interferon's effects against viruses both in the test-tube and the rabbit model. Furthermore, interferon's workings in the laboratory were associated with potential therapeutic effects in the clinic: interferon was presented as a possible new approach to the problem of finding suitable remedies against virus diseases. It was emphasized that hitherto all chemotherapeutic substances which had been found to inhibit virus growth had been toxic to the host, thus implying that this new biological substance, interferon, might offer a non-toxic alternative. In the test-tube, at least, it had been shown to be innocuous.[22]

The *MRC Annual Report*, together with the series of two articles in the *Proceedings of the Royal Society* and further annotations in the *Lancet* and *British Medical Journal*, drew the attention of the science correspondent of the *Daily Express*.[23] The latter immediately phoned Isaacs with a request for additional information. Despite Isaacs's refusal to provide the extra information an article was published at the end of October 1957 in the *Daily Express* on the isolation of an antiviral substance, named 'interferon'.[24] Following this newspaper article, the pharmaceutical companies Glaxo and Benger Laboratories wrote to the MRC, urging for a change of name of Isaacs's substance. Since both had Registered Trade Marks for products with names similar to interferon, they anticipated difficulties if it was to be developed as a drug in the future.[25] However, far from leading to a change of name, these minor events led the MRC, which had been impressed by Isaacs's private demonstration and were closely following the latest developments on interferon, to consider more urgently the patentability of Isaacs's discovery in order to protect the Council's interests.[26] As I argued before, the MRC was on the alert to actively take up research projects with potential links to practical medicine in order to prevent missing out on a penicillin-like medical breakthrough. Furthermore, the MRC wanted to keep up a high profile toward the Treasury, and as a means to this end the MRC was always looking for promising research results produced by one of its research units which would make for a big story.[27]

The MRC consulted their patent adviser at the NRDC to see whether or not the subject matter was patentable, with the result that in May 1958, applications were made for patents in the United States, Canada and Germany. However, they were told by the NRDC that it was impossible to obtain a British patent because more than six months had elapsed since the first public report on interferon had been published. Obviously the MRC had little experience in patent matters.[28]

At about the same time that the MRC and NRDC began to regard interferon as a patentable commodity and started to file patent applications, Isaacs and Burke gave a public demonstration of interferon's workings at the Annual Conversazione of the Royal Society. Probably a similar kind of experimental set-up to the one shown earlier to MRC officers was on display at the Royal Society in London on 15 May 1958 – thereby extending Isaacs's laboratory space beyond the doors of the NIMR. The catalogue of the exhibits to be seen at the Royal Society contained the following description:

INTERFERON: AN INHIBITOR OF VIRUS GROWTH

It has been known for many years that one virus may interfere with the growth of a second. Recently it was found that this interference is mediated through a substance which has been named 'interferon'. Active preparations of interferon can be easily produced by the litre without special equipment, and the activity can be readily concentrated. Such preparations are potent inhibitors of the growth of influenza and related viruses *in vitro*, and preparations which have been concentrated 10-fold inhibit the growth of vaccinia virus on the chick chorioallantoic membrane or in the rabbit skin.[29]

According to a correspondent of the *Daily Telegraph*, a Fellow of the Royal Society had said at a press preview: 'I find enormous excitement in the promise of interferon'.[30] The next day, the *Daily Telegraph*, among other newspapers, reported the discovery of a substance named interferon which might one day play an important part in the fight against viruses. In a comment on the exhibits at the Royal Society by the *British Medical Journal*, 'interferon' was said to break new grounds in medicine as a revolutionary therapeutic tool against viral disease appropriate to the satellite age, 'while the latest and largest sputnik made its first orbit over Great Britain'.[31] Isaacs was even asked to give a demonstration of interferon's workings on the BBC programme 'Eye on Research' (see Figure 12).[32]

The sudden national press coverage of interferon as a kind of new 'antiviral penicillin' in newspapers, magazines, in the scientific press, and on British television had an impact. Translating interferon's viral inhibitory effect at the

Figure 12 Isaacs lecturing to the camera directly from his laboratory to explain his work on interferon to a mass audience in the 1958 BBC programme 'Eye on Research'. Courtesy of Dr S. Isaacs-Elmhirst.

laboratory bench into therapeutic terms brought interferon to the public stage (see Figure 13). Penicillin served as a powerful cultural symbol for interferon's legitimacy as a promising lead toward antiviral therapy. The introduction of 'miracle drugs' like penicillin and cortisone in the 1940s had provided physicians with a curing power never seen before. The spectacular clinical potential of specific and truly curative drugs, so-called 'magic bullets', and the often dramatic portrayals of beneficial effects in the clinic, had raised hopes regarding scientific medicine's ability to come up with yet another 'wonder' for the doctor's bag.[33] It effected a firm belief among doctors, administrators and the public in therapeutic breakthroughs as a dominant feature of a laboratory-supported scientific medicine.[34] Together with the British 'penicillin trauma', or what was generally referred to as 'the American penicillin syndrome', this set the stage for interferon's dramatic reception in Britain and the subsequent efforts to develop interferon as a drug.[35]

Misinterpreton

Interferon mattered in a rather different way to the international audience of animal virologists, thereby yielding a different set of expectations. Instead of immediately assessing interferon's potential benefits to practical medicine, virus researchers judged Isaacs's and Lindenmann's knowledge claims in a different light: first, the association of those claims with the predominant theoretical framework concerning viral interference; second, the expected relevance for research programmes that figured high on national research agendas; third, the practicality of integrating Isaacs's and Lindenmann's experimental features in existing experimental systems in other laboratories; and fourth, the success or failure of replicating these experimental features.

The international virology community was far less receptive than Isaacs had hoped. Despite a series of articles in highly prestigious British journals and the established reputation of both Isaacs and the NIMR, Isaacs nevertheless had to engage in a quest for scientific interest and credibility regarding interferon. In particular, the cool reception within the powerful community of animal virologists in America concerned Isaacs. Although claiming the existence of a factor associated with viral interference, supposedly of viral origin, did no harm to the basic premise that interference was mediated by a virus, some researchers argued that it *did* infringe upon the consensus of viral interference as an interviral competition phenomenon. For others, the mere association with a laboratory phenomenon that had been relegated to the shadows of phenomena with relatively little research priority prevented them from becoming interested: viral interference was just not 'sexy' enough to capture their 'imagination'.[36] In addition, Isaacs's and Lindenmann's publications were criticized for their vagueness as to the nature and mode of production of the novel factor. Moreover, those who tried to implement Isaacs's and Lindenmann's experimental features in their own experimental systems found it extremely difficult to reproduce even part of the material results reported in Isaacs's and

Figure 13 The first human trial of interferon was 'performed' in this Flash Gordon comic strip (dated early 1960).

Lindenmann's publications. Whatever viral inhibitory activity was present in their experimental systems, it could not be linked with the experimental features claimed to be specific to interferon. A growing number of animal virologists in America began to question the validity of Isaacs's and Lindenmann's interferon claims, and a rumour spread that interferon was most likely a noise due to small amounts of inactivated virus coming through during the experiment. For this reason, influential American virus researchers like Harry Rubin and Howard Temin started nicknaming interferon 'misinterpreton'.[37]

With the great expectations building up over interferon in Britain and at the same time the 'misinterpreton' rumours coming in from the USA, Burke witnessed how during the summer of 1958 Isaacs began to appear increasingly stressed. Gradually Isaacs fell victim to doubts about the validity of the interferon findings himself. Part of the assumptions and practices that had been buried in the process of publishing the claims were opened to criticism in the open again. This continued to the point that the claims about interferon as a product and mediator of viral interference were threatened with complete dissolution.

More and more frequently, Isaacs would come into his laboratory in a depressive mood and do nothing but talk with Burke about the possibility that the observed antiviral effects of interferon might after all be due to the presence of traces of inactivated virus or to the absence of one or the other nutrient that prevented the virus from multiplying. Would they not be better off by adding interferon to the list of noteworthy laboratory artefacts? The situation had a paralysing influence on interferon research and a damaging effect on Isaacs's mental health. In the fall of 1958 Isaacs had a nervous breakdown and went on sick leave, suffering from a severe bout of depression which lasted several months. Thus Burke was left in the laboratory on his own.[38]

The crisis of faith in Isaacs's 'interferon kitchen' and the American 'misinterpreton' rumours do not seem to have affected the high expectations in Britain with regard to interferon as a possible new 'antiviral penicillin'. At about the same time the MRC decided to start efforts to develop interferon as a drug and seek collaboration with British drug companies. In addition, drug companies had begun to approach the MRC about interferon.[39]

These circumstances which involved the high hopes regarding interferon as a potential therapeutic breakthrough, widely shared in British society, made MRC's development plans relatively immune to both the late developments in Isaacs's laboratory and the misinterpreton rumours. They were instrumental in creating a niche, in other words a relatively stable and safe environment, for subsequent efforts to develop interferon as an antiviral drug in Britain.[40]

From the brink to revival: meeting the others

The second part of this chapter chronicles a gradual reversal of the balance of forces in Isaacs's quest for scientific credibility and interest regarding interferon. Gradually the significance of experiments and experiences relating to interferon was being transformed from a local into a trans-national issue as part of a new international line of investigation within the field of animal virology. Work on interferon no longer meant work in Isaacs's 'interferon kitchen' only. Through circulation, comparison and combination of written and printed texts (letters, papers, articles, textbooks, charts), research materials and skills, interferon research became increasingly dependent on events outside the NIMR. It was a matter of doing different experiments in a different time/space proportion. Beside Isaacs's 'interferon kitchen', 'others' began to take an active part in describing and defining the nature and meaning of interferon in complementary ways. In analysing this extension process I will consider the material, social and cognitive conditions that were instrumental in shaping the modes of interchange and influence between various researchers working on interferon: between those who claimed to have command of a novel laboratory procedure enabling them to produce a hitherto unknown experimental feature, and those who still had to include and integrate the new feature into their laboratory practices. This shows that the process of validating interferon and forming it into a basis for consensus involved not only evaluating claims in light of the perceived research opportunities and of negotiating consensus, but also a collective process of communicating, manipulating, interpreting and elaborating a common assembly of research objects.

The events in John Enders's laboratory at American's East-coast were key to this reversal and the subsequent formation of an international body of laboratory practices with shared commitments in the pursuit of a common set of research problems and objects. As it was exemplary for the changing relationship between interferon and 'Room 215', the growing interference between Isaacs's 'interferon kitchen' and Enders's laboratory deserves further attention.

Burke recalled the day that Isaacs, who had returned to work in the course of January 1959, received a letter from one of the most prominent American virologists and recent Nobel Prize winner, John Enders, sometime around June 1959, as one of euphoria. Isaacs immediately showed the letter to Burke. Apart from requesting a sample of interferon, Enders informed Isaacs of the fact that together with one of his postdocs, Monto Ho, they had been working for some time now on a so-called 'viral inhibitory factor (VIF)' which seemed to have properties similar to interferon. Here, finally, were American researchers who had found what seemed to be interferon.[41]

It is most likely that sometime during the spring of 1958, Ho and Enders on their part must have become aware of Isaacs's and Lindenmann's articles on interferon. According to Ho, one morning Enders popped into the laboratory and told him that he had discussed their 'VIF' experiments with Werner Henle. Apparently, Henle seemed to have been quite sceptical of Enders's report on VIF, which Henle believed to be rather similar to recent work published by the British animal virologist, Alick Isaacs. The latter claimed to have found a factor called 'interferon', presumably a small viral component which escaped detection by conventional virological techniques and interfered with the growth of viruses. Upon reading the interferon articles, Henle and his associates had almost immediately started efforts to see whether a similar inhibitory factor might be responsible for the observed resistance of chronically virus-infected cell cultures against unrelated viruses.[42] However, despite extensive testing, they had not been able to demonstrate the presence of an interferon-like factor. Henle had concluded that it was most likely that the observed virus resistance, which was said to fulfil the criteria for viral interference, had nothing to do with interferon and he even doubted the validity of Isaacs's experimental claims. These claims did not correspond with Henle's concept of viral interference as an inter-viral competition phenomenon that was not considered to be transmissible. As Enders's report on 'VIF' sounded similar to what Henle had read about interferon, Henle did not think much of it either. Far from discouraging Enders and Ho, the meeting with Henle was an incentive to carry on and to get hold of the English interferon articles.[43]

What were for Henle experimental findings that resisted replication and violated the basic criteria for viral interference, were for Enders, results that not only sounded interestingly familiar but might also be instrumental in furthering work on VIF. It is highly questionable that Enders considered the subsequent situation in these terms, but retrospectively it looks like the transfer of knowledge regarding interferon changed the order of events in Enders's laboratory, rendering the situation different for work on VIF. The subsequent series of experiments that were performed to forge a primary profile of VIF bore a close resemblance to trials Isaacs and Lindenmann had run in the case of interferon, a year earlier.

Ho and Enders realized that they should be careful to rule out the possibility that the observed virus inhibitory effects in their experimental system might after all be due to the direct effect of interfering virus particles. Since interference by virus particles had already been known for years, the presence of virus particles in the VIF preparations would relegate their phenomenon to the limbo of ordinary laboratory events. Hence, all available techniques and procedures that were known to exclude the presence of virus were applied to see whether or not the virus inhibitory activity of VIF would disappear. For example, VIF preparations were centrifuged at a certain speed that was known to separate fluid and virus particles. After removing the 'virus sediment' the presumably virus-free supernatant fluid was tested for its virus inhibitory capacity (result: the supernatant fluid still showed a considerable virus inhibitory activity). In addition to the centrifugation experiment they tried to figure out – like Isaacs and Lindenmann had done before – whether treating the virus specimen with a specific viral antiserum would affect the viral inhibitory activity of their VIF preparations (result: it did not affect the virus inhibitory activity). Furthermore Ho tested the effect of VIF preparations against a number of different viruses, among others vaccinia virus (result: in all cases under investigation a considerable reduction in cytopathogenicity could be monitored) and subjected to various physical manipulations (result: for example heating at 56°C resulted in partial loss of inhibitory capacity).

Moreover, some efforts were devoted to studying the mechanism whereby the inhibition was brought about. The first data indicated that VIF neither acted directly on the virus nor interfered with absorption of virus particles on cells, but presumably affected virus multiplication within the cells. These kinds of trials were run primarily by Ho over a period of eight months, thereby forging a primary profile of the elusive factor which he and Enders thought to be present in their experimental system.[44]

Since there were still many questions concerning the nature of the factor which might be responsible for the inhibitory effects, Enders was in favour of maintaining the term 'viral inhibitory fluid' in their first joint publication which was submitted to the *Proceedings of the National Academy of Science* on 7 January, 1959.[45] In this article they claimed to have discovered that fluids from certain human cell cultures, which were infected with a strain of poliovirus, inhibited the cytopathogenic effect and probably multiplication of several viruses. This property was associated with the presence of a factor that could be separated from the virus and was not inactivated by specific viral antiserum. In addition, the following reference was made to interferon: 'In certain respects this inhibitor is comparable to 'interferon', a factor appearing in chick embryo tissues exposed to influenza virus.'[46]

The rather short reference to interferon in the aforementioned publication did not prevent Ho and Enders from having frequent discussions in Enders's room about the growing list of similarities between VIF and interferon, and from speculating once again on the nature of the relationship between VIF and interferon. It became a matter of sorting out differences and similarities. With a steady growth in the number of perceived similarities between VIF and interferon, the novelty of VIF was clearly challenged. The issue began gradually to stalemate research on VIF.

At the end of May 1959, Enders decided that in order to put an end to all guesswork it would be best to ask Isaacs to send them a sample of interferon in order to test its effects in his Boston laboratory. Enders's decision to ask Isaacs to ship over tubes with interferon in order to arrange for a 'real-time and real-space meeting' of VIF and interferon – in other words, for an experiment to compare the workings of both inhibitory factors in his Boston laboratory – was a final attempt at clearing up the rather confusing and uncertain laboratory situation by practical means.

According to Ho he was more than relieved to find that the specimens of interferon, which Isaacs had arranged for them to receive by air in a special paper bag container, showed no effects at all in their 'VIF system'. After all, more than a year of manipulating and producing at the bench had made Ho develop a major commitment to these VIF studies. Ho considered the test pivotal in the sense that it re-established his belief that they were dealing with something different and novel.[47] It is questionable whether Ho thought of the test in these terms, but I would argue that at least to Ho the testing of Isaacs's interferon alongside VIF was a crucial event. Before that VIF's status as a novel laboratory factor had been in jeopardy. After the experimental event, with Isaacs's interferon having shown no activity whatsoever in Ho's hands, a difference between VIF and interferon seemed self-evident to Ho, thereby reinforcing his commitment to VIF.[48]

In retrospect Ho explained that as Isaacs's interferon preparations had not produced any effects in his experimental system he then strongly believed that VIF and interferon were neither biologically nor chemically the same. Of course there were striking similarities, and some sort of relationship was to be expected, but in essence he considered VIF and interferon to be different.[49]

Debating interferon's mode of operation: the specificity effect

We will proceed by concentrating again on the events and occurrences related to 'Isaacs's interferon kitchen'. At about the time Ho and Enders submitted their second article to *Virology*, Henle sent personal word across the Atlantic to Isaacs about the characterization of what seemed to be an interferon-like factor. Apparently, the same kind of experimental system which had previously resisted all attempts to detect interferon-like substances and had fuelled Henle's scepticism regarding interferon and VIF was now said to produce signs indicating that an observed resistance against viruses was most likely due to the presence of an interferon. It would have been interesting to know more about the sequence of events in Henle's laboratory that tipped the balance in favour of interferon. This was not of great concern to Isaacs, however, who seemed to be more interested in the bare fact that his ideas about the existence of a viral interfering substance were gaining some ground among American virologists.[50]

As I already mentioned, Isaacs was informed about Ho and Enders's research on VIF. However, in contrast to the research team in Boston, the association of VIF with interferon encouraged rather than discouraged Isaacs in his research work. In Isaacs's case, the perceived similarities between interferon and VIF were important. These were considered instrumental in putting an end to all doubts with regard to the isolated character of the interferon phenomenon. From the very first moment that Isaacs had heard from Enders, Isaacs believed VIF to be interferon produced by polio virus in human cell cultures. At once the possibility of producing and manipulating an interferon in any laboratory no longer seemed purely speculative. Neither Ho's failure to produce an effect with Isaacs's interferon specimens, nor Isaacs's failure to prepare interferon in human cell cultures, could reverse the situation. Isaacs also knew that it was possible to produce interferon in *in vitro* systems other than the original chick embryo system, as he and Burke had succeeded in producing interferon in cultures of monkey kidney cells.[51] Concurrent events in the laboratory of his friend and colleague David Tyrrell, who worked as an animal virologist at the Common Cold Unit on the downs to the south of the city of Salisbury, only strengthened Isaacs's belief that VIF was an interferon.

46 *Interferon: The science and selling of a miracle drug*

Figure 14 The Common Cold Unit was housed in a former American army field hospital. Courtesy of Dr D. Tyrrell.

The Common Cold Unit was an extension of Andrewes's Division at the NIMR and was regarded as a sort of field station for research on viruses related to the common cold (see Figure 14).[52] Tyrrell would go over to London quite regularly to discuss research matters and visit the NIMR's library. During one of these visits Tyrrell told Isaacs that he had found indications suggesting the presence of an inhibitory factor resembling interferon in fluids collected from monolayer cultures of calf kidney cells which were infected with virus – a so-called 'Sendai' virus strain of the subgroup of Parainfluenza viruses.[53] Tyrrell emphasized the preliminary status of his observations. He had only just started physical and chemical tests to characterize his virus interfering factor and, moreover, his experimental system was different from the chick embryo membrane system Isaacs used to produce interferon. Tyrrell agreed with Isaacs that it would be an interesting option to take a frozen sample of Isaacs's interferon to the Common Cold Unit and test its effects on his Sendai virus/calf kidney cell system. Communicating the interferon-like experimental features to Isaacs thus led to the transfer of interferon-related expertise and research materials from Isaacs's to Tyrrell's laboratory (see Figure 15).

Back in his laboratory in Salisbury, Tyrrell then started an experiment to see if Isaacs's interferon preparation would induce the same effects as his own viral interfering fluid. Surprisingly it did not. The tube cultures of calf kidney cells which were treated with the interferon sample showed an almost normal level of cytopathogenicity – that was regarded as a measure for virus reproduction – after the introduction of Sendai virus. Conversely Tyrrell could measure

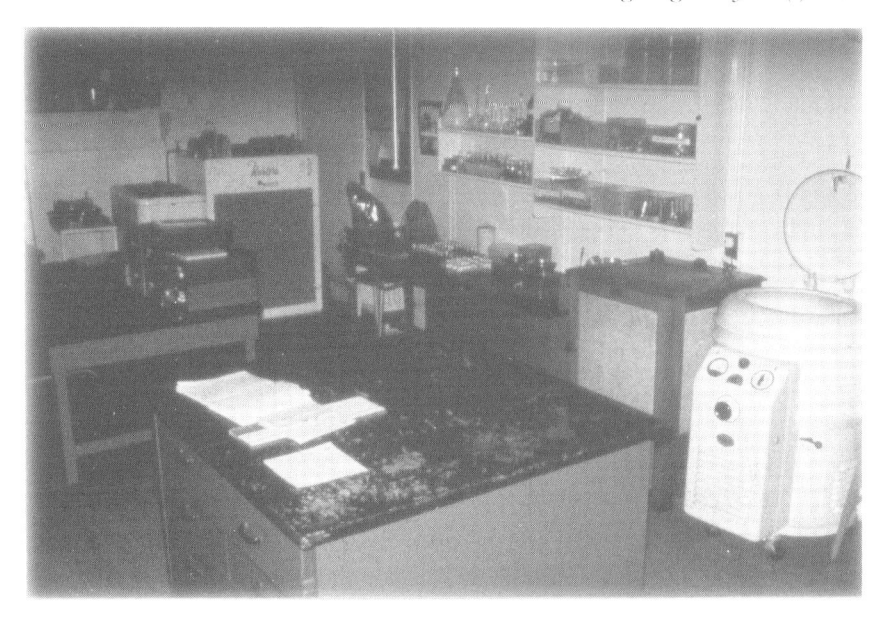

Figure 15 Tyrrell's laboratory was located in one of the central barracks of the Common Cold Unit. Courtesy of Dr D. Tyrrell.

a reduction in cytopathogenicity in cultures which had been treated with his own viral interfering fluid. Tyrrell did not know what to make of the absence of an inhibitory effect in the case of interferon. He immediately initiated another attempt.

However, the second set of data were almost identical. Since Tyrrell had heard from Isaacs that there were frequent problems with the stability of the interferon preparations, he inferred that the interferon sample most likely had lost its virus inhibitory activity over time. In order to prevent this from happening again he considered it necessary to work with fresh interferon preparations in future experiments. The logistics involved in getting a regular supply of fresh interferon preparations from Isaacs's laboratory would be quite demanding. Tyrrell therefore decided first to see if by reproducing Isaacs's interferon system it would be possible to prepare interferon in his own laboratory.

Tyrrell was quite confident that he would succeed in his efforts, as Isaacs's experimental system for the greater part resembled the *in vitro* system Tyrrell had used to study viral interference between influenza viruses as a postdoc at the Rockefeller Institute for Medical Research in New York some years before. Furthermore, on a number of occasions, he had closely watched Isaacs manipulating his interferon system. It took him some time to figure things out, but eventually he was able to prepare a batch of interferon by inoculating tubes containing pieces of chicken embryo membrane and nutrient fluid with inactivated influenza virus.

Remarkably however, his 'fresh' interferon preparation, which clearly showed a virus inhibitory effect in the chicken embryo membrane system, had hardly any antiviral activity when applied to the calf kidney cell system. Likewise the viral inhibitory fluid which had been collected from the calf kidney cell system was much more active in this system than in the chicken embryo membrane system. Yet there did not appear to be a difference in physical and chemical properties between the two preparations. It was a puzzling result.[54]

Tyrrell believed that in both cases he was dealing with more or less the same chemical substance, 'interferon', but there seemed to be a difference in biological activity. According to Tyrrell, the only thing he could think of at that time was that some sort of cellular specificity effect might be involved.[55] In response to virus infection both his calf cell and chick membrane system produced interferon. However, in one way or another, in both cases, the interferon shared specific features with the cell type from which it was formed, such that the chick cell interferon was less active in calf cells and the same was true for calf cell interferon in chick cells.[56]

Sometime early in 1959, Tyrrell talked about his findings with Isaacs and Burke. The first reaction was one of disbelief: 'Well it had never occurred to us at all that interferon might show host cell specificity. Viruses don't, antibodies don't and so there was no reason to believe that interferon would'.[57]

However, they both knew Tyrrell as a careful experimenter, and the data seemed to be quite clear. It was Isaacs who first acted on Tyrrell's idea that interferon might share specific features with the cell type from which it was formed. Tyrrell's experimental results seemed to offer a possible explanation for the inconsistencies Isaacs had experienced in testing the effect of both interferon and inactivated influenza virus on laboratory rabbits more than a year ago. At that time Isaacs had been surprised to see that inactivated influenza virus regularly protected rabbits against infection of the skin with vaccinia virus whereas rather variable results were obtained with interferon. This was thought to be inconsistent with his idea that the protective effect of inactivated influenza virus was mediated through interferon. Now, with Tyrrell's experimental results at hand, the rabbit experiments no longer seemed problematic to Isaacs. The perceived inconsistency could be explained away. The most likely conclusion was that, following injection with inactivated influenza virus into the skin of laboratory rabbits, interferon was locally produced. Upon taking into account a cellular specificity effect it was only to be expected that this rabbit-skin interferon would protect rabbits more effectively against infection with vaccinia virus than was the case with rabbits that had received an injection with chick interferon. Isaacs, Tyrrell and Burke agreed that it would be worthwhile to examine this cellular specificity effect more closely.

With regard to Isaacs's laboratory the specificity issue had both practical and theoretical consequences, and resulted in a rather substantial reshaping of research on interferon. The production of interferon as a 'blind alley' of virus production no longer seemed to make sense, once the specificity effect had been confirmed in Isaacs's laboratory. The host specificity was thought to indicate differences between interferons from cells of different animal species. One and the same virus could induce different interferons depending on the type of host cell, and this simply did not seem to fit the idea of interferon as an abortive product of virus multiplication. Apparently the process of virus multiplication and the mode of production of interferon had to be independent. Isaacs began to associate the production of interferon with a cellular defence mechanism against viruses. According to Burke, it was no longer the virus, but the cell, that was at the centre of their thoughts.[58] Isaacs confirmed this at a symposium on virus diseases in September 1959 in London, where he proposed the following line of thought with regard to the nature of the interferon phenomenon:

> It may be that interferon is concerned in a natural mechanism of cellular resistance to virus infections. For example, the fact that people recover from colds in the absence of any clear evidence of specific immunity [resisting infection by the presence of antibodies in the blood] suggests that local cellular immunity may play a part.[59]

From this point of view interferon was considered to be a product of the cell, which might be involved in the natural defence of cells against viruses.

As the viral point of view gave way to a cellular research perspective, there was a shift in emphasis from experimental virology to a more interdisciplinary, molecular biology type of research. It meant employing whatever tools and techniques the problem at hand seemed to demand to explain interferon's role as part of the cellular defence mechanism in terms of fundamental biochemical and molecular processes. Changes in the experimental system and research perspective not only meant doing different experiments but also reshaping interferon's identity as a research object. The ideas regarding interferon as a deviant product of the cell closely associated with virus multiplication were abandoned in favour of interferon as an essential product of the process of cellular resistance to viral infections. The events in Ho and Enders's and Tyrrell's laboratories played an active part in this transformation process, as did the practical context in which knowledge of interferon was claimed by Isaacs to be applicable to the problem of antiviral therapy. In other words meeting the 'others' resulted in 'working forward' with the 'others'.

As a result interferon emerged as a central concept to a network of laboratory practices within the field of animal virology. These practices were particular and local in nature, each with its own ways of doing things and of presenting the manipulations at the laboratory bench. Through the interaction of their activities, the heterogeneous material and conceptual resources of these particular practices were fine-tuned, leading to new ways of doing and representing things, independent of the immediate local experimental situations. We will see that out of this adaptive behaviour a new subfield of research emerged in which a group of experimentalists learned to manipulate, communicate about and work on a common set of research problems and laboratory phenomena.

50   *Interferon: The science and selling of a miracle drug*

What's in a name?

In the fall of 1959 additional foreign reports reached Isaacs's desk concerning work on substances which seemed to have properties similar to interferon. Among these was a French paper by two Japanese animal virologists, Yasuiti Nagano and Yasuhiko Kojima from the Kitasato Institute in Tokyo. In their article they described the production by vaccinia virus of a virus inhibitory factor, named 'facteur inhibiteur', which was separable from the virus and was able to inhibit vaccinia virus infection in the rabbit skin. Their factor was said to be relatively heat resistant, unlike vaccinia virus not sedimented by centrifugation at 100,000 g for two hours and resistant to ultraviolet radiation. However, no further pronouncements were made as to the nature of their factor.[60] Upon reading the article Isaacs became convinced that the Japanese had been working with interferon. 'I have just seen a paper by two Japanese . . . who have found interferon in vaccinia virus but don't know it'.[61]

Isaacs then wrote a letter to Nagano to communicate his findings. Nagano answered by return post that this 'facteur inhibiteur' was indeed very much like interferon, and included a 1954 article indicating that they had already pointed in their experimental system to the presence of a virus inhibiting factor, years before Isaacs and Lindenmann published their interferon data. Since Isaacs thought Nagano's 1954 article even less clear regarding the nature and effect of this factor, he did not pay much attention to Nagano's remark about the early date of his research. For the moment he let things go at that and instead focused on yet another American report concerning an interferon-like substance.[62]

Isaacs had received information through Andrewes that an American colleague of theirs at Johns Hopkins Medical School, Robert Wagner, who in the early 1950s had worked as a research fellow in Andrewes's department for one year, had an interferon-like substance under investigation in his virus laboratory. Apparently the substance was referred to as 'interferon B' and was said to be distinct from Isaacs's interferon. As with Nagano, Isaacs immediately got in touch with Wagner.[63]

He learned that Wagner had become seriously interested in the interferon phenomenon when he repeatedly managed to induce an inhibitory effect on the growth in chick cell cultures of a highly pathogenic virus, a so-called Eastern Equine Encephalitis (EEE) virus under investigation in his laboratory.[64] However, the experimental data indicated that Wagner's preparation was more stable in relation to heat than the interferon preparation described by Isaacs. As in the case of Enders and Ho, Wagner let the differences outweigh the similarities and called his virus inhibitory factor 'interferon B'. In response Isaacs argued that the differences between interferon and interferon B did not seem to be significant at all.

Upon assessing Isaacs's latest data Wagner saw no further need to make a difference between interferon and interferon B, and decided to withdraw the latter term. According to Wagner this decision was not based on any tests with

Isaacs's interferon in his laboratory at John Hopkins. It should be emphasized that Wagner's 'conversion to interferon' does not necessarily mean that Isaacs and Wagner both identified interferon B as interferon in a similar way. At least they agreed on identifying interferon B with interferon biologically, but it is far from clear if Wagner shared Isaacs's belief that both were chemically identical too.[65]

In the process of discussing the nature of their respective virus inhibiting factors both researchers informally agreed on a provisional set of experimental criteria as to what laboratory material they both would call an interferon. The production of this set of criteria added up to listing the biological, chemical and physical properties the individual research objects had in common, and choosing from this list the most exceptional ones – exceptional in the sense that it would distinguish their factor from other known biological factors. For instance, one of the informal criteria was pH resistance as both factors shared a resistance to acid conditions. 'Fairly early on we had criteria among ourselves as to what we would call an interferon. One of the very important things was pH-resistance. There are not too many proteins with that property'.[66]

In the meantime, word got around within the animal virology community that an increasing number of laboratories were reporting work on interferon-like factors. Even fierce critics like the American virologist Harry Rubin began to pay some attention to these 'interferon stories', although the issue still remained controversial. The published experimental data were still considered circumstantial, as none of the laboratories involved could present more than biological activities and a limited number of physicochemical properties. For instance, the claim that one was dealing with a protein was based on little more than experiments showing that the virus inhibitory activity of preparations was destroyed by trypsin, which was known to inactivate proteins, and precipitated by saturation with ammonium sulphate like most other proteins. Was it really a protein though? One of the things that disturbed people was that hardly any other proteins were known to resist both heat and acid conditions in a similar way. Moreover, assuming that it was a protein, some wondered how you could know that it was a cellular protein and not a viral protein? Of course, different viruses reportedly induced the same kind of inhibitory activity in one particular type of cells, whereas one and the same virus could induce different kinds of interfering principles in cells from different animal species. But what other proof was there to say that one was actually dealing with a primary product of the cell? Repeatedly, researchers like Ho, Enders and Wagner had to respond to such questions from their fellow scientists.[67]

While attending a meeting of the American Cancer Society in New York in November 1959, Andrewes noticed the growing interest in interferon. He was even approached by one of the members of the Program Committee of the prestigious Gustav Stern Symposium on Perspectives in Virology to be held early January 1960 with a last minute request to present a paper on interferon. On his return to Mill Hill, it was agreed that Isaacs would deliver a paper with his latest data and ideas regarding the nature and function of interferon.[68] The latter immediately dropped Lindenmann a note of the news: 'I am going to America early in January to speak at a meeting on Perspectives in Virology and will take the opportunity to meet all the interferonologists there, viz. Schlesinger, Henle, Wagner and the senior interferonologist, Enders.'[69]

In response, Lindenmann, who was aware of the growing number of claims regarding interferon-like substances, congratulated Isaacs on his success, but at the same time expressed his concerns about a possible confusion in the field. He thought it important to try to give at least some direction as to what antiviral factors should be called interferon:

> It is probably premature to undertake this, but on the other hand there is the danger that as more workers pop into the field different names might be proposed for the same thing. It would be a good policy to reach a provisional agreement as to what points are to be considered relevant for inclusion of a new antiviral factor in the interferon group.[70]

Isaacs, for his part, thought that Lindenmann went a bit too far in his suggestions, and told him that they could not stop 'other people from describing whatever they like and calling it whatever they like'.[71]

Isaacs was in favour of a more subtle approach to the problem. He believed that it might be possible to prevent people from using different names for substances similar to interferon in the future by adopting a fairly broad description of interferon. To persuade other researchers to adopt the term 'interferon' for biological factors similar to interferon, Isaacs proposed to use the following 'tailor-made' definition at the 'Perspectives in Virology' meeting in January 1960:

> A protein, slightly smaller than antibody globulin, produced by cells of different animal species following inoculation with inactivated or live virus of many different kinds and capable of inhibiting intracellularly the growth of a variety of viruses in cells of the same animal species, in doses which are not obviously toxic for cells.[72]

Isaacs told Lindenmann that he did not think it easy for them to be dogmatic in saying that a substance with these properties should be called an interferon. 'It is usually much better to let someone go his own course and see what happens in time . . . I have already transformed Wagner and hope to work on Enders in January'.[73] Isaacs's strategy appeared to be successful. At the Perspectives in Virology meeting, Enders, who acted as chairman of Isaacs's session, made a public statement, saying in effect that VIF might be considered an interferon:

> The more recent data that Dr Isaacs and his colleagues have assembled, together with those of others, interest me because we have been working with an analogous factor, which we have called viral inhibitory factor

(VIF). We have noted certain differences between interferon and the principle we have dealt with, but these more recent observations have, I think, explained most if not all these differences, and I now believe that we are all dealing with fundamentally similar, closely related factors.[74]

Enders made his announcement at the end of the discussion following Isaacs's paper presentation on the nature and function of interferon. In adopting the term 'interferon' for his and Ho's viral inhibitory factor, Enders, like Wagner before him, seemed to agree with Isaacs on what kind of biological factors would be qualified to bear the name interferon. Enders's formal announcement had an impact among those present and was instrumental in gradually attuning the research activities of a growing number of laboratories in America, Britain, Japan and on the Continent, to one another.

The blossoming of a new subfield

The process of validating interferon and forming it into a basis for consensus involved, besides the recurrent elements of evaluating claims in terms of the perceived research opportunities and of negotiating consensus, a collective process of engineering mutual ways of communicating, manipulating, interpreting and elaborating on a common assembly of research objects. The lead on interferon in the May 1961 issue of the *Scientific American* symbolized a turnabout with regard to the scientific status of interferon (see Figure 16).[75]

The controversy over the significance of the original knowledge claims regarding interferon had died down. This was consonant with a shift in status of scientific statements regarding interferon. Whereas up to 1960 journal articles contained modalities with reference to interferon such as 'The data presented are too scanty to allow even the crudest guess as to the nature of . . .' or 'It therefore seems reasonable to postulate that . . .' or 'It is necessary to state from the outset that the evidence is purely circumstantial . . .', by 1962 these kind of modalities were mostly dropped.[76] Statements such as 'In 1957 Isaacs and Lindenmann observed the formation of a virus interfering substance, to which they gave the name interferon . . .' or 'The most notable breakthrough in the general area of host resistance to viruses was the discovery in '57 by Isaacs and Lindenmann of a substance called interferon . . .' became commonplace.[77] As such, the provisional state of interferon was transformed into scientific fact. Interferon became widely regarded as a product of the process of cellular/host resistance to viral infection.[78] At least from a biological point of view, the existence of interferon as a specific biological entity was beyond dispute. However, in chemical terms interferon was still more a putative substance, the active fraction of an impure biological preparation, or what often was dubbed a 'protein soup', which showed some activity in a particular bioassay.[79] Without a chemical structure, interferon's chemical 'reality' was still up for debate.

In the process, 'misinterpreton' rumours gave way to talking about the pros and cons of joining what some individuals now jokingly called an 'interferon

54   *Interferon: The science and selling of a miracle drug*

Figure 16 Cover of the *Scientific American* May, 1961. Photograph of plaque assay for measuring the activity of interferon preparations. Reproduced with kind permission from the publishers.

bandwagon'.[80] The number of laboratories and researchers working on research projects associated with the term 'interferon' was indeed on the rise. However, for assessment of Table 1, which shows the number of papers relating to interferon appearing in any of the given categories of scientific journals, it is necessary to take into account the fact that the expansion of interest in interferon studies after 1960 was associated with a rapid growth of biomedical research in general.[81] Seen in this perspective, the growth of scientific interest then attached to interferon as a research topic and the investment of resources in it was less spectacular than the numbers suggest.[82] By the end of 1963, about 30 laboratories and 80 researchers worldwide had become involved in interferon research, but far from all of them pursued interferon research to the exclusion of all other research problems. The researchers who contributed to the study of interferon were mostly animal virologists with a training in medicine and, in a few cases, biochemists with a training in science.

Gradually the outlines of a new subfield of research emerged, which was reflected by the organization of an official symposium on interference and interferon at the 8th International Congress for Microbiology in Montreal in 1962. And subsequently in 1964 the first formal interferon gathering was held at the Home of Scientists in the wire-tapped castle of Smolenice in the woods near the town of Bratislava behind the Iron Curtain (see Figure 17). Out of fear of being tapped by the secret police, all informal talking occurred

Table 1 The number of interferon-related papers published worldwide over a ten-year period (1957–1967)

1957–1967	'57	'58	'59	'60	'61	'62	'63	'64	'65	'66	'67
Nature		1	2	1	3	2	9	8	7	2	7
Science							3	4	1	2	4
Lancet			2	3		2	2	1	2		
N. Eng. J. Med.						1		1	3		1
Virology		1	1		1	8	15	1	6	8	8	5
J. Immunology							2	1	1	2	4	6
PSEBM				1		4	1	6	8	10	20	14
Other publications: Britain	3	3	2	1	3	4	1	5	3	5	6
Other publications: USA, Canada		1	1	2	3	6	16	15	13	17	31
Other publications: Europe, Russia				4	6	3	5	8	15	16	21
Other publications: Asia, Aust.		1		2	1				7		2
Total	3	7	9	14	28	36	54	57	71	73	97
Number of different authors or groups of authors	1	3	5	8	12	21	31	38	45	51	50

in the surrounding woods. Word has it that the flight of the virologist Jan Vilcek to the West was the most important achievement of the four day meeting. Apparently he was smuggled out of the country in the car of two of the participants to the conference. Due to Vilcek's flight to the West, the publication of the proceedings of the meeting in the form of a monograph by the Publishing House of the Czechoslovak Academy of Sciences was banned.[83]

Finally, it is interesting to see that once interferon was gaining some momentum in the international biomedical research community, informal priority disputes concerning the discovery of interferon began to surface. In the corridor it was increasingly rumoured either that the renowned American polio-researcher Hilary Koprowski had worked with interferon long before Isaacs and Lindenmann, or that the Japanese researcher Nagano had first discovered interferon, what he called 'facteur inhibiteur'. According to Burke, finishing off these rumours was one of the primary incentives for Isaacs to put his mind to writing a review on interferon. The introductory section of this review paper, published in *Advances in Virus Research* in 1963, shows that Isaacs wanted to set things straight regarding who deserved priority over the very first interferon claims:

> Interferon derived its name from virus interference, since it was first isolated and characterized during a study of this phenomenon (Isaacs and

Figure 17 From left to the right, Jan Vilcek, Jacqueline De Maeyer Guignard and Edward De Maeyer in the corridors of the Smolenice meeting. Courtesy of Dr J. Vilcek.

Lindenmann, 1957). However, similar substances were previously observed, although they were not characterized.[84]

By placing the work of others like Koprowski, and Nagano in a particular order of relations with one another he succeeded in making a convincing argument for justifying his and Lindenmann's priority position. In doing so he superimposed new meaning on past scientific work, thereby providing directions in research activities both retrospectively and prospectively.[85]

The priority dispute provides a clear illustration of the fact that the knowledge claims regarding interferon had been weaned from their matrix of origin and gained acceptance among fellow scientists. Isaacs's laboratory and interferon were no longer inseparable. Though in essence Isaacs's laboratory had not changed that much, Isaacs's research world had become rather different from the one he had inhabited when submitting his first interferon papers. The scientific notion of interferon, which matured in conjunction with the work of Isaacs and 'the others' had undergone a transformation. From a *viral interfering factor* which was provisionally considered a *deviant viral entity*, interferon came to refer to *active antiviral substances* in particular culture fluids – as a generic term – which complied with a specific set of experimental criteria and which were considered *essential products of the cell* that played a part in resistance to viral infection.[86] However interesting this may be from a cognitive point of view, we should keep in mind that the scientists under survey did not regard the situation in terms of a shift of world view. They manipulated, produced and interpreted a body of laboratory phenomena in what seemed to them the most appropriate way – as a means to the end of increasing the range of manipulative options at the bench and of promoting the blossoming of a new, cutting-edge field. What in 1957 was still a laboratory phenomenon of local significance that had yet to pass its referees had not only become the organizing principle for a promising new line of investigation but as we will see in the next chapter had also become a serious possibility for the development of an antiviral drug.

3 Interferon on trial[1]

For a good understanding of the drug testing culture in the 1960s in Britain it is important to realize that there was no national system for controls regulating the testing of potential remedies. Neither the Dangerous Drug Act 1951 nor the Therapeutic Substances Act 1956 provided a regulatory context for the testing of medicinal substances in human subjects. And there were no ethical committees to sanction (or otherwise) experiments on human volunteers or formal rules on how to carry out a clinical trial.[2] The control over testing of new pharmaceutical substances for toxicity was left to the good sense of the researchers and pharmaceutical companies involved.

Except for the routine testing of immediate toxic effects on at least two species of laboratory animals for several months, no standard procedures existed to evaluate the clinical safety and efficacy of new therapeutic substances. Usually after completion of the acute toxicity studies in laboratory animals British academic and commercial laboratories began with studies on healthy volunteers.[3] This earliest trial of a new drug on humans involved administering a single dose or a small number of repeated doses of the pharmacological substance to a small group of about twenty volunteers. In most cases the volunteers participating in these early trials were employees of the pharmaceutical company or institute involved in the development of a specific therapeutic substance. Volunteering came with the job.[4]

The British Ministries of Health argued that there was no need for stricter regulations as long as the manufacturers continued to discharge their responsibilities effectively within the limits of contemporary knowledge of methods of testing. Of course the tragedy of thalidomide, of gross malformations appearing in children of mothers who had taken this sedative during pregnancy, brought home to the public and hence politicians the inadequacy of the control of medicines in the UK. However, a new drug regulatory legislation which would provide a framework for the testing for new drugs had to wait until 1971 when the Medicines Act of 1968 came into force.[5]

In this chapter I will show how in the absence of a standard system for the testing of potential therapeutic drugs in the 1960s in Britain, interferon researchers, government administrators and entrepreneurs from several drug companies participated in the evaluation of interferon as an antiviral agent.

I shall describe and analyse how interferon resisted attempts at being integrated into what counted as front-line drug research and development programmes and state-of-the-art testing practices. And how the various parties reacted through a change of expectations and research agendas to these resistances. To what extent did they succeed or fail to cooperate across the domains which those involved routinely demarcated as science, policy and industry? As I will show their dilemma was to prevent choices and futures becoming either merely scientific, political or commercial beyond the grasp and thus control of the other. The intention here is to point out how in the process of testing the effectiveness and safety of interferon not only research agendas and commitments underwent changes but also the modes of perceiving interferon changed and differentiated.

A partnership originating in national interests

Following the extensive press coverage of interferon in which it was presented as a possible 'magic bullet' against viral disease in Britain in May 1958, the MRC received a letter from the pharmaceutical company Johnson & Johnson. The American drug firm had noted the publicity surrounding interferon in the UK and was most interested in receiving additional information on this possible new lead to antiviral therapy. In particular they wanted to know whether any results from the administration of interferon to humans were available.[6] The American request was passed on to Isaacs for reply. In response Isaacs provided Johnson & Johnson with references of the latest publications on interferon while emphasizing that the NIMR and the MRC had not yet 'contemplated any experiments in man and even if present experiments go favourably it is unlikely that we would do so for some time'.[7] Isaacs responded in a similar way when contacted informally by Maurice Hilleman, the head of Virus and Cell Biology Research of the American drug firm Merck Sharpe & Dohme ('Merck').[8]

At about the same time, the Secretary of the MRC, Sir Harold Himsworth, was informally approached by Glaxo's top executive, Sir Henry Tizard, about the patent position on interferon.[9] Together with the American requests for research information, this event provided an important incentive for the Secretary of the MRC to consult the director of the NIMR, Sir Charles Harington, about further action on interferon. Both parties realized that in order to develop interferon as a drug some kind of collaboration with the British drug industry would be needed. Harington indicated that he would welcome cooperation with one of the UK's leaders in biologicals if it took the form of supplying interferon in large amounts, but he would wait for Glaxo to take the initiative. However, Himsworth was in favour of a different approach.

The MRC's reluctance and restraint in coordinating things with regard to the industrial development of penicillin during the Second World War and the subsequent 'British failure' was still fresh in everybody's memory.[10] It had an important influence on the MRC's attitude toward interferon.[11] If interferon proved to be the new 'antiviral penicillin', the opportunity should not be lost to develop it in Britain.[12] Taking into consideration the American interest and a possible drift of knowledge on interferon across the Atlantic where it might be developed into a commercially successful product, Himsworth was in favour of taking determined policy steps.[13] He realized that the pharmaceutical industry alone had the expertise and capacity to produce large amounts of biologicals for use in humans as well as the financial resources needed to develop interferon as a therapeutic drug. The MRC should therefore nurture the links with the pharmaceutical companies. Moreover, by actively seeking collaboration with British commercial firms only, they could best serve the national interest.[14]

Under the specific circumstances of British post-war science policy with the 'penicillin syndrome' as one of its guiding principles, the National Research Development Corporation (NRDC) was entrusted with the task of making proposals for a form of cooperation. As a result of the British failure to exploit penicillin, the NRDC was set up under the British Industries Act of 1948 to secure the development of inventions resulting from public research and to obtain patents for these inventions. One of the first research projects in the field of biomedicine the NRDC took an interest in was the development of the antibiotic cephalosporin C. In accordance with its terms of reference the NRDC had been steadily patenting research results ever since the first results on cephalosporin had been published in 1951 by an Oxford research group. By the time its chemical structure had been established in 1955, the NRDC had initiated efforts to call in the help of the British industry. In exchange for certain rights over use of the developed product, several British pharmaceutical companies agreed to look at the problem. However it never really came to a formal collaboration agreement and the NRDC experienced a rapid loss in interest when progress turned out to be slower than expected. The cephalosporin experience influenced NRDC's preference for a formal multi-party collaboration arrangement in the case of interferon.[15]

In consultation with the NRDC the MRC decided to formally invite Glaxo as well as other major British pharmaceutical firms to cooperate, by invitation to the Association of British Pharmaceutical Manufacturers (ABPM).[16] In general, caution prevailed among British drug company executives. Without Isaacs's claim about interferon's potential as a therapeutic drug being substantiated by more extensive laboratory studies they decided to wait and see. A reserved attitude toward interferon prevailed even within companies like Glaxo and Wellcome with a history of producing drugs which originated in the preparation and purification of biological material, so-called 'biologicals' (e.g. vaccines and hormones). Few company scientists seemed to know what to make of the 'iffy' status of interferon as an undefined biological substance – what essentially was not more than a mere protein soup with a claimed biological activity. The perception of biologicals – whose potency and identity could only be determined by cumbersome and inaccurate bioassays – as profitable but troublesome and high-risk commodities played an important role in preventing the drug companies from throwing themselves into interferon research. In being primarily oriented toward medical chemistry and pharmacology the drug companies preferred to deal with the relatively straightforward chemicals – synthetic compounds that are made of source materials whose potency and identity can be adequately tested by chemical or physical means. As far as the pharmaceutical industry was concerned most biologicals had a problematic developmental track record. They were tricky to produce – requiring elaborate and expensive production, safety testing and standardization procedures – and difficult to quantify and store.[17]

At the same time British drug company executives were quite familiar with the industrial potential of substances of biological origin: the unparalleled clinical success record of the biosynthetic penicillin-like antibiotics was all too visible. Their spectacular clinical activity in eradicating bacterial infections had created a firm belief in the 'Ehrlichean' chemotherapeutic research approach. Paul Ehrlich's ideal of generating chemical agents which, like 'magic bullets' seek out and destroy the enemy and injure nothing else, seemed to have materialized.[18] Except for viruses most disease inducing microorganisms were known to be susceptible to antibiotics. The common view was that the chemotherapeutic research approach would eventually also provide them with some kind of effective and clinically useful antiviral therapeutic drug. One might say that in a clinical sense there was already an established frame of thought for chemotherapeutic control of viral disease. This despite the fact that only costly research failures had been reported as a means of developing a serious lead on viral chemotherapy. Most promising leads in the test-tube had proven to be too toxic *in vivo*. As a result increasing effort was being given to the control of viral disease by time-proven vaccination ('immunoprophylactic') principles.[19]

Gradually, in reaction to the surge in public enthusiasm in Britain, and in relation to the perceived clinical and economic promise of interferon as an 'antiviral penicillin' the caution of British drug firms subsided. The promise of interferon as a potential specific (that is truly curative) and innocuous drug which seemed to fit into the conventional concept of chemotherapy as a sort of 'magic bullet' against virus infections was rather tempting. Moreover, from industry's point of view, risk-sharing with the MRC who offered their unique expertise in interferon research and the eventual rights in the developed product made work on interferon look more feasible. Joining in with the MRC-led interferon project, without having to start from scratch a private high-risk developmental programme, reduced the liability and as such made it easier to accept the many imponderables.[20]

Nearly all major British drug companies decided to accept the invitation and sent representatives to the meeting with the MRC and NIMR. On 22 April 1959 in one of the NIMR's conference rooms at Mill Hill, London, representatives of the companies were informed about the current state of affairs concerning interferon. Sir Charles Harington opened the meeting by pointing out American interest in interferon, but emphasized MRC's preference to first

62   *Interferon: The science and selling of a miracle drug*

seek collaboration with British firms. Harington told the pharmaceutical companies that the NIMR needed their help in developing this possibly interesting biological. The NIMR simply lacked the expertise to tackle the production problem. In exchange he offered them eventual rights in the developed product. Subsequently, both technical and patent matters were discussed. A patent expert from the National Research Development Corporation (NRDC) helped to clarify the patent position on interferon whereas Harington would call upon Alick Isaacs, to answer technical questions.[21] Isaacs was highly regarded by everyone as the discoverer of interferon. Except for Harington nobody was aware that Isaacs had only recently resumed all his duties at the World Influenza Centre (NIMR) after recovering from a severe bout of depression and that at times he could behave in a rather manic way.[22]

Most experimental results were argued to point in favour of the possible therapeutic value of interferon. Enthusiastically Isaacs told the audience that in his laboratory interferon had been shown to inhibit the growth of a wide range of viruses *in vitro*. He emphasized the fact that apart from these *in vitro* studies, early animal experiments had also been carried out which showed that interferon protected animals against virus infection. Moreover, so far all the indications were that interferon was innocuous without any obvious ill-effects in animals. Of course, Isaacs argued, his audience should take into consideration that systematic and large scale investigations had yet to start and that there were still a couple of experimental problems to solve. A major difficulty was the production of large enough amounts of active material to start systematic investigations of both interferon's effectiveness against viral infection and its toxicity *in vivo*. However, Isaacs expressed his confidence that with the help and expertise of some of the companies present they would be able to find satisfactory solutions to these practical problems.[23]

Just as Isaacs drew legitimation from the use of his knowledge claims in MRC's deliberations, so MRC and drug company officials were better able to legitimate their decision to participate in efforts to develop interferon as a drug by attaching to it the authority of Isaacs's expertise. Thus, the MRC, the drug companies and Isaacs had good reasons to get close to each other, but not too close. Their dilemma was to prevent choices and futures becoming either merely 'scientific', 'political' or 'commercial' beyond the grasp and thus control of the other. In order to fit their own programmatic goals and maintain the integrity of mutual interests there was a continual need to maintain accountability to one another with each party working hard to reinterpret and translate the concerns of the others.

Subsequently, with the help of the NRDC's patent agent, Harington succeeded in playing down as a mere technicality the fact that the information available was considered insufficient by the American patent examiner. Mainly because of a lack of evidence on the drug's utility, the American examiner had decided that the patent application for interferon as an antiviral agent had to be rejected until further research results could be made available.[24] At the end of the meeting it was agreed that each firm represented at the meeting would let

Harington know as soon as possible whether or not they wished to collaborate on the development of interferon.

In the weeks following the meeting the two British leaders in biologicals Burroughs-Wellcome (Wellcome) and Glaxo Laboratories (Glaxo), and Imperial Chemical Industries-Pharmaceuticals (ICI) informed the MRC of their willingness to cooperate.[25] Significant was the fact that Sir Harry Jephcott, the chairman and managing director of Glaxo, announced that Wellcome and Glaxo were prepared to act as one, hoping thereby to exclude ICI (who as yet had not been active in the field of biologicals) from collaboration with the MRC.[26]

The following two factors played a most obvious role in this respect. First of all Wellcome and Glaxo regarded biologicals as their core business and tried to defend their position as British leaders in biologicals against a newcomer like ICI (ICI's pharmaceutical group was an offshoot of its dyestuff division and as such had built up a strong position in chemicals). Moreover, they both had had bad experiences with ICI as an unreliable collaborator on penicillin. During the Second World War a consortium of British pharmaceutical companies (including Wellcome, Glaxo and ICI), incorporated as the Therapeutic Research Corporation (TRC), had collaborated on a penicillin development programme.[27] However, to the unpleasant surprise of the others, soon after the war had ended ICI had prematurely pulled out of the penicillin programme.[28]

The matter was eventually settled at an informal meeting when the MRC made it clear to Jephcott that for public policy reasons ICI could not be arbitrarily excluded if the company wished to participate.[29] Anxious not to miss the boat on what might turn out to be a promising innovative lead toward antiviral therapy, the drug companies were eventually prepared to put up with the multi-party collaboration arrangement. As will turn out the MRC's intervention could not prevent the issue from being an early burden on the success of the collaboration.

Further meetings were organized a couple of months later with all parties (NIMR, MRC, NRDC, Glaxo, ICI and Wellcome) to discuss in more detail the proposed collaboration and settle the outlines of an agreement for a collaborative programme of further work on interferon.[30] At the joint meeting which was held on 3 July 1959, the NRDC proposed to set up an Executive Body to hold property, grant licences and administer the arrangement between all the parties involved.[31] They suggested that NRDC's Patent Holdings Company, which had been founded in 1953 with the aim of carrying out future collaborations between British government institutions and private industry, might be used for this purpose. In addition they proposed to set up a Scientific Committee on Interferon, which would come under the Executive Body and provide a platform for the exchange of technical information and know-how.

Following further talks between the MRC, the NRDC and the three drug firms, all parties in principle agreed to these proposals.[32] The MRC patents on interferon which were currently being filed in the USA, Canada and West Germany, as well as any further Council patents regarding interferon, were made available to the NRDC's Patent Holdings Company. The Executive Body would be run by senior managers of all the parties involved and the Scientific Committee was to consist of research workers of the industrial firms and the MRC, under the chairmanship of Isaacs. This committee was not only meant to serve as a clearing house for information but also as a kind of scientific steering committee to allocate research work to the various parties and to advise the Executive Body on the scientific aspects of further work on interferon.[33]

The parties concerned only agreed to the main points of an agreement for collaboration. Further discussions and negotiations were needed to determine the exact terms of a formal agreement between the NRDC, MRC and the three pharmaceutical firms relating to interferon. In particular the differences in opinion on the issue of publication of research results would require quite a bit of talking through before they could be settled. Harington and Isaacs maintained that the results of all research work should be made available for scientific publication without limitations in the normal way in scientific journals. MRC workers were used to freedom of publication as means to the end of advancing scientific reputations and to 'follow their nose' in exploring research problems. Only in order to ensure adequate patent protection of any discoveries would they allow some delay to occur in publication. The representatives of the drug companies, however, emphasized the need for stringent control over publication. If necessary it should be possible to reject publication on grounds that harm might be done to current and future patent positions.

Ultimately all parties would agree to the NRDC approving all papers before they were submitted: every first draft of a paper would be submitted to the NRDC, who had agreed to give their approval within 14 days.[34] However, this could not prevent the parties from having occasional skirmishes over the exchange of research information. The majority regarded the issue of publication as the one point on which the whole programme of collaboration could easily break down. As I will show much effort had to be invested in preventing this from happening.[35]

Establishing a collaborative programme for research on interferon

While negotiations on the precise terms of a Collaboration Agreement were dragging on, Isaacs was informally given the go-ahead to start discussing a programme of research as chairman of the Scientific Committee on Interferon.[36] In showing their unconditional belief in the authority of Isaacs's scientific expertise the MRC and the drug companies had virtually given Isaacs carte blanche to pursue his personal research agenda regarding interferon. Isaacs emphasized that ultimately all parties hoped to develop research on interferon to a point where it could be used therapeutically in humans and animals.[37]

This would require both basic research and research on scaling-up from laboratory to manufacturing procedures. At the present stage of work it seemed that more scaling-up than basic work was needed. Hence, Isaacs was very much in favour of a provisional programme for research that would focus on the development of methods for the large-scale production, purification and storage of interferon. He also emphasized the need to examine thoroughly the question of whether or not interferon was a species-specific biological substance.

Despite the fact that this species-specific effect did not seem to be absolute, it could have far-reaching consequences for the testing trajectory. If it were to be substantiated by further experiments, tests on the virus-protective effect of interferon in animals, and later in humans subjects, would require samples of interferon that had been produced in cells of the same or a closely related species. The species-specificity issue most notably bothered the drug company scientists. They immediately realized that this would complicate working in line with the conventional industrial new drug development trajectories with preliminary toxicity studies on at least two species of animals.[38] It meant that testing interferon in mice would require the production of mouse interferon, testing it in rabbits would require the production of rabbit interferon and so on. In addition, comparing test results in different animal species would become far more difficult.

Until the spring of 1960, research on interferon was carried out, more or less, along the lines of this provisional programme, which closely resembled the initial stage in the industrial development of a new drug involving a division of work and interplay between various workers and disciplines (biologists, biochemists and physicians).[39] The MRC's researchers mainly worked on the development of a purification method, the problem of the species specificity, and the mechanism of action. Meanwhile, researchers from the collaborating firms focused on the development of methods for the large-scale production of interferon.[40]

As work at the bench and policy-making co-evolved, research agendas underwent changes. Initially the plans for research closely resembled the initial stage of conventional industrial new drug development trajectories with an emphasis on animal testing and the development of production techniques. However, from the beginning the difference between the restraint of the company scientists and the drive of the academic scientist Isaacs to bridge the worlds of the laboratory and the clinic comes to the fore. In succeeding to define interferon in terms of a natural non toxic-agent, which was different from most other existing pharmaceutical substances, Isaacs created latitude to follow an alternative testing trajectory. He received unwitting support in his rush to human testing from the American patent examiner, who rejected interferon's patent application on utility grounds.

After eight months of collaborative research, the Scientific Committee suddenly revised the provisional research programme quite radically.[41] During the April 1960 meeting research workers from the MRC reported the following two research results. First, when testing the specificity of interferon prepared in a variety of animal tissue cultures, monkey interferon had shown activity against a range of human and simian cells. In particular, the observation that monkey interferon appeared to be active against a number of respiratory viruses in human tissue cultures aroused interest. It was agreed that, although extrapolation from individual tissues to the complete organism was hazardous, monkey interferon might be fit for future use in humans. Moreover, Wellcome's and Glaxo's experience of many years with the production of poliomyelitis and measles vaccine in large-scale cultures of monkey kidney cells was argued to be helpful in overcoming problems with the production and safety of monkey interferon. Second, the purification work on interferon had reportedly resulted in a pure product.

The prospect of producing sufficient quantities of pure monkey interferon had a profound impact on the planning of further work on interferon by the Scientific Committee. Immediately, four suggestions were made for future experiments with purified interferon in monkeys and humans:[42]

1 Since the MRC was in the process of testing a measles vaccine, facilities to test interferon's effect against measles would be widely available. It therefore seemed a feasible option to test its protective effect against measles infection in volunteers. However, objections were immediately raised. As measles was a systemic infection a measles trial would require considerable amounts of interferon and extensive safety tests. Furthermore, interferon's effect on measles had neither been tested *in vitro* nor *in vivo*.
2 Interferon had shown a clear-cut inhibitory effect against vaccinia virus ('smallpox') *in vivo* and *in vitro* and it was proposed to test the protective effect of interferon against local infection with attenuated smallpox in volunteers. There would be enough recruits available who had not yet been vaccinated against smallpox and were willing to volunteer in such a trial. Moreover, as a local infection it would need much less interferon than a systemic infection and require fewer safety tests.
3 The idea to test the protective effect of interferon on respiratory infections in volunteers originated from a combination of experimental results and logistics. Interferon had shown a protective effect against respiratory viruses in the laboratory and the MRC had an official site for common cold trials in the form of the Common Cold Unit in Salisbury. Moreover the common cold was a local infection which would require only small amounts of interferon and relatively limited safety testing.
4 Quite recently Andrewes's Division had succeeded in culturing trachoma virus, which caused local but potential debilitating infections in the eye and the MRC seemed interested in finding a therapy against this virus disease.[43] Testing the protective effect of interferon on trachoma virus infection in monkeys was considered an excellent experiment to do first as a final check before proceeding to experiments in humans. As a local infection it would require only small amounts of interferon.

Isaacs, in particular, seems to have been rather keen on having an early demonstration of interferon's effect in humans. He felt that it would help to keep people interested and maintain the necessary pressure, because the negotiations on the Collaboration Agreement were not completed yet.[44] The company scientists were less enthusiastic about deviating from the conventional drug testing trajectory. However, they too were aware of the fact that the American patent examiner had rejected interferon's patent application on utility grounds. If interferon could be shown to have an effect on humans, even in an experimental infection, this decision might be reversed and hence safeguard their companies' future market position. Moreover, Isaacs convincingly argued that interferon was different from most other biologically active compounds in showing unusually few toxic side-effects in human cell cultures. According to Isaacs any substance being tried out in humans always presented hazards, but in the case of interferon it was a great comfort to realize that it was not so much a matter of administering a foreign substance as supplementing a natural mechanism of resistance to virus infection. For this reason there was little chance of interferon producing any severe adverse effects when administered to humans. Isaacs met the company scientists part of the way by indicating that for safety reasons they would carry out the trachoma experiment in monkeys first, followed by the vaccinia and common cold trial. A measles virus trial was just too demanding to carry out at the present time. All agreed that in order to be able to carry out the trial programme an early start had to be made with work on the large-scale production, purification and storage of monkey interferon.[45]

Zooming in on the question whether or not interferon could induce a clinical effect in humans did not result in a wholesale shift in research work but in a reshuffling of research priorities. There was no such thing as laboratory oriented research giving way to clinically oriented research – as either or alternatives. As can be expected from 'exploring uncharted territory' the scientists under survey in this chapter worked in a 'zigzag' fashion; as soon as any progress was perceived toward tackling one problem, they would simultaneously begin work on the next problem regardless of whether this involved a transition from *in vitro* to humans or from *in vivo* to *in vitro* and back to humans. Recall that when members of the Scientific Committee reported the observation that monkey interferon showed an activity against viruses in human cells in the test-tube, it was agreed that extrapolation from individual cells to the complete organism was problematic. Yet it was decided to switch attention to trials in monkeys and humans without further pursuing the legitimate question to what extent extrapolation from one organism to the other was justified. However this did not mean that in the preparations for and performance of the clinical trials there was no guidance of any sort. Whenever possible those involved tried to create landmarks by linking up their trial efforts with already existing testing practices.

In the absence of a standard system for testing new therapeutic substances in Britain, the members of the Scientific Committee tried to fall back on existing testing practices for biologicals and other therapeutic drugs within the companies and the MRC.[46] Due to the non-compliant nature of the subject matter or resistance arising from the continuous interplay between the various parties, adjustments had to be made over and again. Deciding on how to go about problematic and uncertain situations and cope with the 'mangle' of practice involved ad hoc judgements and assumptions. This is not to say that they necessarily proceeded arbitrarily, but that given the specific problem context they collectively weighed the perceived pros and cons of proposed solutions derived from the material, social and cognitive resources at hand.[47]

Preparations for an early trial in volunteers

In preparing for clinical trials at its 1960 meetings, the Scientific Committee on Interferon started with general discussions about the nature and design of testing procedures. Eventually, on the basis of these discussions, preliminary safety requirements were drafted for the production of interferon to be used in clinical trials. This included an early virus identification test as was used for poliomyelitis vaccine, subsequent treatment with both acid and formalin, and a series of *in vitro* and *in vivo* control tests. Moreover, an early trial in volunteers would be organized first before any large-scale clinical trial was projected. In addition, Isaacs proposed to cancel the trachoma experiment and concentrate on the vaccinia and common cold experiments. The latest information on trachoma was that it responded, unlike any viral disease, to some antibiotics and this casted doubt on the virus aetiology of this disease. The agent responsible for trachoma might as well be of bacterial origin and hence insensitive to the actions of interferon. Most likely, a trachoma trial in monkeys would only complicate research matters instead of contributing to the understanding of interferon's actions *in vivo*.[48] All agreed with Isaacs that taking into account the innocuous nature of interferon and the proposed extensive safety testing procedure it was justified to start with volunteer studies right away. Since a vaccinia trial seemed to require less preparatory time than a common cold trial, the former was chosen as a first experiment to test the antiviral effect of interferon on virus infections in human subjects. All agreed that the first volunteers should be recruited from the staffs of the NIMR, Glaxo, Wellcome and ICI.

Apart from the nature and design of the human experiments, and the recurrent safety debate, the need for standardization was discussed at length in preparing for trials with interferon. From the very beginning the variability of assay results had been on the Scientific Committee's research agenda. Researchers reported considerable variations in the potency of interferon preparations from day to day, week to week, from researcher to researcher and laboratory to laboratory. Each time the matter came up for discussion, pleas could be heard for the standardization of assay methods and the creation of standard interferon preparations for common use. However, apart from individual efforts to set some interferon preparations aside to be used as internal standards in one's own laboratory, little was done to establish common interferon standards until the issue began to hamper the preparations for human trials.

Without sharing a standard it appeared already difficult to agree on the activity of a particular preparation, let alone justifying the decision to discard a costly batch of clinical trial interferon because of its poor biological activity. Moreover, it was hardly possible to produce sound answers to important questions – like 'Does the interferon remain stable over time?', and, 'What dose of interferon is required to produce a detectable antiviral effect in humans?' – in the absence of a tool to correlate and compare the potency of interferon preparations.[49]

All these considerations underlined the immediate need for establishing a reference standard preparation for monkey interferon, to which an agreed potency would be assigned. This time things went beyond the discussion stage. The Scientific Committee decided to start efforts to establish a provisional standard for monkey interferon against which all monkey interferon samples could be compared. As the NIMR's Division of Biological Standards was responsible for establishing and supplying all British biological standards it was thought most appropriate to leave the task of keeping and distributing a provisional standard to this Division.[50]

The Executive Body was far less enthusiastic about the efforts to establish national interferon standards. The free circulation of standards, however practical and profitable in research terms, was perceived as troublesome in management circles. Senior company executives had just signed the final Agreement for Collaboration on Interferon and the plans with regard to establishing a provisional British standard was giving rise to concern among the Board of the Patent Holdings Company.[51] It was of particular concern to them that the standard preparation would most likely be made available to workers outside the collaboration. They strongly believed that this would inevitably lead to the disclosure of valuable research information and hence to the commercial disadvantage of the collaboration.

The concerns over standardization coincided with rumours within the Executive Body that Isaacs had conveyed secret information to an outsider. Reportedly he had told the American virologist Enders about his latest research results concerning the stimulation ('induction') of interferon production in cell cultures by the administration of so-called 'non-viral (foreign) nucleic acid fractions' – preparations of nucleic acid (either DNA or RNA) derived from cells not infected with viruses, such as RNA from yeast cells. Moreover, reference to this highly confidential subject matter also appeared to have been made in an article that was published by Isaacs in the *Scientific American*.[52] It was generally felt that Isaacs's unauthorized disclosures were putting efforts in jeopardy to establish a strong patent position for inventions that might emerge from work on the use of these foreign nucleic acids as interferon inducers.[53]

The affair gave rise to substantial disquiet within MRC quarters. Himsworth, who wanted at all costs to make a success out of the first formal post-war collaboration between the Government and the British pharmaceutical industry, personally demanded that Harington should provide an explanation.[54] Harington immediately consulted Isaacs, who unaware of having caused resentment in MRC circles, was enjoying his recent appointment as head of the Division of Bacteriology and Virus Research. Isaacs told Harington that he could not remember leaking significant information to Enders or the *Scientific American*.[55] Isaacs was equally surprised to hear about the fuss that was made over the standardization issue. As far as he was concerned an experimental standard had been prepared for research purposes only, and it had so far only been distributed to members of the Scientific Committee.[56]

The whole affair clearly shows the enormous energy that continued to go into bridging the differences in cultures. On the one hand, you had the government researchers like Isaacs who continued to regard the free exchange of information and laboratory materials as a precondition for fruitful research. On the other hand, virtually opposed to this position, the drug company people stuck to their perception of the same working practice as a threat to the commercial development of therapeutic drugs. But at least for the time being they were able to craft compromises.

Handling production and safety problems

Meanwhile, problems were forming relating to the preparations for tests in humans. The fact that the production of a large batch of monkey interferon for clinical trials took much longer and resulted in less material than was anticipated began to put the collaboration within the Scientific Committee under strain. Isaacs, in particular, was disappointed by the slow rate of progress. He was surprised to see that the scaling-up of production and purification procedures took so long. The latest in a row of setbacks was the news from Glaxo Laboratories about the loss of a large batch of satisfactory monkey interferon due to one or another filtration error. Isaacs wondered whether something could be done about this rather frustrating situation and what steps were needed to speed up research work. Far too often there had been duplication of research work and, in his view, the contribution of the scientific staff of the collaborating firms could have been more substantial. The representatives of the firms agreed that progress was slow, but this was only to be expected at this stage and certainly not due to a lack of commitment. Confronted with an unexpectedly stiff opposition Isaacs gradually toned down his criticism and showed his good will by promising to pay special attention as chairman to a more efficient distribution of research work.[57]

Almost all of Isaacs's spare time went into keeping the wider public interested in the interferon cause. Both in lead articles in the *Scientific American* and the *New Scientist*, and on the BBC television programme 'Achievement 1961' Isaacs made reference to imminent tests of interferon in humans which would show whether interferon fulfilled its promise as medicine's first really effective cure for virus infections.[58] He also became one of the first academic investigators to deliver a lecture at the annual British Pharmaceutical Conference (see Figure 18).

Addressing the audience of mainly pharmacists and representatives of the British Pharmaceutical Industry he called interferon 'a round unvarnish'd tale', an exciting example of how investigation in an academic field of research can lead to the practical prospect of treating virus infections in humans. Apart from informing his audience on the general state of the art of interferon research Isaacs emphasized the fact that success in reaching the stage of clinical use of interferon in humans largely depended on the ability of the MRC and three pharmaceutical firms to develop a productive research partnership.[59] In publicly linking the success of the research venture to the success of the Collaboration Isaacs cleverly put all parties under an obligation. At the same time, by presenting the Collaboration as a crucial test-case that might be extended in the future to many other fields of research, Isaacs limited his own freedom of speech and action: neither the others nor he was supposed to prejudice its possible success.

In November 1961, two months after the disruptive notice that a large batch of monkey interferon had been lost due to a technical failure, news came from the Wellcome Laboratories that the first three litres of clinical trial interferon and control material were ready and had proved satisfactory in both activity and sterility tests (see Figure 19). Subsequently the clinical trial interferon and control material had been subjected to toxicity tests in mice and guinea pigs. The toxicity test in mice went satisfactorily. However, this was not the case when the guinea pigs were used. One guinea pig out of two died three to four days after subcutaneous inoculation. Confronted with the high mortality rate in guinea pigs the Wellcome scientists immediately had decided in favour of additional tests in dogs and monkeys. The same interferon and control material had then been given to a couple of two-year-old dogs subcutaneously and to two rhesus monkeys intradermally, but in both cases no sign of toxicity was monitored. It was altogether a puzzling affair. What was the meaning of the high guinea pig mortality after interferon inoculation?[60]

Consequently, a debate arose within the Scientific Committee over the safety of interferon. According to the Wellcome scientists guinea pigs were described in the literature as being extremely sensitive to penicillin. Since penicillin was used to prevent bacterial infection in the large-scale culture of monkey kidney cells in which interferon was produced there might be enough penicillin in the interferon material to be fatal to guinea pigs. After much discussion the majority agreed that although further guinea pig tests were required to confirm their singular sensitivity to penicillin no harm would be done by proceeding at the same time with toxicity tests in human volunteers. The members of the Scientific Committee would perform the toxicity tests on themselves and volunteering colleagues. As a precautionary safety measure penicillin-sensitive and allergic individuals would be excluded. The proposed course of action showed the strong belief in interferon as a non-toxic natural antiviral agent in humans.

Figure 18 A round unvarnish'd tale. Courtesy of Dr S. Isaacs-Elmhirst.

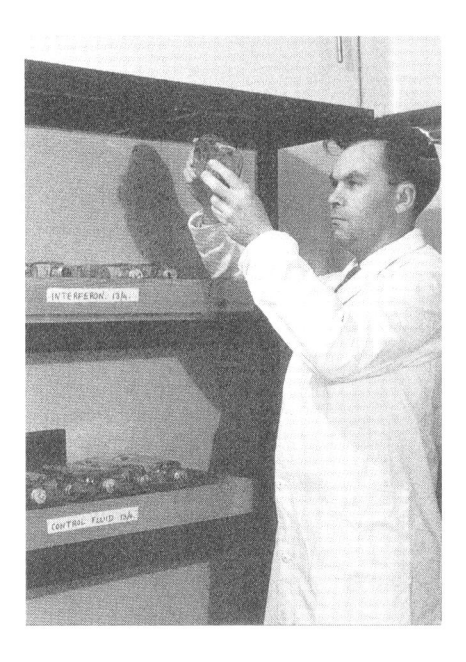

Figure 19 Dr R. F. Sellers supervises the production of monkey interferon at Wellcome Laboratories (1961). Used with the permission of Glaxo Wellcome plc.

Instead of waiting for the results of additional tests in guinea pigs the members of the Scientific Committee volunteered for self-testing.[61]

At the subsequent meeting of the Scientific Committee the misgivings of the past months seemed to be forgiven and forgotten with the preparations for a trial in humans moving into gear. Isaacs reported that of five volunteers inoculated at the NIMR only two showed an abnormal flushing of the skin after 24 hours, but only in one case was the reaction so severe as to prevent the person from being enlisted in the ultimate trial. Tyrrell had inoculated himself and three other subjects and only he himself had had an abnormal flushing of the skin. Of the five volunteers inoculated at Glaxo the members of Scientific Committee, John Beale and Robert Andrews, showed a reaction to the interferon inoculation consisting of a substantial swelling of the skin of the arm and pain but no reaction to the control. Beale argued that in having experienced a rather severe inflammatory reaction to the interferon inoculation himself, he was far less worried about the safety aspect than about the possibility that such an inflammation due to interferon might interfere with the interpretation of the experiment. In assessing the provisional results all agreed that since no alarming reactions had occurred it was justified to go ahead with the trial and start discussions on the trial design. With no standard rules on how to perform a clinical trial the trial design was left to the wisdom of those involved.

Most common at the time was a so-called 'open' experiment in which the investigator was held responsible for administering the pharmaceutical substance to the volunteers or patients and monitoring their condition. On the basis of the volunteer or patient records the clinical investigator then formed impressions of the potential toxicity and the probable therapeutic value of the substance and sent a summary of the experiment together with his conclusion to the sponsor of the trial. However, the practice of testing therapeutic drugs in humans was slowly beginning to change in Britain. Convinced that the methods of the statistically controlled trial or so-called 'randomized controlled trial' would provide a more reliable, more objective basis for evaluating medical research and directing clinical practice, the MRC actively promoted the use of randomization in testing new therapies.[62]

In the most advanced type of randomized clinical trial the new agent was tested against a lookalike placebo – a harmless, inert preparation – under circumstances where neither the patient nor the doctor and nurses knew whether the remedy or the placebo was administered to particular subjects. This type of controlled experiment was usually referred to as a 'double blind' experiment. The double-blind clinical trial was the most complicated to run. Every dose of both trial medication and placebo had to be coded, and the codes had to be cryptic enough not to be cracked by the ingenious minds of the people involved. An independent third party had to keep careful records of who was given what and keep those records secret until after the trial monitoring data had been gathered, counted, compared and evaluated. Among biostatisticians the double-blind trial was considered an absolute requirement if you did not want to throw the statistical validity of your test right out of the window. For instance, in a single-blind study, the 'blind' subjects might pester their physician who kept records of who received what, to tell them what they did or did not receive, or the doctor might react to the condition of his patients according to their expectations about the new agent. In double-blind placebo-controlled studies, subjects and investigators alike were believed to be much less susceptible to psychosomatic or other personal factors which threatened to undermine the statistical validity and hence assessment of clinical trial data.[63]

Taking into account the central role of the MRC in the Collaboration it may not come as a surprise that the members of the Scientific Committee decided in favour of the state-of-the-art double-blind placebo-controlled trial design. The trial would be carried out on a limited number of about forty non-allergic and unvaccinated volunteers, who would be recruited within the various collaborating laboratories. The trial procedure went as follows.[64] Two sites 5 cm apart on the upper part of the arm of each volunteer were to be inoculated with either interferon or the control material. The ampules with interferon and control preparations would be labelled either X or Y and distributed 'blind' together with special forms by a researcher from NIMR's Immunological Products Control Laboratory, who would be the only one to know the code. A so-called form A with the name of the volunteer should indicate, as decided by a random coding procedure, whether the inoculator was supposed to inoculate X or Y in the front or back of the arm. Subsequently the inoculated areas would be marked with indelible pencil to make it possible to carry out the vaccination with attenuated cowpox virus 24 hours later on the

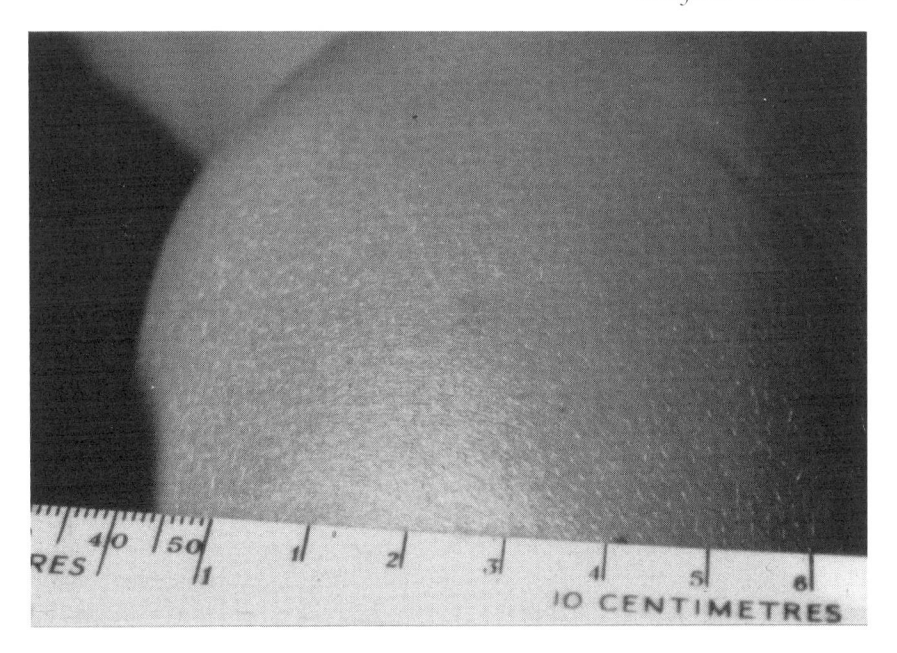

Figure 20 The arm of a human volunteer before vaccination. Courtesy of Dr D. Tyrrell.

exact site of each of the previous inoculations. A different person, the observer, then received a so-called form B with only the name of the volunteer on it and spaces for daily observations for a period of 14 days to ensure that reaction to the vaccine would be read 'blind'. Any possible sign of a lesion developing at the injection site would be monitored as closely as possible as inflammation due to interferon might in one way or the other affect the development of virus lesions (see Figures 20, 21 and 22).[65]

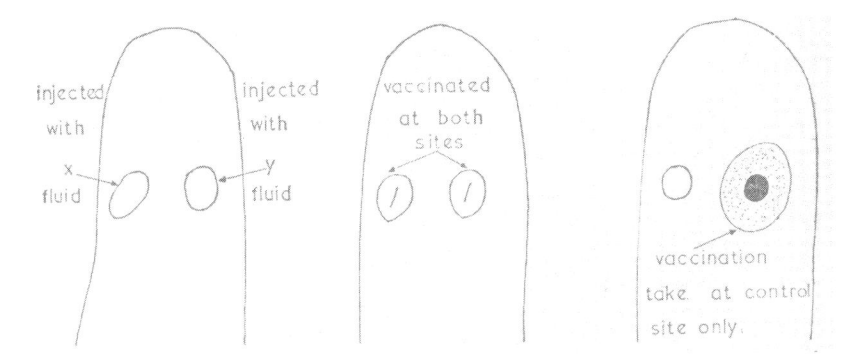

Figure 21 Principle of the vaccination experiment with interferon. The arm is injected with two fluids, 'x' being interferon and 'y' an inactive control material for comparison. Subsequently smallpox vaccinations are made at both sites. Courtesy of Dr D. Tyrrell.

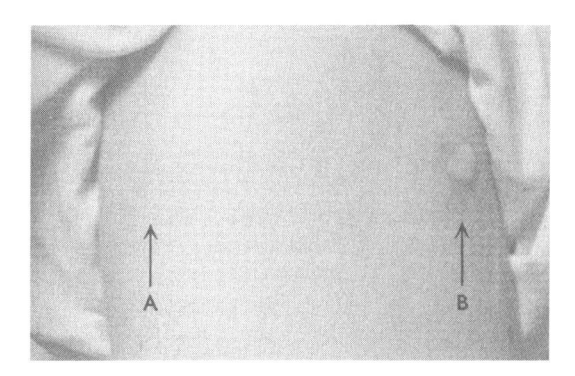

Figure 22 The arm of a human volunteer after vaccination. The vaccine has clearly 'taken' in site B (no interferon), but not in the other (interferon). Courtesy of Dr D. Tyrrell.

On Thursday 22 February 1962 the great day had come for Isaacs and the other members of the Scientific Committee on Interferon to exchange and discuss the results of the first trial of its kind to see whether interferon prepared in the laboratory could prevent the establishment of a virus infection in humans. Unusually large delegations of all the parties involved were assembled for the occasion in one of the conference rooms of the NIMR and waited anxiously for the chairman to open the meeting. Isaacs cheerfully announced that the trial with interferon had yielded promising results.[66]

At the NIMR 17 out of 23 volunteers showed a clear protection at the interferon site against cowpox infection and the development of vaccinial lesions. As far as interferon's protective effect was concerned the NIMR volunteer group had been exceptionally successful. At Wellcome Laboratories 7 out of 19, at Glaxo Laboratories 2 out of 4, at MRC Hampstead 2 out of 7 and at the Common Cold Unit only 1 out of 5 volunteers showed clear protection against virus infection. Despite the considerable deviation in test results between the various volunteer groups the overall protection rate of 60 per cent against virus infection was considered a definite success.[67]

All agreed that the results of the clinical trial were certainly qualified for publication in either the *Lancet* or the *British Medical Journal*. Isaacs immediately offered his services to prepare a first draft. This time even the commercial side to the Collaboration encouraged him to do so. All parties realized that with such a publication they would have a strong case in support of the American patent application. For once the interests with regard to the dissemination of information of both government researchers and company executives seemed to correspond.

The *Lancet* report

As promised, Isaacs prepared a rough draft of the trial results and circulated it to the members of the Scientific Committee and to the MRC before finalizing it as an official report to the MRC from the Scientific Committee on Interferon, and submitting it for publication to the *Lancet*. During a routine check of the trial data in Isaacs's report, the MRC's medical officer who had been present at the presentation of the preliminary results came across oddities in the figures from the Wellcome laboratories. As far as he could remember more than the reportedly three volunteers had participated in the Wellcome trial group. Furthermore he could not avoid the impression that apart from the fact that the vaccinator and the observer had been one and the same person, this individual also seemed to have taken part as a volunteer in the trial. The medical officer immediately informed Isaacs, who confirmed that his suspicions were correct: 'Dr Isaacs telephoned me this morning to say that the query I had raised in my letter to him of 2nd April about the results from the Wellcome laboratories had been very astute (more so than I had suspected)'.[68]

Isaacs told him that after the meeting of the Scientific Committee he found out to his dismay that the Wellcome scientists had made a mess of things. How on earth could the monitoring of the volunteers have been blind if there was no difference between the vaccinator and the observer? They even appeared to have forgotten about the agreed upon single arm procedure and inoculated volunteers in both arms. By tampering with procedures the Wellcome laboratories obviously had broken the rules of the trial which was supposed to be double blind. Isaacs indicated that he had tried to cover up for Wellcome's sloppy data in his report by including only 3 out of the 19 volunteers tested at the Wellcome Laboratories. The issue worried him and he asked the MRC officer for advice on what to do next. Would it be best to make no reference to the Wellcome data at all as suggested most recently by David Edward of Wellcome, who had been away in the United States during the time of the trial?

In response the medical officer urged Isaacs to discuss the matter with Sir Charles Harington. Eventually, in consultation with the research director of the Wellcome Laboratories, Colonel Mulligan, Harington decided that the Wellcome laboratory figures should be deleted from the text of the report.[69] In the meantime the original report had been submitted to the *Lancet* but was being held up until the matter had been settled. With the 'delete decision' in mind, it is surprising then to see that in the final *Lancet* report the three Wellcome volunteers can still be found. Furthermore, both in the case of the Common Cold and Hampstead figures, two volunteers are missing (see Table 2).[70]

According to the report in the *Lancet* – 'A report to the Medical Research Council from the Scientific Committee on Interferon' – the clinical trial with interferon to study the effect of interferon on vaccination in volunteers was carried out as a result of unambiguous experiments with an antiviral substance in the laboratory. The transition from experiments in the test-tube and in laboratory animals to experiments with humans was depicted as unproblematic. In addition, the trial was said to have proceeded smoothly and unmistakably showed that interferon had an antiviral effect in humans.[71]

Obviously the contingencies and uncertainties which the committee members had faced in preparing for and performing the vaccination trial and the simultaneous management of uncertainty were deleted from the public

78 *Interferon: The science and selling of a miracle drug*

Table 2 Results of individual vaccinations in the vaccination trial of the Scientific Committee on Interferon

Laboratory	Volunteer	Site treated with Interferon	Control	Interferon at
National Institute for Medical Research	D.H.	0	++	Front
	K.R.P.	0	++	Front
	J.S.C.	++	0	Back
Inoculations:	B.B.	0	++	Back
Dr G.W. Bissett and	F.C.F.	0	++	Front
Dr D.H. Sproull	S.S.	0	++	Front
	K.M.S.	0	++	Back
Observations:	W.J.B.	+	++	Front
Dr A. Isaacs	N.G.B.	0	++	Front
	J.M.G.	+	++	Front
	A.C.	0	++	Back
	R.C.	0	++	Back
	S.D.D.	+	++	Back
	B.D.	0	++	Front
	M.H.G.	0	++	Front
	T.G.J.	0	++	Back
	G.A.	0	++	Back
	S.C.	0	++	Front
	J.E.	0	++	Back
	M.G.	0	++	Back
	V.H.	++	++	Back
	J.J.S.	+	++	Front
	W.F.W.	0	++	Back
Common Cold Research Unit, Salisbury	V.M.B.	+	++	Back
	A.E.M.G.H.	0	++	Back
	P.K.B.	+	++	Front
Inoculations: Dr D.A.J. Tyrrell				
Observations: Dr M.L. Bynoe				
Immunological Products Control, Hampstead	J.W.	++	++	Back
	D.J.	0	++	Back
	H.H.	0	++	Front
Inoculations:	A.P.	+	++	Front
Dr M. Clarke	B.H.	0	++	Front
Observations: Dr M.M. Winter				

continued on facing page

Glaxo Laboratories	P.L.	0	++	Back
	R.H.	0	++	Front
Inoculations:	A.C.	++	++	Back
Dr R.J. Andrews	R.C.	++	++	Back
Observations:				
Dr A.O. Hagger				
Wellcome Laboratories	Mrs. G.	0	++	Front
	Mr. W.	++	++	Front
Inoculations:	T.M.P.	+	++	Back
Dr T.M. Pollock and				
Dr W.T.W. Lawson				
Observations:				
Dr T.M. Pollock				

Notes:
++ = Definite take
+ = Take, but distinctly smaller than the take at the other site.
0 = no take

record. While Isaacs proclaimed the trial a success and wrote in the *New Scientist* that work on interferon had 'just passed a critical stage' by producing statistically highly significant results which showed that interferon protected human volunteers against infection with the vaccinia virus, the managing boards of the companies judged the efficacy data differently.[72] In their opinion the resources invested in the trial efforts were in no proportion to the experimental results: interferon had only shown a protective effect in a clinically insignificant viral infection under controlled circumstances. The trial was not regarded as particularly informative as to whether or not interferon deserved the label therapeutically-interesting with regard to everyday practice. They were concerned about the already high costs of producing minute amounts of semi-purified monkey interferon, not to mention the possible production costs of human interferon.

Going by the information from their scientific staff the production process seemed to be bedevilled by technical problems too and much more basic laboratory work would be needed than expected to figure out interferon's clinical potential. The idea began to settle within management circles that it might take years before the large-scale commercial production of material suitable for clinical use might even be taken into consideration. With the growing doubts about whether interferon was indeed worth the big effort that was being put into it, the commitment of the three drug companies to interferon came under pressure.[73]

My analysis of the 1962 *Lancet* report shows that the consistency of an experimental study is influenced by more than the rigour of its experimental design and the statistical elaboration of the trial data; this can be seen by the way test subjects are recruited, by the compliance of those involved with the agreed-upon trial procedures, and by the way in which decisions for starting, stopping and reporting a study are made. Moreover, in closely paralleling what one might call a treatment's scientific basis, the published efficacy data appeared to be of little value to those primarily interested in assessing interferon's therapeutic potential in everyday practice ('effectiveness'). Following the publication of the first trial report a gradual reversal of commitments and objectives of those involved would come about. Questions were raised about whether or not the development of interferon as a therapeutic drug constituted a worthwhile pursuit, and about the appropriate research approach among those working at the bench and administering the collaboration.

The dissensions which arose in the process reflected the conflicting time-horizons and conflicting ways of handling and judging research data of the various parties involved in the Collaboration. The academic scientists were interested primarily in creating new opportunities for research, advancing their scientific careers and the prestige of the institute or research organization. They regarded interferon as a somewhat unruly but promising biological substance that represented a major new stratagem of defence against infection by viruses. Sooner or later interferon would prove of use to medicine in one way or another – if not as an end in itself than as a means for stimulating people to make their own interferon. However difficult and time-consuming, this was not believed to be beyond the powers of modern pharmaceutical technology available through the pharmaceutical industry.

The executives of the drug companies, on the other hand, were trained in evaluating research projects rather sooner than later in terms of their market potential, and the likely time and expenses involved in translating laboratory data and processes into practical and commercially viable medical technologies. They took note of the published efficacy data but reacted to what they judged and perceived to be therapeutically-interesting remedies. They actively created their own profile of the treatment's effectiveness, attaching much importance to commercial viability (research expenses compared to market potential) and social acceptability (whether or not a provisional everyday therapeutic rationale existed for its use). Their negative assessment of interferon's therapeutic potential was accompanied by mounting pressures for accountability.

Interferon losing its momentum

Company scientists working on interferon sensed a shift of research priorities away from interferon. Unlike before, their research plans and efforts were assessed with the utmost rigour in competition for company research resources. It is unlikely that its development as a therapeutic drug would have been pursued much longer on the ground of purely economic motives. It is in this context that the 'American penicillin syndrome' continued to play a most vital role in sustaining interest.[74]

All senior company executives were aware of the fact that one of the US leaders in biologicals, the large drug company Merck Sharpe & Dohme, was putting considerable resources into interferon research. While the British Interferon Collaboration invested heavily in efforts to test the effect of interferon in humans, the American drug firm focused on biochemistry: making sure that interferon could be processed to a high degree of purity in a cost-effective way.[75] In the most recent informal talk with one of the members of the Board of the Interferon Collaboration, Max Tishler, the president of Merck Sharpe & Dohme Research Laboratories at Rahway, New Jersey, had indicated that it would only be a matter of time before interferon would be obtained in pure crystalline form.[76] This kind of competitive rhetoric from the other side of the Atlantic kept the various research projects dealing with interferon within the three drug companies going, though on a much more modest level.

It did not last long before the lower rating of interferon within drug company quarters made itself felt within the Scientific Committee. Isaacs's ambitious plans to step up immediately the production of clinical trial interferon to proceed to further clinical trials met with increasing opposition from Glaxo, Wellcome and ICI scientists. Confronted with a decline in research resources and increasingly critical managements they questioned more and more frequently the aim and relevance of the Committee's research programme. Instead of stepping up the costly and troublesome production of monkey interferon and expanding the clinical trial programme, the representatives of the drug firms argued in favour of doing more animal work and more chemical studies on purification. As the most outspoken critic of a continued clinical trial approach, the ICI researcher Norman Finter was on a collision course with Isaacs.[77]

The mutual dissatisfaction over research policies began to manifest itself when the material consequences of the eroding support for the clinical trial approach were making themselves felt through a shortage of clinical trial interferon. By the end of May 1962 Isaacs was fed up with the situation. The fact that the collaborating firms continued to fail to meet the agreed-upon production requirements originally seemed to confirm his earlier fears that the firms were dragging their feet. In consultation with the MRC, who told him not to beat the big drum, Isaacs decided that he would first challenge the firms in the Scientific Committee. As chairman he felt it would be relatively easy to put the problematic supply position of interferon on the agenda.[78]

At the last Committee meeting before the 1962 summer holidays Isaacs brought the matter up for discussion. In accordance with previous plans, the next step should be for Tyrrell to perform a trial at the Common Cold Unit in Salisbury in order to test the effect of interferon on the common cold in volunteers. The idea was that if the human nose and pharynx was treated repeatedly with interferon by means of nasal drops or nasal spray it might become protected against infection with viruses capable of causing colds like rhinovirus.[79] The trial would be carried out in line with the general experimental routine at the Common Cold Unit – fortnightly trials with volunteers kept under strict isolation at the special volunteer flats, daily clinical examinations, a short quarantine period to exclude intercurrent infections, and a double-blind assessment of symptoms.[80] Isaacs emphasized that Tyrrell did not envisage having difficulties in finding appropriate volunteers for the interferon trial. The Common Cold Unit had a regular supply of volunteers from the general public, mostly students and civil servants, who were willing to participate in the Unit's ongoing common cold studies programme with a nationwide reputation. The interferon trial would be run side by side the regular research on common colds.[81]

However, Isaacs sneered, with the current problematic supply there would be hardly any need for volunteers.[82] The available clinical trial interferon only allowed for 3 instead of the planned number of 20 volunteers per trial. With less than a handful of trial subjects one could not expect to generate statistically significant data. Isaacs was willing to accept that at Wellcome Laboratories the production of interferon was affected by organizational problems. The two scientists who had been undertaking the production and purification work at Wellcome had been transferred, and it was only to be expected that the production would fall or stop in consequence until substitutes had been found. However, the other parties had no valid excuse for their shortcomings.

Glaxo's representatives on the Scientific Committee time and again promised half a litre of concentrated interferon, the last time within a month. When at last he decided to contact someone at Glaxo more immediately concerned with the matter, Isaacs learned that no such goal was set and that there were still numerous problems to overcome. The situation at ICI was even worse. Isaacs's impression was that ICI not only gave no sign of producing any clinical trial interferon, but did not even seem to be prepared to undertake any initiatives in that direction.

Finter rather boldly confirmed this position. He said that he could not envisage ICI making interferon for use in the common cold trial from (primary) monkey kidney cells and, to be honest, he thought that there was no ultimate hope for the production of interferon from any primary tissue culture. At ICI they preferred to concentrate on animal work, on the mode of action of interferon and on the production and purification of interferon. He thought it absolutely necessary to develop a more practical way of making interferon since the current procedures were commercially unacceptable.

It cost quite some time to ease the tension between Finter and Isaacs, but towards the end of the meeting the situation was looking up a little. Without succeeding to recapture his full authority as chairman of the Scientific Committee, Isaacs settled the differences by compromise: they would await the results of the common cold trial before entering into further debates about the research programme.

Isaacs's disillusionment was great when at the end of September 1962 Tyrrell revealed that the trial results had been disappointingly negative. After testing interferon against three different common cold viruses in 20 volunteers there was no clinical evidence of a protective effect (see Figure 23). However negative his trial results looked, Tyrrell indicated that he had not given up hope. Personally he did not consider discontinuing this line of work as it would

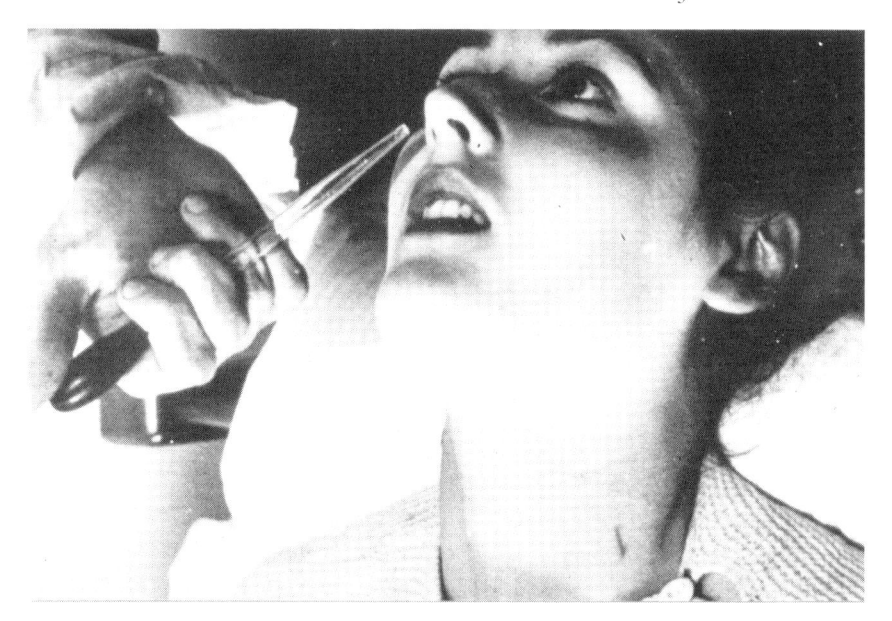

Figure 23 After having inhaled an interferon preparation the woman volunteer at the Common Cold Unit at Salisbury is infected with common cold virus. Courtesy of Dr D. Tyrrell.

most likely only be a matter of increasing the dose of interferon.[83] Tyrrell's reassurance could not prevent the company scientists from reopening the discussion on the targets for future work on interferon. The company scientists seized the opportunity with both hands to bring the research programme more in line with the standard research trajectory for biologically active compounds in the pharmaceutical industry – focusing on animal testing and production processes first.[84]

Outside the private rooms of the collaborators few seemed to know that the Interferon Collaboration went through hard times. A clear indication was the following story entitled 'Whitehall men join battle on common cold: This new drug may be the answer', which appeared in the Evening Standard on Friday 5 October:

> Scores of top Civil Servants are being asked to take 10 days off to act as 'guinea pigs' in the first large-scale trial of a British medical discovery which doctors hope can be used to beat many virus diseases including the common cold . . . scientists will inject them with a newly-discovered substance called interferon and at the same time try to infect them with thousands of cold germs. The hope is that colds will not break out . . . A Medical Research Council spokesman told me today: . . . 'Not everybody will be given interferon: some will get water or a salty solution but

nobody will know who has which. We have to satisfy the statisticians that any effect on colds is not due to chance.'[85]

The final assessment 'indoors' of the carefully controlled common cold trial in December 1962 doubtless must have satisfied the statisticians, but it was disastrous for the clinical trial programme. The number of volunteers who either developed disease symptoms or improved in the interferon-treated group was said to differ insignificantly from the control group. It seemed pointless to pursue these clinical studies any further until more information had been obtained on dose–effect relationships and the method of administration. Even Isaacs agreed that the trials should not be proceeded with for the present, knowing that none of the collaborators was willing to continue the costly production of monkey interferon to perform further clinical trials.

Isaacs reconciled himself to the situation and decided in favour of more animal experimental work in the hope of finding a better test model. Moreover, confronted with the negative sentiments regarding the testing of the clinical potential of 'exogenous' interferon, Isaacs chose to ride his latest hobbyhorse. He told the other members of the Scientific Committee that after all he was becoming more and more convinced that interferon injected or otherwise given to patients might prove less useful than developing a means for stimulating people to make their own interferon. He pointed out that he soon hoped to clarify the use of foreign nucleic acids as a means of stimulating 'endogenous' interferon production through tests in laboratory mice.[86] In his view this would make a most promising new approach for future work of the Committee. However Isaacs had lost too much political capital to get a hearing for what the company scientists already had dubbed 'Isaacs's ludicrous nucleic acid story'.[87] By reacting rather half-heartedly to his proposed change of research priorities the fellow members of the Scientific Committee made him feel more and more isolated and exposed.

The December meeting of the Scientific Committee not only proved to be somewhat of a landmark inasmuch as the decision had been taken to stop the clinical trial programme but also with regard to Isaacs's fragile mental health. Isaacs's continuous concern about the working of the Collaboration, the disappointing trial results, and the growing opposition to his research guidance within the Scientific Committee began to take their toll. The succession of problematic events was instrumental in provoking a second bout of depression that was more severe in nature than the first three years earlier. He felt extremely restless and irritable most of the time with little self-confidence. This time Isaacs had to go into a mental hospital where he received extensive medical treatment for a couple of months.[88]

Meanwhile across the Atlantic Ocean at the Merck Institute for Therapeutic Research, a paper had been cleared for publication through Merck Sharp & Dohme's (MSD) patent office that would play a major role in the reordering of the British research agenda concerning interferon. Sections of this paper on the purification and characterization of interferon had already been leaked purposely to MSD's British rivals.[89] In making a tentative approach early in 1962 to join the Collaboration on Interferon (which was ignored by the British) Max Tishler, the President of Merck Sharp & Dohme Research Laboratories revealed that his researchers had prepared interferon material that was far superior to anything the British had thus far produced. Almost simultaneously Maurice Hilleman MSD's head of Virus and Cell Biology Research – a renowned scientist with a rather strong ego who was not averse to competition and throwing his competitors into confusion – let Isaacs informally know that they had prepared chick interferon that not only was 200 hundred times more pure than the British interferon but also had a different molecular weight.[90]

While these American news leaks mainly had the effect of stimulating the sense of rivalry on the British side without really affecting work on interferon, the formal publication of MSD's studies on interferon in the prominent American journal *Proceedings of the Society of Experimental Biology and Medicine* had a more profound impact.[91] Reading in black and white that for years British researchers had studied interferon which essentially contained only minute amounts of the interferon agent and consisted largely of extraneous material was quite a different cup of tea. Hilleman's research team claimed to have purified interferon 4500 times with respect to initial protein content – an unprecedented degree of purity of interferon material.[92] In the article the previous failures of purification were articulated by juxtaposing purity with impurity: emphasizing the marked difference in properties between their highly purified interferon and the allegedly crude interferon studied heretofore by other researchers, most notably Isaacs and his British colleagues.

Although the disagreement was not complete it was substantial enough to throw serious doubts on the British research achievements so far and to further weaken the British patent position. To the same extent that it lent credibility to MSD's efforts, the American publication threw discredit on the work of the British collaborators. In doing so it would have an additional discouraging effect on senior British company executives to invest considerably in interferon, thereby reshaping the balance of forces both within and outside the Interferon Collaboration.

No 'magic bullet'

By the time Isaacs had recovered from the severe manic-depressive episode at the end of April 1963, deputy chairman David Tyrrell had brought the American article to the attention of most members of the Scientific Committee on Interferon. Finter was first to react.[93] In two letters to Isaacs, he mentioned having read the studies on interferon published by Hilleman's research group and indicated that on the face of it these appeared to be an advance over anything which had been achieved within the Collaboration on Interferon. In Finter's opinion the collaboration's lack of success in purifying interferon remained a major stumbling block to further progress in the development of interferon as a practical chemotherapeutic agent. 'From the scientific point of view there might therefore be very real advantages to be gained from liaison with Merck Sharp & Dohme if this would give us access to their "know-how" in this field (since in all probability they have data available other than those published).'[94] Finter then proposed to reopen the question of whether or not MSD be invited to join the collaboration, assuming that they still wished to do so, and discuss it at the next meeting of the Scientific Committee.

Isaacs, however, showed himself rather reluctant to bring the matter up for discussion.[95] He didn't think much of a liaison with yet another party and certainly not if it meant collaborating with Hilleman whom he suspected of being driven solely by self-interest. At least in the article, Hilleman had done everything to make the work of Isaacs, Burke and the Committee look rather silly. It was only because of Finter's persistence that eventually the question of liaison with MSD was discussed. Finter must have been disappointed to find out that the other members of the Committee did not think much of the idea either and decided to hand over the matter to the Board. However, rejecting the proposal to collaborate across the Atlantic ocean did not mean ignoring the American interferon studies.[96]

Isaacs rapidly got the better of his initial embarrassment about losing the initiative to Hilleman's research group in a field of which he considered himself the founding father. More determined than ever before, he devoted himself to steering the Scientific Committee, pursuing his interferon studies and promoting his ideas.[97] He decided to turn the published American purification and characterization studies to his advantage. Should interferon indeed be powerfully adsorbed onto glass, as claimed in the American report, and should as a result interferon preparations stored in glass bottles lose their antiviral activity in time, then this put the outcome of the clinical trials in another perspective. Isaacs therefore thought it worthwhile to assay some interferon bottles left over from the trials. When it indeed turned out that the antiviral activity of the clinical trial interferon remnants had dropped considerably compared with the original material, Isaacs took the matter up with the Scientific Committee. Isaacs's story was considered very convincing, and it was unanimously agreed that since there was no guarantee that the interferon samples used were still potent at the time of the trials, the assessment of the clinical trials of interferon should be suspended – a private report should be sent to the Council but not published.[98]

So far so good, it seemed. However, much to Isaacs's regret, the reassessment did not take away the general reluctance to plan further trials. Even more disappointing in Isaacs's opinion was the subsequent lack of support for his proposal to discontinue the obviously disastrous purification studies and instead concentrate on the radically different approach of exploring ways to exploit the potential interferon productive capacity within a living organism. If it were possible to find ways to fight off viruses by stimulating the production of interferon inside the body ('endogenous interferon') by non-viral means, Isaacs believed that there would be no further need to continue the costly and problematic efforts to produce large amounts of exogenous interferon.

His proposal for a change of course in the work on interferon resembled to some extent Hilleman's simultaneous shift in attention from interferon administration to interferon induction as the most promising lead toward antiviral therapy, although Isaacs did not share Hilleman's radical ideas about exogenous interferon as a dead end.[99]

After almost four years of interferon studies, Hilleman had come to believe that despite its remarkable non-toxicity, administering exogenous interferon had no real potential for clinical usefulness in antiviral therapy. Interferon's prophylactic rather than therapeutic action and the expected high costs and technical problems involved in producing large amounts of high quality interferon material were instrumental in bringing about a change in research perspective. Instead of regarding interferon as a potential chemotherapeutic agent in itself, Hilleman began to see and promote interferon in terms of a research tool to study resistance to and control of viral infection. Eventually this new approach was thought to open the door to the development of chemical means for achieving viral chemotherapy. This would also fit in well with the majority of chemistry-based drug development programmes within MSD. Consequently Hilleman decided to give up his purification studies. Instead he turned his research resources to assessing the possibility of engineering chemical compounds which are potent inducers of 'endogenous' interferon production and can be used to limit, and preferably prevent, viral infection in a host organism. Hilleman's new endogenous approach closely resembled Isaacs's ideas about the non-viral stimulation of the production of interferon in living organisms.[100]

Taking into consideration Hilleman's motivations to change course it comes as a surprise to see that the 'endogenous' approach did not receive the backing of the Glaxo, Wellcome and ICI scientists within the Scientific Committee.[101] The wonder grows if we add to this Isaacs's accusation that the attitude of the company scientists within the Scientific Committee seemed to be conditioned for the greater part by what he called a 'chemical habit of thought'. Apparently the mere association of the endogenous approach with Isaacs's interferon inducer studies sufficed to relegate the new concept to the circular file within the Scientific Committee. As such it marked the credibility crisis concerning Isaacs's abilities to steer and coordinate research on interferon.[102]

Increasingly frustrated in his efforts to direct research as chairman of the Scientific Committee, Isaacs eventually decided to plead his case through other channels. In July 1963, during the yearly visit of medical and science correspondents to the National Institutes for Medical Research, Isaacs saw his opportunity both to promote his new line of research on interferon inducers and to show his dissatisfaction with the Interferon Collaboration. While showing the journalists around in his laboratory, Isaacs told them that he believed the UK's effort on interferon to be inadequate and particularly so when compared with what was going on in the USA.[103] About the same time Isaacs sent a formal letter with a similar vote of no-confidence to the new director of the NIMR, the immunologist and recent Nobel prize winner Peter Medawar.[104]

88   *Interferon: The science and selling of a miracle drug*

Medawar, was informally – through his predecessor Harington – already well aware of Isaacs's dissatisfaction with the cooperation they had received from the consortium of three drug companies. And he immediately passed the letter on to Sir Harold Himsworth. The letter did not take Himsworth by surprise either. Himsworth already knew for some time that Isaacs was dissatisfied with the efforts of the industrial collaborators. At first Himsworth thought that Isaacs was too soft to be chairman of the Scientific Committee, tending to discuss possible lines of action rather than requesting that certain work should be done. Later on however he came to believe that this was not the case. The main reason for the firms' 'lethargic' effort was a collective action problem in terms of the logistics involved in communication between so many different parties and the growing feeling amongst them that there was no practical value in interferon.[105] After extensive consultation with his own staff and Medawar, Himsworth eventually decided that Medawar should meet the Board of the Patents Holding Company ('Executive body') to put forward and discuss all the reasons for concern about the working of the Interferon Collaboration. Medawar met the Board at their 20 November meeting. The whole atmosphere was charged with an anxiety as to whether the collaboration should be continued.

Whereas Medawar was disappointed with the contribution of the industrial collaborators, the commercial partners concentrated on the negative assessment of the possible practical and commercial value of interferon. They generally felt that the prospects of interferon becoming a saleable therapeutic agent had not improved since the start of the collaboration. Despite its imaginative action against a wide range of viruses, interferon did not live up to the hopes of an Ehrlich sort of 'poison-arrow' specifically aimed at the virus invader.[106] Worse yet, far from being the 'magic bullet' that was hoped for, interferon was known as a chemically undefined biological substance that was part of a poorly understood natural mechanism of resistance to virus infections. As such it was regarded as alien to the industry's Ehrlichean chemotherapeutic programme.

Medawar's interpretation of the backwardness of the enterprise as a lack of engagement and effort of the three drug firms met with severe criticism from the commercial collaborators. Though lack of progress was not denied, the effort of the companies was said to be 'genuine and wholehearted'.[107] The chairman of the Board of Directors of the Patent Holdings Company indicated that the industrial partners had spent about twice as much as Merck Sharpe & Dohme – over 250,000 pounds so far. (On a yearly basis this meant that about 0.75 per cent of the annual research budget of all British-based pharmaceutical firms was spent on interferon.) He clearly wondered how much further they should go.[108] So little progress had been made to date that there were serious doubts as to the wisdom of continuing the Collaboration. At the same time, the companies were reluctant to write off all their efforts over the past few years. None of the parties appeared to be willing to take the initiative in bringing the collaboration to an end without giving it another trial period.

The fact that a marketable product was no nearer than when the Collaboration started was attributed, if not to the 'cussedness of nature' and to the incompetence of the scientists engaged on the work, then to the shortcomings of Isaacs as chairman of the Scientific Committee and research coordinator.[109] It was generally felt that Isaacs was too deeply involved in interferon, scientifically and emotionally to continue carrying the responsibility for the scientific policy of the Collaboration.

Faced with a lack of imaginative research results, not only drug company executives but also the British medical community grew impatient with interferon as is succinctly illustrated by the following excerpts from a 1964 editorial in the *British Medical Journal*:

> Immense practical problems attend any future use of interferon in man . . . , and even if they can be overcome the problem still remains of ensuring that the interferon reaches the cell in time to produce any therapeutic response. The use of it for controlling virus infections in man and other animals is likely, therefore, to be limited.[110]

Through this kind of high-profile vote of no confidence in the clinical potential of interferon as a therapeutic drug, interest plummeted and interferon became widely regarded as just another medical 'discovery' which had not lived up to its promises.[111] The waning of public enthusiasm and the new arrangement for scientific policy within the Interferon Collaboration made itself felt in the work of the Scientific Committee. In response to the steady decline in research resources and new research directives from the Scientific Steering Committee, plans for clinical trials were pushed into the background. Attention shifted to studies of the physico-chemical properties and mode of action of interferon and to assessing through animal experiments the possible therapeutic value of interferon.[112]

The political jousting over his position and the change of research priorities passed Isaacs by for the greater part. On New Year's Day 1964 Isaacs suffered from a sub-arachnoid haemorrhage which kept him away from work – owing to a slight paralysis and blurring of his eyesight – for more than three months. The haemorrhage was said to be related to an abnormal tumorous blood vessel which compressed the surface of the brain, as revealed by angiography. Apparently, from its position, the tumour was out of reach of the surgeons.[113] On return to the Institute Isaacs found himself relieved from his post as head of the Division of Bacteriology and Virus Research and instead he was appointed head of a small group for research on interferon – consisting of Isaacs, Joseph Sonnabend and his laboratory technician Dennis Busby – that was officially named 'The Laboratory for Research on Interferon'.

Isaacs's lamentable condition was symbolic of the paralytic state of affairs not only of British interferon research but of interferon research in general. Despite a short-lived excitement that accompanied Hilleman's research group's purification studies, the enthusiasm for interferon within the international scientific community had definitely waned. The powerful group of quantitative, molecular-oriented virologists in America (with people like Delbrück, Dulbecco, Rubin and Huebner), who had tolerated what they considered as one of those short-lived 'research fashions' as long as it lasted, had an important hand in marginalizing interferon research.[114]

By openly displaying their serious doubts about interferon as a troublesome biological substance, which after more than five years of research still resisted chemical characterization and purification, they gave interferon researchers a hard time. Interferon became widely perceived as an odd laboratory substance that had never got much further than an assumed protein and had not lived up to its promise as an important lead toward viral chemotherapy – and that was unlikely to be ever of much practical value. The fact that the scientists interested in problems of interferon admitted that the agreed-upon criteria for identifying a biological factor as an interferon were not exclusive enough to determine authoritatively whether or not one was dealing with an interferon only complicated matters.[115]

In the corridor of scientific conferences and meetings, questions like 'Do you really think that interferon exists?', and, 'How do you know that the phenomenon which you observed in your laboratory was due to interferon and not to an impurity?' were repeated.[116] Unlike in the early 1960s, not 'everything' on interferon was publishable anymore by the end of 1964. Edward de Maeyer, for instance, received rejection letters from both *Virology* and *Science* on the grounds that the material under study ('interferon') was ill-defined and impure, and therefore of no great scientific interest.[117] Never before had this been a reason for rejection.

Considering interferon's low scientific image as an ill-defined laboratory substance of peripheral interest, for most biomedical researchers, entering the area of interferon research just did not seem worth the effort. Why join a marginal subfield of virus research that was considered to lie well outside the mainstream of virology (viral genetics and animal-tumour virology)?[118] Apart from image related problems there were other more technically related matters that 'locked people out'.[119]

If you wanted to start working on interferon as an outsider to the subfield of interferon research you first had to acquire specific skills to set up your own production and assay system. This was rather time consuming. Practically it meant that one had to learn through hands-on apprenticeship the production and assay techniques from established interferon researchers. This was far from a straightforward process as the 'interferonologists' themselves were still in the process of getting to grips with their experimental systems and the biological substances that went by the name interferon.[120] As we will see in the next chapter a great deal of their efforts went into managing differences.

4 Managing differences[1]

> We have finished with the romantic period of interferon and we now have to start on the scientific period.[2]

From the very moment the British started a collaborative research programme in September 1959, as a means to the end of developing interferon as an antiviral drug, the issue of mastering differences had been on the agenda. Making comparisons between the results obtained on different occasions within any one of the collaborating laboratories was already quite demanding; far greater difficulties arose when attempts were made to compare results obtained in different laboratories. Again and again interferon researchers reported considerable variations in experimental results from day to day, week to week, from laboratory to laboratory. For the greater part, this was thought to be due to the wide variety of methods and materials employed in the assay of interferon. It was generally believed that standardization might be helpful in order to ensure the reliability and credibility of the products of their research work, both locally and transnationally.

The need for standardization was nourished by the mere involvement of the pharmaceutical industry, which had practical as well as strategical interests in preventing the existence of a variety of different arbitrary units for expressing the potency of one and the same drug. Performing clinical trials meant passing judgements on the actions of experimental substances within certain limits related to human health. Due to the high level of public responsibility that was involved in experiments with humans, there was special concern for quantitative control and rigour with regard to issues like effectiveness, toxicity and stability. Establishing and using standards was considered a necessary operation to give authority to trial data, which could be appealed to in the future.

However, without some extra encouragement in the form of a clinical trial programme, there was little inclination to use standards at the bench and rigorously define common units of measurement. With an uncertain return on investment, most chose in favour of the proven informal exchange mechanism. The researchers involved accomplished much by informally agreeing on exchange mechanisms.

I shall show how, amongst other factors, the intrusion of powerful spheres, such as the pharmaceutical industry and national centres for biological standardization, into the relatively private domain of 'laboratory life' was instrumental in breaking down informal practices of measurement and in creating a demand for formal types of standardization.[3] My claim is that the standardization discourse in itself played an instrumental role in disciplining interferon researchers into specific routines and actions, and thereby coordinated their research practices. Thus discussing standards was as important to the emerging field of interferon research and the vicissitudes of interferon as their actual establishment and use.

Portrait of a 'gift culture'

In general, researchers working on interferon operated with a high degree of informality in a relatively small and flexible network of personal contacts that was built and maintained through conferences, colloquia, sabbatical visits, post-doctoral fellowships and extensive correspondence. Basically it was a 'gift culture' based on the regular exchange of laboratory samples, techniques and skills.[4] This gift culture was instrumental in facilitating the comparison of research results and the management of differences between laboratories and researchers in the field of interferon research. A case in point is the relationship of many years which existed between the Finnish virologist Kari Cantell and the American virologist Kurt Paucker. When, on his return to the Department of Virology at the State Serum Institute in Helsinki at the end of the summer of 1962, after a two-year stay as a post-doctoral fellow at the Virology Section of the Research Department of the Children's Hospital of Philadelphia, Cantell experienced difficulties in establishing an experimental system similar to that he had worked with in America (the transfer of skills and materials had only been partly successful), he immediately fell back on his former research 'pal', Paucker. As we can see in the following excerpt from a letter by Cantell to Paucker, there was more at stake than an occasional call for assistance to reproduce a particular production system for chick interferon:

> Our work on interferon is in slow progress here (*Helsinki*). We produce interferon in eggs with mumps virus, but have not yet found suitable conditions to get very high titers. This is why I would like to have WS (*strain of influenza virus*), if we can not improve the yields with mumps virus. The interferon is assayed by a similar plaque inhibition test we had there (*Philadelphia*) and it works nicely. We get the Falcon plastic dishes (*assay dishes*) from California. Our purpose is to start soon producing interferon on a rather large scale and to purify the stuff to some extent. One of the things I would like to do with it (*the purified interferon*) is to prepare good anti-IF sera. I wonder whether we could not establish some co-work or co-operation between your lab and my lab. We would supply you with

antiserum against chick interferon and you us with antiserum against mouse interferon. What do you think about this? (Italics are mine).[5]

By emphasizing the profitability of mutual assistance Cantell obviously tried to establish a permanent connection between his Helsinki laboratory and Paucker's laboratory at the Children's Hospital of Philadelphia. As it turned out Paucker too thought it profitable to set up a kind of permanent barter with Cantell. During Cantell's stay in Philadelphia both men had learned to respect each other's abilities as laboratory researchers and they knew that they could count on each other to produce high quality data and research materials which would be useful in pursuit of one another's research.[6]

Both Cantell and Paucker were well aware of the problems involved in exchanging data and research materials between laboratories. They knew by experience that even when working together in the same laboratory it could be quite demanding to make comparisons between each other's data. According to Cantell, different researchers could make research materials perform differently and these differences could be even greater in the case of experiments involving biological specimens. Collaboration between the same pair of researchers in different laboratories would only complicate matters.[7] Apart from occasional difficulties with customs over the contents of packages and occasional loss of material owing to damage to containers during the journey or to contamination, most problems originated from the changeable behaviour of biological specimens.[8] However skillfully one packaged the tubes with biological materials (mostly in containers filled with dry ice and labelled as 'Biological specimen, no commercial value') for shipment, the exchange of biological specimens often met with resistance.[9] More often than not, strains of cells or viruses which were sent from one laboratory to the other behaved differently on arrival and it required some manipulation to make them 'adapt' to the local situation. At times, for no obvious reason at all, a virus strain or cell line lost some of its original properties on transfer to a foreign laboratory, or even stopped growing at all.[10] In general, however, the differences were manageable and laboratory organisms could be made to work in a similar fashion in both Cantell's and Paucker's laboratory.[11]

The regular exchange of research materials turned out to be profitable for both parties. Whenever either Cantell or Paucker experienced difficulties in working along similar lines of research, they could appeal to the other for help. Paucker called it 'the easy way out'.[12] Cantell indicated that the exchange mechanism was instrumental in addressing problems of uncertainty and variability.[13] As long as researchers like Cantell restricted their dealings to fellow scientists like Paucker with whom they could make frequent recourse to shared, often tacit knowledge, this made perfect sense. However, when following the first formal interferon gathering in Czechoslovakia, in 1964, interferon researchers began to expand their networks, the informal form of measurement began to show flaws. This is clearly illustrated by the problems involved in an exchange of samples of human interferon between Cantell and the Head of the Division of Infectious Diseases of Stanford University School of Medicine (CA), Thomas Merigan.

Creating a breeding ground for standardization

After noticing in the literature in the fall of 1965 that Merigan was also involved in setting up a production system for human interferon, Cantell immediately contacted Merigan.[14] With most interferon researchers using an assay system of their own, it was difficult to compare the assay data in the literature and to judge which production system was more promising in terms of the yields of human interferon. In order to get some idea about how his production and assay system compared with Merigan's, Cantell therefore proposed to exchange interferon probes. Merigan, who was keen to improve on his laboratory results with an eye toward producing a preparation suitable for limited trial in man, was receptive to the suggestion. Subsequently Cantell sent Merigan a batch of his human interferon. As per the specification the human interferon had been prepared by incubating cultures of human white blood cells ('leucocytes') with Sendai virus. The potency ('titre') of the material was said to be '1:500' or 500 units.[15]

Whereas a similar kind of specification made perfect sense to Paucker, in the case of Merigan it gave cause for confusion. In Merigan's hands, Cantell's interferon had measured a titre of about 1:2600 or 2600 units, but Merigan did not know what to make of the fivefold difference between his and Cantell's results.[16] If confirmed, it would imply on the one hand that his assay system was definitely more sensitive, but on the other hand, and less positively, it would imply that Cantell's production system for human interferon was superior to the one he employed (based on virus-infected cultures of neonatal fibroblast cells). However the results were far from clear yet. Much depended on what Cantell meant exactly by indicating that the titre of the Finnish interferon was 500 units. In a letter to Cantell, Merigan therefore put the matter up for discussion.

> I am not clear when you say 500 units whether or not you mean per ml or per 4 ml as you use as an absorption volume in your assay on 15 × 100 mm plastic plates or the equivalent. If you measure your interferon in 4 ml and divide it by 4 to express it on a per 1 ml basis then our results on your WBC (white blood cell) interferon would be quite similar.[17]

Merigan emphasized that in order to figure things out and enable a more accurate comparison of assay data, it would be necessary to know as precisely as possible the assay procedure each of them employed, as well as the mode of expressing interferon units. In his view, it was most important to know the volume the units were measured in.[18]

Judging by a follow-up letter from Merigan to Cantell, Cantell disagreed with Merigan on the relative importance of the volume aspect, but he shared Merigan's concern for more rigour in the exchange of information. They were ultimately able to settle, at least for the time being, on a fivefold difference in sensitivity of their assay systems. However, before extending this sort of comparison to a larger number of laboratories they were both in favour of establishing a common standard for human interferon. Merigan indicated that he would bring the standardization issue up for discussion at the next international interferon meeting that was planned for the spring of 1967 in London.[19]

For a couple of years Merigan had vocalized his support for the establishment of interferon standards among the group of American scientists interested in this biological agent. In order to assess the therapeutic potential of interferon, Merigan had chosen to work along the lines of exploring in a quantitative way the conditions needed for optimal protection against experimental virus infections both *in vitro* and *in vivo*. Merigan had not been discouraged by the inconclusive results of the British experiments with interferon in humans. The various published data on the broad antiviral spectrum and non-toxic nature of interferon *in vitro* were too promising to be put off by a single unsuccessful human trial. In his view, the British scientists had been too hasty.[20] They should have at least waited until they had investigated the quantitative aspects of the interferon phenomenon more thoroughly. It did not seem to make much sense to start clinical trials without knowing more about how much interferon was needed to protect laboratory animals against experimental virus infections.

However, he learned that for a majority of workers in the field standardization was not a major issue. A large portion of the work done was qualitatively descriptive in nature and usually considered to stand on its own merits. In general these researchers felt less need to be able to correlate published results from different laboratories. As long as the variation in research results did not make a qualitative difference, the matter was not taken very seriously.[21] As a subject, standardization was considered rather boring.[22] According to Finter, as far as most interferon researchers were concerned, if they got an answer then that was the answer. Should somebody else then come with a tenfold difference in research results, for instance, mostly the reaction would be in the following terms: 'Okay if you say so, most likely this means that your assay is more sensitive than mine'.[23]

Merigan was pleased to hear during an informal gathering of American interferon researchers at the 1964 meeting of the American Society for Microbiology in Washington, DC, that the established interferon researchers Sam Baron and Monto Ho shared his concerns about the need for common units of measure. They agreed to request financial and material support from the National Institutes of Health (NIH) for establishing and distributing reference standard preparations for chick and mouse interferon, which were most widely used among American interferon researchers.[24]

An experiment in scientific communication

Without much trouble, the NIH granted the request. In his capacity of Medical Director of the National Institute of Allergy and Infectious Diseases (NIAID) – one of the institutes that make up the NIH – and as former Head of the Virology Section of the Division of Biological Standards, Baron knew his way around the NIH bureaucracy. Mainly through Baron's offices the NIH administrators gave the undertaking to establish reference reagents for both mouse and chick interferon.[25]

While the preparations for research standards were under way, Baron learnt through his NIH network that NIH administrators were looking for research areas that might benefit from experimenting with a new form of communication, the so-called 'information exchange group'. In studying ways to speed up the development of the biomedical sciences, the NIH had engaged in an experiment in communication which aimed at improving and speeding up communication between scientists. It was argued that because of the inevitably long delays in publication and the distaste of editors for polemics, there was no longer any space left for real discussion in journals. Setting up information exchange groups with the means of dispatching communications, without any editorial restrictions, within a matter of days, free of any charge to all the members of a group, was expected to restore the role of argument as a public instrument of scientific progress. As Baron understood it, the working of this new form of information exchange would enable anyone working on interferon to communicate research findings or other scientific information in record time to all the others in the same area of research throughout the world.

This seemed just the kind of boost the embryonic field of interferon research needed. Baron decided overnight to take the initiative and ask the NIH Information Exchange Group Office for support to set up an information exchange group to provide better communication among scientists actively interested in problems of interferon.[26]

> I went to the administrator and said, 'Here is an area that is not fully accepted with a few struggling laboratories around the world. If we can keep them in constant communication they are likely to accelerate the development of this biological substance that will be of ultimate use as a therapeutic drug.' I indicated that it would be catalytic to set up an information exchange. Would they be willing to try it and they said 'yes'. I wrote to Isaacs and he agreed too. In fact he was chairman.[27]

Whereas most interferon researchers in America and overseas welcomed the initiative and immediately applied for membership of the NIH-sponsored Information Exchange Group no. 6 ('IEG 6') for research on interferon, its reception in Britain was mixed. The letter of invitation which was signed by both Baron and Isaacs caused some unease.[28] Participation in the exchange group was thought to conflict with the British policy of stringent control

Managing differences 97

over publication of work on interferon.[29] The industrial collaborators, in particular, were horrified by the thought of the free exchange of information with anyone actively working in the field. The American invitation letter spoke about communications that would include not only manuscripts but also discussions, unpublished research findings, technical suggestions, unofficial notes of meetings or any other original communications. The drug company executives took the view that to accept this would be to destroy the basis of collaboration.[30] Eventually, it was decided that on the condition that only manuscripts would be circulated that were already intended for publication and had been cleared by the patent adviser, members of the Interferon Collaboration were allowed to participate in the exchange group.[31]

By the end of January 1965, 109 biomedical researchers from 17 countries had been enlisted in IEG 6.[32] In its first two years the interferon exchange group distributed over 220 different communications of its members.[33] For the greater part these consisted of preprints of articles in press and reprints of talks given at symposia, but also included unofficial notes of meetings, technical suggestions, notes of unpublished research findings, and discussions of research problems such as difficulties in reaching some decision on what one ought to call an 'interferon', or difficulties with correlating research results from different laboratories.[34] Generally the information communicated by the IEG 6 preceded journal publication by six to nine months. It was this gain in time that appealed most to the members of the exchange group. According to Baron once you had your laboratory results processed you could have them out in two weeks time. 'This allowed for the following kind of dialogue: in response to so and so's item of last month I want to say that I did the same experiment but it did not come out quite the same way and I therefore think that the interpretation should be different'.[35]

However valued by its members the new communication system began to face increasing opposition from the editors of the biochemical journals. The editors heavily criticized the fact that a large part of the communications appeared to consist of preprints of full articles. As such the communications were in effect publications and could no longer be considered to be an equivalent to a society newsletter. They threatened to ban from publication anything that had first gone through an IEG. In order to meet the objections of the journals IEG 6 was modified and transformed into 'Interferon Scientific Memorandum' in the course of 1967. However, the change of name and sponsorship – from the NIH IEG office to the NIAID Policy Council – and the new rule that only abstracts of work being prepared for publication, plus the tables and figures would be accepted for circulation, did not affect the principle of the communication mechanism.

Like IEG 6 before it, the Interferon Scientific Memorandum functioned as a central clearing house for communications on interferon.[36] Sharing a formal mechanism for communication that was meant to accelerate the dissemination of information required sharing and shaping common denominators that allowed for dialogue and enabled participants to share research problems and cope with differences. As such the NIH-sponsored international information exchange system was instrumental in creating needs for standards. At the same time, by bringing out difficulties with correlating research results obtained in different laboratories, the IEG and Scientific Memorandum contributed to a growing awareness of and attention to this issue. The gradual rise in interest in the standardization issue manifested itself in American and subsequently British efforts to establish common units of measurement.

Safeguarding British interests

The news that the NIH was about to establish national interferon standards stirred a renewed interest in the standardization question in Britain. Before the news about the existence of American interferon standards, there had been occasional calls for more attention to the use of standards by the NIMR's Division of Biological Standards, but without much effect.[37] However, the fact that the Americans seemed to treat the matter very seriously led to an immediate reappraisal of the subject.[38] How should British interferon researchers correlate their interferon data with those of American researchers? How should they judge recent reports by American workers that high levels of interferon-like activity could be induced in cell cultures by adding a fungal extract known as Statolon?[39] Did this new American approach justify opening up a new British line of research? Because of the differences between the various circulating interferon standards, it seemed rather difficult to make a comparison between this new endogenous approach, and the conventional exogenous interferon approach.[40]

While the need for international standards was recognized as a matter of national interest, word came from Isaacs – who after a series of manic-depressive episodes with periodic hospitalizations and occasional medication was having one of his rare good periods[41] – that the virologist and interferon researcher Charles Chany in his laboratory at the Hôpital St. Vincent-de-Paul, Paris, had succeeded in setting up a production system that seemed feasible for the eventual 'mass production' of human interferon. Chany claimed to have produced the largest batch of partially purified human interferon available anywhere in the world. The news struck home as two months earlier it had still been difficult to envisage the production of large amounts of human interferon that would be needed to perform clinical trials.[42]

At the October 1966 session of the British Scientific Committee on Interferon Isaacs raised the question as to whether or not efforts should be made to join forces with the French in instituting clinical trials. Isaacs argued that establishing such a liaison would be in the interest of both parties. As the three participating firms (Glaxo, ICI and Wellcome) were clearly lagging behind in the production of human interferon, he believed that the Committee might take advantage of Chany's expertise with the large-scale production of human interferon. While giving Chany and his collaborators the credits for their production skills, there was reserve with regard to French clinical methodology: the

French were generally considered hopeless at arranging clinical trials. Judging by Chany's interest in supplying human interferon for a clinical trial in the UK, the French, in Isaacs opinion, were well aware that they could benefit from the Committee's experience of conducting clinical trials with interferon and from the availability of testing facilities at the Common Cold Research Unit. Isaacs then asked if the other members of the Committee thought the idea of arranging a clinical test of human interferon acceptable.[43]

Isaacs's question stirred anew a discussion within the Scientific Committee about the desirability of clinical trials as opposed to more basic *in vitro* and animal studies of interferon. Eventually the majority agreed that they should give it another try four years after the negatively assessed clinical trial programme had been aborted.[44] However, it was emphasized that before any arrangement was made with Chany the work which might be involved must be carefully considered. The chairman should therefore send a letter to Chany asking him for a quantity of human interferon and only suggesting the possibility of a joint trial.[45] ICI researcher Norman Finter thought it terribly important first to correlate the units used for human interferon in different laboratories. Without correspondence between units used by different workers, it was difficult to compare Chany's interferon with human interferon that was produced in a similar fashion by Cantell in Helsinki and Merigan at Stanford. Finter indicated that he would be able to discuss the idea for a comparative study in detail with all three workers at the Second International Symposium of Applied Virology and Medicine held in Fort Lauderdale (Florida) in December 1966.[46]

In the corridors of the Florida meeting there were rumours buzzing that interest in interferon was picking up momentum. No less than two international interferon meetings were scheduled for 1967 and after years of radio silence the field of interferon research seemed to produce results that were newsworthy. Beside news about French as well as Finnish and Russian investigators having developed production methods for human interferon and plans to use this interferon in clinical trials, word spread about promising laboratory results with synthetic interferon inducers.[47]

The news that Maurice Hilleman at the Merck Institute for Therapeutic Research at Rahway, New Jersey, and William Regelson at the Medical College of Virginia had both successfully explored the possibility of preparing synthetic materials of known composition in their laboratories which were capable of stimulating endogenous interferon activity was snapped up. It was most exciting that endogenous interferon production could be stimulated and studied *in vivo* by the use of easy to make and inexpensive chemicals without having to worry about impurities. Basically, working along these lines meant that interferon was no longer perceived in terms of a practical therapeutic agent but as a research tool to develop preferably chemical means – in other words synthesizing chemical compounds – for achieving viral chemotherapy. The dominant view was that if interferon was ever going to come to something clinically useful, the synthetic interferon inducers showed most promise.[48]

Isaacs was not to get a great deal of pleasure out of the reappraisal of the field of interferon research. Following a fatal haemorrhage Isaacs died on 26 January 1967 in London University Hospital, at the age 45. However shocking the news of his untimely death was, it hardly affected the course of events. Due to his illness he had been for quite some time a mere shadow of his former self.[49]

Through the rapidly expanding work on interferon inducers new needs were being created for common denominators. The research question that began to dominate the research agenda was whether the endogenous approach offered a fruitful alternative to the exogenous approach. Gradually, a division emerged between those who favoured working along the lines of the endogenous approach – the 'inducer people' – and others opting for holding on to the exogenous approach – the 'interferon people'.[50] However different their modes of perceiving interferon as an object for research, both groups shared a growing interest in standardization and establishing common yardsticks as a means to the end of comparing and criticizing research data from different laboratories.

The higher rating of the subject found expression in the fact that during the general discussion of the April 1967 Interferon meeting in London, special attention was paid to ongoing attempts to prepare standard interferon preparations and two months later at a follow-up conference in the beautiful setting of Siena (Italy), Finter – to his own amazement – was invited to give a lecture on standardization.[51] According to Finter, he did not think at the time that his presentation made very much impact, but it was instrumental in emphasizing once again the need for standards and in stimulating discussions in the corridors concerning British and American efforts to set up interferon standards.[52]

The NIMR's Division of Biological Standards had enlisted on behalf of the World Health Organization (WHO) seven laboratories in Britain, the US and Finland which were prepared to participate in an international collaborative assay of chick interferons. The participants were told that the assay was to be carried out in order to assign potencies to both a new British and American chick interferon standard in terms of the existing British Research Standard. This was said to enable researchers in the future to assign potencies on a common unitage basis.[53] What the British did not tell the participants was that they hoped that the Americans would agree to assay their standard in terms of the British standard.[54]

The NIMR's Division of Biological Standards had also hoped to start with an international collaborative assay of human interferon. This assay was to be carried out with the primary intention of investigating the suitability of Chany's material to serve as a research standard to be used in clinical trials. If successful, the MRC would be the first to offer a standard for human interferon, and it was hoped that such a standard would eventually prove suitable for submission to the WHO for consideration as an international standard.[55]

However, from the very beginning events were against the establishment of an international human interferon standard. The complications began with the presentation of the results of the comparative study of human interferon from four different laboratories at the March 1967 meeting of the British Scientific

Committee on Interferon.[56] Rather unexpectedly, the results of the comparative assays showed that Chany's French interferon material was not particularly potent. It turned out that Cantell's Finnish interferon was approximately 130 times more potent than Chany's sample which was of average potency. The members of the Scientific Committee were in a scrape, because they already had officially arranged for a meeting with Chany during the London Interferon Symposium to discuss a proposal to carry out a clinical trial with Chany's interferon in MRC facilities. They did not feel like putting the friendly relationship with the influential French interferon researcher Chany on the line. That could only put at risk the growing coherence in the field of interferon research, and that was in nobody's interest.

Eventually it was decided to tell Chany that the Scientific Committee could only perform clinical trials with the most potent material available – Cantell's interferon, but that Chany's material could well serve the purpose of setting up a standard. It was a political compromise that would leave its mark over the years. Although, initially, everything went smoothly. Chany accepted the offer and at the beginning of May 1967 sent five litres of human interferon preparation to the NIMR's Division of Biological Standards.[57]

At the June 1967 meeting of the British Scientific Committee it was reported that more than 3000 1-ml ampoules (labelled with the code number 67/87) filled with freeze-dried human interferon material were in store. The proposed British research standard 67/87 was currently undergoing stability tests, and everyone hoped that they could soon start arranging an international collaborative study in which this material would be compared with other preparations of human interferon. During the same meeting plans for clinical trials with Cantell's interferon were discussed and it was decided that the action to be undertaken first was to organize an inspection visit of Cantell's laboratory.[58]

A Finnish detour

On the afternoon of Sunday 6 August 1967 at Helsinki airport Kari Cantell nervously awaited the arrival of Norman Finter. It had been more than half a year since he had decided, during informal talks with Finter at the Fort Lauderdale conference, to start a collaboration with the British Scientific Committee on Interferon Research. Over a six month period Cantell had supplied the Glaxo biochemist Karl Fantes with several batches of his human interferon produced in white blood cell (leucocyte) cultures for purification and potency studies. Apart from the development of a precipitation method by which the interferon could be concentrated, little had been achieved in terms of the purification of interferon. He had therefore been pleasantly surprised early in July to receive an official request from the Scientific Committee to intensify their collaboration. The British asked for additional amounts of his interferon to carry out work which would lead to a trial in humans.

However eager to participate in an ambitious international project such as this, Cantell was somewhat taken aback by the fact that he had to allow for an inspection of his laboratory by foreign researchers. At first he felt irritated by the apparent lack of confidence of the British in the safety of his production method: the production protocols in his laboratory were in accordance with the same stringent safety standards that were operative at the Finnish Red Cross Blood Transfusion Service nearby. Finter's personal assurances that laboratory inspections were made a compulsory part of British research projects that involved making biological material for use in humans helped to overcome his negative feelings. In talking over the matter with his Swedish research assistant, Hans Strander – who was trained in medicine and immunology at the Karolinska Institute in Stockholm and had arranged to pursue his doctoral thesis at the State Serum Institute in Helsinki – he was further persuaded to cooperate.[59]

Hadn't they been concerned too about the risks involved in using a product stemming from leucocytes from blood donors in a trial in humans? Apart from the possibility of allergic side-reactions, the final product might be contaminated with hepatitis or other pathogenic viruses originating from the blood of infected donors. It was therefore understandable that the Scientific Committee like they themselves wanted to proceed with the utmost caution in this matter. Moreover, they both regarded the inspection visit as a sign that the British took the project very seriously. Cantell also discussed the likelihood of virus infection, in particular the hepatitis problem, with Harri Nevanlinna, the director of the Red Cross Blood Transfusion Service in Helsinki. The information he gathered was less promising than expected. Cantell learned that there could be no absolute answers. The only thing that Nevanlinna could be sure about was that so far his service's plasma products had caused very few cases of hepatitis. It made him feel a little bit uneasy about the inspection visit.

Cantell was relieved to find out that Finter would arrive one day in advance of the other members of the British inspection team and stay overnight at his private home. This provided them ample opportunity to have preliminary discussions about the ins and outs of the collaborative project and about the state of the art of interferon research. Over dinner it turned out that they both looked upon the rapidly growing popularity of interferon inducer research with envy and felt themselves increasingly marginalized by the 'inducer people' under the virtual command of the powerful and charismatic Maurice Hilleman. According to Hilleman there was no future for exogenous interferon in medical therapeutic terms. This made Cantell and Finter feel that they were fighting for a common cause: researching the actions of exogenous interferon. They felt encouraged by reports from Vladimir Soloviev of the Virology Department of the Russian Academy of Medical Sciences in Moscow – despite a shared scepticism about the state of the art of Russian interferon research – who had performed several successful trials with human leucocyte interferon in large numbers of volunteers infected with influenza virus. The best protection seemed to have been obtained when the interferon was given to humans 2–24 hours before being challenged by the virus.

By the time the other members of the inspection team arrived, Cantell had already familiarized Finter with his work as head of the Virus Department at the State Serum Institute in Helsinki. Finter learned that optimizing the production of interferon by human leucocytes was a laborious and time consuming research activity governed by trial and error. The production process of interferon appeared very sensitive to disturbances and changes – for instance they experienced that for uncertain reasons using flat instead of round bottles for culturing the white blood cells appeared to affect the interferon yields in a dramatic way.

Keeping up the production levels already required hard and meticulous work – let alone engineering an increase in interferon yields and to scale up the production system. So far the most rapid and abundant production of interferon was achieved by incubating purified cultures of white blood cells obtained from blood donors with large doses of Sendai virus.[60] After storage overnight the supernatant fluid which contained interferon was harvested and underwent a basic purification procedure. Since both Cantell and Strander lacked the biochemical skills thought to be necessary for purifying the interferon preparations to a degree acceptable for use in humans it was quickly agreed to leave further purification and concentration work to Glaxo laboratories.

The British inspection team was impressed by the amount of work that went into setting up and maintaining an experimental production unit for human interferon, and by the efficiency and skilfulness with which the production of interferon was dealt with. Interferon was produced in a special room that was isolated from public health work and was occasionally used for the production of vaccines. The small laboratory space for the production of interferon was continuously supplied with sterile air. And before entering the production area the British researchers had to change clothing in order to prevent contamination and to maintain sterile conditions during the production process (see Figure 24). This resembled very closely the stringent British vaccine production procedures.

They were also shown around the transfusion service where they learned that the preparation of interferon was not done at the expense of blood that was needed for blood transfusions. Since separate fractions of blood (plasma, packed red cells, antihaemophiliac factor and platelets) were increasingly used for transfusions instead of whole blood, the transfusion service was in the process of fractioning more and more of its donor blood. Of the 250,000 blood donations collected each year about 100,000 were currently centrifuged to separate the various parts of blood. It appeared to require only a little extra work to collect the leucocytes routinely during this procedure as a 'by-product' that was to be sent to Cantell's laboratory for further processing.[61]

However short the actual tour through the Red Cross building, the visit itself was drawn out. Lengthy discussions took place concerning the risk of hepatitis infection of blood products. In an attempt to reassure his visitors with regard to the hepatitis problem, Cantell emphasized that Finland compared favourably with other European countries with its low incidence of infectious hepatitis

Figure 24 The production of leucocyte interferon in Helsinki (late 1960s/early 1970s). Courtesy of Dr K. Cantell.

and that he would as soon as possible send them the latest official figures. It certainly meant that safeguarding the clinical safety of the final product would be a costly affair.

Wrestling standards

About the same time as the British visit to Cantell's human interferon production facility, industrial interest in interferon rose to a new height. Rumours started to buzz around that at Merck and the NIH experiments were already in progress which showed that endogenous interferon stimulation through treatment with specific synthetic substances could cure acute viral infection in animals at a non-toxic dosage level.[62] Drug company executives in both America and Britain were drawn to the synthetic interferon inducers. These compounds were easy to make chemically – fitting perfectly well their chemically oriented drug development trajectories – inexpensive and induced large quantities of interferon, whereas the production of useful, in other words safe, potent and pure enough preparations of exogenous interferon appeared to be difficult and costly.[63] The new industrial drive to capitalize on the clinical promise of the interferon-stimulating agents only added to the growing interest in establishing interferon standards. Since little was known of how exactly interferon acted or what its precise nature was, common yardsticks at least were needed to facilitate the assessment of the relative merits of employing interferon inducers versus exogenous interferon.[64]

The parcel of Cantell's frozen leucocyte interferon arrived with flight AY 831 from Helsinki on the 29 January 1969, 9.05 am local time at Heathrow airport, and was safely transferred to the deep freeze at the NIMR's Division of Biological Standardization.[65] About two months later the material was thawed, filtered, distributed into glass ampoules, coded 69/19 and freeze-dried. It had taken two years to reach this stage, but except for the availability of two different proposed research standards for human interferon little progress was made. Establishing the 'Finnish made' 69/19 in conjunction with the 'French made' 67/87 even created new disparities and therefore could only be expected to complicate the job of managing differences.

The problems of coordinating standardization efforts at an international level were a constant source of irritation between the British, Americans and the French. In an effort to bring the parties closer to one another the Permanent Section of Microbiological Standardization of the WHO started preparations for an international conference on the subject of standardizing interferons.[66]

In his conference address the chairman and former head of the NIMR's Division of Biological Standardization, David Evans, made an appeal to the 84 delegates from 13 countries to direct their attention to reaching agreement on setting up international interferon standards. In Evans's opinion this was the only way to overcome the major stumbling block in the interferon field – 'namely, the inability of workers to express the activity of interferons in meaningful terms'.[67] At the same time, in an apparent effort to reassure those who were afraid of the fuss that was felt to be associated with the standardization efforts, Evans tried as hard as he could to play down the requirements for providing all laboratories with the same yardsticks. However, setting criteria, assigning units and choosing suitable reference materials would require far greater efforts from the participants than Evans tried to make himself and the others believe.

This is aptly illustrated by the numerous and lengthy discussions which were needed to achieve consensus about the ins and outs of interferon preparations that would make most suitable research standards. For instance, Evans indicated that the standard should be distributed in a form which would ensure its stability for many years. Among other things, it should be kept at a low temperature. However, what was for some a 'frozen state' that would guarantee stability was for others a freezing temperature that would still allow for degradation. It was eventually agreed that the reference standard should ideally be freeze-dried and kept at $-70°C$, but that the more practical temperature of $-20°C$ was acceptable too.[68] In addition, Evans's criterion that ampoules containing standards should be sealed by fusion with glass threatened to exclude the United States, where it was common practice to use rubber stoppered vials, from contributing standard preparations. The issue was amicably settled by a recommendation to siliconize the rubber stoppers, which would remove most likely contaminants such as oxygen and water. Furthermore, should interferon standards be crude or purified? Crude interferon preparations were known to produce easily misleading results because of the presence of loads of impurities. While highly purified interferons would have the advantage of giving those involved less cause to rack their brains over the nasty question: 'Is this phenomenon which I observe in my laboratory due to the interferon, or is it due to an impurity?', they were claimed to be rather unstable.[69] The compromise was to purify them in such a way that would rid them of as many biologically active interfering impurities as possible.[70]

In the ensuing final discussion of the conference the adoption was recommended of four research reference standards. While it was decided that the research standards for mouse (6000 units per ampoule) and rabbit interferon (3000 units per ampoule) would be made available through the Reference Reagents Branch of the NIH, they agreed to adopt as the research standards for chick and human interferon the preparations that were stored at the NIMR's Division of Biological Standards.[71]

I do not know what went on behind the scenes in the weeks after the meeting, but what I do know is that the recommendations had changed by 18 November, when the conclusions of the Standardization Conference were sent to all participants.[72] The British Conference Secretary, Frank Perkins, indicated in an accompanying letter that in consultation with Evans he had decided to include in the recommendations that the more potent preparation 69/19 of Finnish origin instead of the 'French made' preparation 67/87 should be adopted as the research standard for human interferon.

'Having to haul down one's colours'

On receipt of Perkins's letter and the enclosed conclusions of the Symposium, Charles Chany was, to say the least, surprised to see the changes and immediately sent back a letter to Perkins to tell him that he had misquoted the recommendations.[73] Critical reactions from American and other quarters followed swiftly, and within ten days of the first letter Perkins was forced to send around a revised statement on the section concerning human interferon.[74] It said that the Conference agreed to adopt 67/87 as the research standard for human interferon and that the Conference was also informed that a more potent preparation 69/19 would be made available to serve as a working standard for special use in insensitive assay systems. The section ended by giving the advice – which eventually would not be published in the Conference Proceedings – that 67/87 should be replaced with 69/19 sooner rather than later. Perkins had done his best to find a compromise that was thought to suit both sides, but in doing so he all but helped to clarify the situation.

The summary report which was sent to all the participants in the collaborative studies caused quite some upheaval in French quarters. Chany was baffled by the fact that, regardless of the admitted inadequacies in the test results, a decision seemed to have been made regarding the suitability of his and Cantell's material to serve as a research standard. Had not 67/87 and 69/19 both been found sufficiently stable to be employed for longer periods? How could the Finnish material be established as a research standard without being able to produce a reliable estimate of 69/19 in terms of 67/87? As far as Chany knew it had been agreed upon at the Standardization Conference that all future reference materials would be adjusted to 67/87.[75] The British not only did not seem to feel bound to abide by the recommendations of the Standardization Conference, but on top of that they proposed to fly the British flag over what was supposed to function as an international research standard.[76]

Chany immediately decided to take action. He began by sending his grievances in diplomatic terms both to Anderson and the participants in the collaborative studies.[77] He realized that he had the odds stacked against him. In the January 1971 issue of the Interferon Scientific Memorandum a joint statement from the NIMR and the NIH had been published to formally acquaint the workers in the field with some modifications of the recommendations of the Standardization Conference. The position regarding the provision of reference materials was said to be that 67/87 was no longer regarded as the research standard for human interferon but that, depending on their suitability, either 67/87 or 69/19 would soon become established as a research standard.[78] In Chany's view this was already a first formal step toward abandoning 67/87, and unless he succeeded in gaining the support of the Americans he was fighting for a lost cause. So, when Perkins did not show any response to his adverse comments Chany chose a more aggressive approach.[79] In January 1972 Chany paid a visit to the NIH in Bethesda, Maryland, to win support from those directly involved in the standardization efforts.[80] At the same time he requested that the British return to him the material that he had given to them as a possible interferon standard. He told them that he was planning to distribute it on his own initiative as the original interferon standard.

Eventually, by the end of May 1972, through the intervention of the Americans, a compromise was reached that offered a way out of the confused situation. Basically it meant that 67/87 was retained as the original reference reagent available only on special request and used as the basis for assigning unitage to any subsequent reference material, while 69/19 would be distributed routinely as the standard for general use. Furthermore, to satisfy the American demand that 'nationalism be removed from the reagents', the NIMR's Division of Biological Standards eliminated the word 'British' from the label and renamed 67/87 as 'Research Standard A' and 69/19 as 'Research Standard B' for human interferon.[81]

The political jousting hardly affected the work at the laboratory bench. Insofar as researchers had been involved in the collaborative studies of human interferon they (apart from Chany) already showed a strong preference for using the more potent preparation 69/19 as a working standard. Eventually this preparation became regarded as the base line for comparing work on human interferon (see Figure 25).[82]

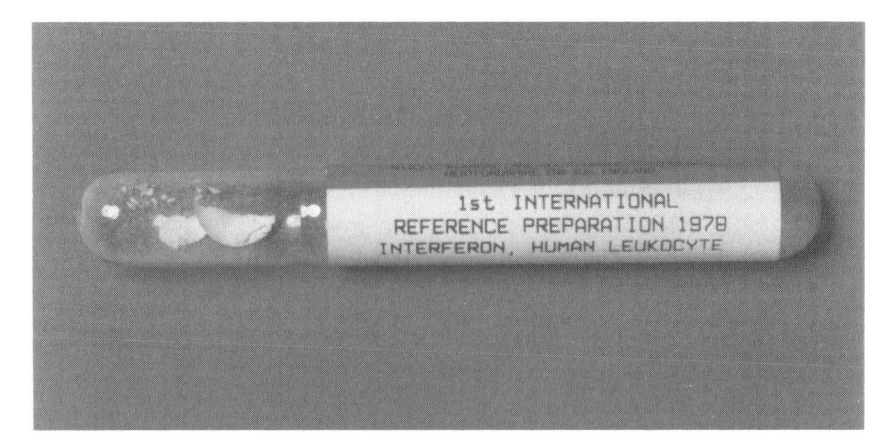

108   *Interferon: The science and selling of a miracle drug*

Figure 25 The Expert Committee on Biological Standardization of the WHO considered data relating to 69/19 and finally in 1978 established part of the stock of preparation 69/19 (MRC Research Standard B) as the International Reference Preparation of Interferon, Human Leukocyte. Courtesy of NBSB.

However little this chemically putative natural substance still meant to most British drug makers, who apart from Burroughs Wellcome had abandoned work on interferon, interferon had managed to establish itself as an object of laboratory study for an international field of researchers.[83] They mainly worked with chick, mouse and rabbit interferons. By 1972 most major laboratories were in the process of including the internationally recognized research standards into their laboratory routines by calibrating their internal laboratory standard against the appropriate research standard obtained free of charge from the NIMR's Division of Biological Standards or from the Reference Reagents Branch of the NIH. Increasingly, editors of pertinent scientific journals requested that papers dealing with interferons should include results in terms of the internationally accepted unitage, but also in informal contacts a growing number of researchers started to communicate in terms of the appropriate MRC or NIH unit.[84]

The researchers involved were still in the process of learning how to handle common units of measurement and disciplining themselves into specific routines and actions. The local variances were still considerable, and time and again problems arose with interpreting unexpectedly high variation in laboratory results. Had the activity of the standard gone off, was it related to flaws in the standardization procedure, was one dealing with yet unknown biological characteristics of interferon or even with a different type of interferon?[85]

By the time a second international workshop on standardization was being held at Woodstock, in September 1978, the use of specific research standards had become embodied in daily laboratory practice in the field of interferon research. It was widely considered to constitute a necessary condition for performing research on interferons. The worldwide distribution and use of a fixed set of research standards not only was generally thought to have made the comparison of results from different laboratories more interpretable but also to have played an instrumental role in redefining the nature and functions of interferons.[86] It was no longer a question whether interferon was heterogeneous in nature but what the nature and function of the newly established differences were. Not only different species but also the same species produced different interferons – two different types of human interferon were officially distinguished: so-called 'classical' (acid-stable) type I interferon and 'immune' (acid-labile) type II interferon. Thus, in attempting to cope with and master differences, interferon researchers had created and as I will show in the next chapter would continue to create new differences, albeit of a different nature.

What makes a lasting impression is the large time span (more than 15 years) between the first discussions about standardization and the routine use of at least three different international research standards as stable and unquestioned parts of interferon research. Apparently the perceived need for standardization was dependent on the nature and purpose of research, and on institutional and national as well as field interests. As long as the researchers were able to narrow down the margins of interpretation to workable levels by informal means there was hardly any incentive for formalized forms of standardization.

Not until the informal form of measurement began to show flaws, with the expansion of the field of interferon research and with the intrusion of a higher order of interests (company, governmental and public interests) into the relatively private domain of laboratory life, did a more general and permanent need for formal types of standardization begin to manifest themselves. The value and power of interferon standards as common frames of reference or 'laboratory yardsticks' – which in being mobile, and conveying relatively stable information, transform individual laboratory practices into universal practices – appeared largely dependent on the extent to which they became embodied in actual research practice.

Ultimately as a result of handing around and learning to use standards, the interferon researchers achieved far more than resolving research problems. In shaping and defining the outcome of experiments, interferon standards came to serve as a constitutive part of interferon research. As such, the use of standards not only helped researchers to discriminate between and compare outcomes of experiments, but was also instrumental in creating differences and redefining problems of research: in a collective effort to master differences those involved had established new differences of such a nature that these multiplied and changed their options for research. In the next chapters we will see how the creation of new options between bench, bedside and the public sphere in the 1970s and 1980s resulted in new career opportunities for interferon.

5 About mice, malignancies and experimental therapies[1]

In the USA in early 1970, Mary Lasker the philanthropist and notorious lobbyist of American medical research had assembled a panel of consultants. This group, largely consisting of friends and influential associates also known as the 'Laskerites or Mary's little lambs', was to advise the American Senate on legislation involving a 'moonshot' approach for cancer that held out the promise of major progress in the officially declared 'war on cancer'.[2] One of the scientific panel members was Mathilde Krim, a Swiss-born geneticist and virologist at New York's Memorial Sloan Kettering Cancer Center, whose husband Arthur Krim, a media tycoon, was influential in the Democratic party. Krim helped draft working papers on the progress of cancer research, which became part of the technical portion of the panel's report entitled *National Program for the Conquest of Cancer*.[3]

Krim did her share of the work in searching the literature for promising areas of cancer research that deserved more attention. In doing so she ran across the interferon literature and learned that studies had been published claiming that interferon and interferon inducers, by means of their antiviral activity, had an inhibitory effect on tumour viruses and virus-induced tumours in mice and rats. In accordance with the mainstream in virology, Krim firmly believed that tumour viruses played a major role in the aetiology of malignancies not only in animals but also in humans. Unlike most of her scientific colleagues, however, she qualified interferon as a potentially interesting anti-tumour agent for use in humans. Krim managed to include interferon research in the panel's report as a promising area that needed intensive further study.[4]

After presentation of the report to the Senate, Krim remained closely involved with the prelegislative agenda-setting activities of Mrs Lasker up to the enactment by President Richard Nixon of the National Cancer Act in December 1971. This event marked the start of a much expanded National Cancer Program and was the outcome of more than a year of political struggle and compromise in which Mathilde Krim became familiar with the ins and outs of American cancer politics. According to Krim it was Lasker who taught her the political groundwork of science lobbying and the persistence needed to succeed as a self-appointed lobbyist.[5] In the case of interferon more than a little endurance would be required.

However promising in the eyes of Krim, within the context of American cancer research interferon was virtually non-existent or at the best considered part of the history of dead-ends in the search for antiviral therapies. Even for researchers with an open mind for the new and unexpected the idea that the action of this antiviral substance might be pleiotropic sounded questionable. Neither did it seem to fit the dominant 'biomedical model' of specific aetiology of disease, nor did it fit the notion of specific therapy on which most therapeutic drug research programmes were based. Moreover, there were concerns about funding research with a poorly defined and impure substance for which a production and purification technology had yet to be developed.

The question that will be addressed here is how in working between bench, bedside and the public sphere French and Scandinavian laboratory researchers and clinicians with the help of the self-appointed American lobbyist Mathilde Krim succeeded in reshaping interferon's profile; from a dime a dozen failure in the search for antiviral therapies into a promising lead in the search for anti-cancer therapies. Before following up Krim's wheeling and dealing with interferon, I will first describe how the notion of the anti-tumour effect of interferon came into being, primarily in France, Finland and Sweden.

Interfering with cancer

The action of interferon on tumour viruses had first been singled out for study and speculation by two French virus researchers, Charles Chany and Pascu Atanasiu in the early 1960s. They claimed that interferon inhibited the growth of polyoma virus, a specific strain of tumour virus, in hamsters.[6] Initially scepticism prevailed among fellow scientists, but after a couple of British and American publications confirmed that interferon seemed to have an inhibitory effect on certain strains of tumour viruses *in vivo* the subject matter was taken more seriously.[7] At about the same time several articles were published including one in *Science*, suggesting that leukaemia in humans was a viral disease.[8] Since viruses which were considered to induce leukaemia had already been isolated in mice and chickens, the claims were in general taken seriously by biomedical scientists.[9] Consequently the 'hunt' for a human leukaemia virus was intensified and a growing number of biomedical researchers became involved in the field of leukaemia research. As did the American virologist Ion Gresser.

Having accepted Chany's invitation to set up his own laboratory of viral oncology at the *Institut de Recherches Scientifiques sur le Cancer* in Villejuif in the outskirts of Paris, Gresser was looking for research projects that would satisfy his interests in interferon research, fit into the research profile of a cancer research institute, and would be promising and daring at the same time.[10] To meet these conditions he chose to organize his research around the problem of leukaemia and viruses.[11] To Gresser's best knowledge there was no published report of interferon's value or lack of activity in mouse leukaemia.[12] Furthermore, he knew of only a small number of other interferon researchers who thought it worthwhile to invest their resources in studying interferon's effects *in vivo*.

The consensus among interferon researchers was that with the available techniques you could never produce enough interferon to make a difference in animals, let alone man.[13] Moreover interferon had only been shown to prevent the development of virus diseases (as a 'prophylactic') and was without effect once the infection had established itself. Gresser, however, did not want to take this for granted as long as nobody had made a serious attempt at administering high doses of interferon in virus-infected animals for longer periods of time.[14]

Gresser could afford to plunge himself into a problem which entailed a break from mainstream interferon as well as tumour virus research. The research expenses in his Paris laboratory were for the greater part covered by a generous unconditional grant provided for by the Nestor of American Cancer Research, Sidney Farber, who worked as Director of the wealthy Children's Cancer Research Foundation in Boston, the largest clinical centre for the study and treatment of child leukaemia in the USA.[15] This enabled him to take a relatively independent research position and choose an area of research which most fellow researchers shunned.

Gresser's first objective was to find a method to produce large amounts of highly active mouse interferon. From the literature Gresser knew that Norman Finter had developed a production method – obtaining interferon from the brains of mice after infection with so-called 'West Nile' virus. However, the methodology was rather demanding because of the large number of mice needed and the laborious isolation procedure.[16] But, with a limitless supply of laboratory mice at his disposal and the help offered by Chany and his team – who were also interested in interferon as a possible approach to anti-leukaemia therapy – Gresser decided to give it a try.[17]

By the end of November on Mondays they routinely injected about 500 mice with West Nile virus and on Fridays it was 'butcher's day'. Standing all together along large tables they mechanically opened and emptied the hundreds of mouse skulls and threw the brains in ice chilled beakers.[18] At the end of Friday afternoon the brains were mashed up in an electric mixer and the resulting mouse brain suspension was taken for further biochemical processing, including centrifugation. The final product was the equivalent of soup broth after all solids have been strained out and was designated 'mouse interferon'.[19]

These preparations of mouse interferon were employed to figure out what effect interferon had on the pathogenesis of murine leukaemia. A fixed number of mice were treated with mouse interferon and subsequently inoculated with Friend mouse leukaemia virus ('Friend virus') which he had obtained from the American tumour virologist Charlotte Friend. Daily administration of interferon was continued for 2 to 5 weeks. Simultaneously a control group was only inoculated with the virus. The course of the leukaemic disease process in the infected interferon-treated and infected untreated mice was compared. This entailed daily monitoring of the condition of the mice until the animals were sacrificed between the second and fifth week.[20]

For more than four months they seemed to go absolutely nowhere. Analysis of data in the lab notebooks was disappointing since no statistically significant differences could be determined between treated and control group. Over and over again the experimental conditions were changed until they were able to demonstrate by May 1966 that interferon, provided that extra concentrated preparations were given, showed an virus inhibitory effect in mice. While immediately initiating further experiments Gresser put himself to the job of writing up the first results for publication. Ultimately in August 1966 two articles were submitted to the *Proceedings of the Society of Experimental Biology and Medicine*.[21]

The referee reports were received with mixed feelings by Gresser and his team. They were credited for showing that mouse leukaemia could be inhibited by biological means, but the data indicating that interferon was responsible for the inhibitory effects were considered unconvincing. The identification of an antiviral substance as an interferon was known to be fraught with pitfalls implicit in reliance on bioassay procedures. This was certainly the case when using almost crude preparations. The inhibitory action might as well be due to residual interfering virus or to mechanisms as yet undetermined. They insisted on Gresser employing the term 'preparations of interferon' instead of interferon, since there was not sufficient proof that the inhibitory effect was mediated by interferon. It was sheer pragmatism that made Gresser, who himself was convinced that interferon was the decisive element in his experiments, adapt his text. At least the incident underlined the importance of investing part of his resources in attempts to further purify and concentrate his interferon preparations.[22] The data showing that purified preparations of mouse interferon were as effective as crude interferon preparations, were used instrumentally in subsequent publications and seminars in both Europe and America to argue in favour of interferon being responsible for the inhibitory effects.[23]

At about the same time the discussions during the Friday 'slaughter' sessions in Gresser's laboratory centred around the mechanism of the observed inhibitory effects. Gresser would usually bring the topic up for discussion in the following way:

> Given the dogma in the field of interferon research that interferon inhibits virus growth we automatically seem to accept that it acts by repressing viral multiplication and thus protecting normal cells from infection and subsequent malignant behavior. But what evidence is there to exclude the theoretical possibility that also some kind of direct inhibitory action of interferon on virus-transformed malignant cells is involved?[24]

Started as a kind of funny thought experiment, it gradually became one of Gresser's favourite topics for discussion both in and outside his laboratory.

Most interferon researchers laughed off his theoretical speculations about interferon having additional biological effects.[25] They were not prepared to give up on the generally accepted concept of interferon as an essential part of a particular cellular defence mechanism against viruses. The fact that no other research group had succeeded in inhibiting Friend leukaemia disease in mice by the administration of interferon did not add to Gresser's credibility.[26]

Far from discouraging him the critics made Gresser more stubborn about his mouse leukaemia studies.[27] In the process he was able to show that daily administration of mouse interferon was also effective if treatments were started 2 and even as long as 7 days after the mice had been infected with Friend virus. To his knowledge, these results represented the first demonstration that interferon exerted a significant antiviral effect after infection had been well established in the animal host.

At the end of 1967 Gresser learnt that a fellow researcher in his institute happened to be in the process of establishing a so-called 'RC19' tumour cell line which originated from mice infected with Rauscher leukaemia virus. The available tumour cell line suggested the opportunity to test the idea about interferon having a direct effect on mouse leukaemia that was different in nature from inhibition of viral multiplication.[28] It did not seem much extra work to do some experiments with the 'RC19' tumour cells, since a couple of changes in the experimental design of the Rauscher virus experiment would suffice. Gresser left the bench work to his experienced laboratory technician Chantal Bourali, who started with the actual mouse trials early in 1968. Instead of inoculating mice with Rauscher virus she injected Rauscher leukaemic cells and waited for 24 hours for the leukaemia to establish itself. Then she split the animals up into two groups. One control group received nothing and the other the usual daily administration of interferon. Bourali ear-tagged all the animals with a metal ring, wrote down the number of the mouse and the treatment it received. When after two weeks Bourali told Gresser that the first mice in the control were dying while the interferon-treated mice were only having minor disease manifestations he responded coolly that it was best to wait and see what happened next.[29]

Two weeks later in reaction to Bourali's presentation of the preliminary results of her experiment Gresser was much more attentive. Whereas most of the control mice were reported dead more than 90 per cent of the interferon mice were still alive. Though initially excited he feared the dramatic result might not be repeatable. He read Bourali's laboratory notes over and over again. He suspected Bourali of having made a major mistake, but he could not figure out what had gone wrong except for the fact that she had forgotten to inject the tumour cells into the interferon group. He asked her to repeat the experiment. When by the end of April Bourali again reported similar results Gresser was perplexed. He was still reluctant to accept the results at face value and opted for yet another repeat experiment, this time together with Bourali. However, the student riots in Paris – bringing public life to a standstill for more than two months – put their patience to the test.[30]

Mice, malignancies and experimental therapies 115

By the time the international interferon conference was held in Lyon early in January 1969, Gresser had enough confidence in the consistency and persuasiveness of the laboratory data to come out with the claim that continued administration of mouse interferon increased the survival of mice inoculated with tumour cells (see Figure 26). However, most interferon researchers continued to smile at the suggestion that interferon might have a direct inhibitory effect on the tumour cells – implicating that interferon was a multi-functional agent. They argued that the anti-tumour effect was most likely due to non-interferon contaminants in Gresser's preparations.[31] Moreover, another paper presentation by the NIH researcher Hilton Levy, who worked along similar lines using synthetic interferon inducers instead of mouse interferon, was not particularly supportive of Gresser's claim.[32] Levy indicated that he did not think that the observed anti-tumour action of his chemical compound was directly related to its interferon inducing capacity, particularly since only two of the tumours tested were of viral origin. He believed that something different was involved in the chemical's anti-tumour action and this made him very doubtful about the possibility that he and Gresser were dealing with the same phenomenon. In Levy's opinion interferon was in neither case at the basis of the observed effects.[33]

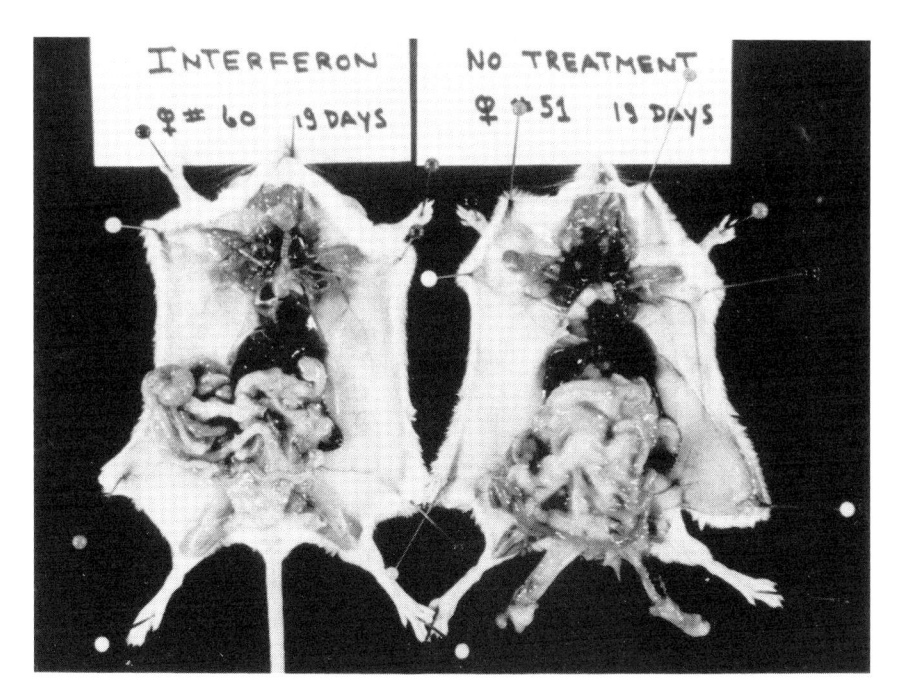

Figure 26 Gresser's conference slide with interferon treated and untreated female (Balb/c) mice inoculated with (2,000 RC19) tumour cells and sacrificed on the 19th day. Note intraperitoneal tumour masses in untreated mouse on right-hand side. Courtesy of Dr I. Gresser.

Despite the controversy about the nature of the reported anti-tumour actions, there was general consensus about the practical relevance of both studies. Supporters and critics alike agreed that Gresser's and Levy's animal data were fascinating in their own right by suggesting new therapeutic possibilities in the treatment of cancer.[34]

Hoping for patients to behave like laboratory mice

Like most other interferon researchers present at the Lyon meeting, the Finnish interferon researcher Kari Cantell had never paid special attention to the series of articles from Gresser's research group showing that interferon was effective in treating virus-induced leukaemias in mice. It never occurred to Cantell that Gresser was exploring 'different territory'.[35] His experiments seemed to confirm interferon's potential as a broad spectrum antiviral agent but this never made Cantell think about a serious therapeutic link between interferon and cancer. However, hearing Gresser talk about increased survival of mice inoculated with tumour cells and treated with interferon preparations made a difference to the physician Cantell. Whatever the mechanism of the effects the exciting thing in his view was that interferon seemed to show definite anti-tumour activity.[36]

Gresser's animal studies in combination with the interest of the National Institutes of Health (through its Antiviral Substances Program that was established in 1969) in his interferon production facilities encouraged Cantell to initiate trials in humans on his own.[37] Collaboration with the Americans was attractive, because of the high payments they offered for the production of human leucocyte interferon.[38] This enabled Cantell to employ more staff and obtain more sophisticated equipment needed to expand his production capacity as well as research activities.[39]

While visiting one of the paediatricians at the University Hospital in Helsinki to discuss plans for a leukaemia trial in children in October 1969, Cantell learned that a 15-year-old boy was suffering from a rare complication of a measles infection ('SSPE') – a severe and ultimately fatal subacute sclerosing panencephalitis (inflammation of the brain) – without any hope for recovery. Apart from an immediate professional interest in the case as a virologist, Cantell became personally touched by seeing the young patient fighting a losing battle against the virus disease. In consultation with the responsible paediatrician and the parents, who were willing to try anything that might help, Cantell decided to give it a try. Treating the boy seemed better than doing nothing and at least it provided a good opportunity to learn more about the pharmacology (dose–effect relationships) of his interferon preparations.[40]

It was a big disappointment – although no one knew what to expect – to see that the intravenous injections with the interferon material were followed by serious side-effects. In spite of the fact that a couple of hours after the injections the clinical condition seemed to improve considerably, the paediatrician and Cantell decided to discontinue the treatment. They would not resume interferon treatment until Cantell succeeded in producing a more purified interferon preparation that was believed to cause less toxicity problems.

Two months passed before Cantell felt confident enough about his purification procedure to give it another try. As a precautionary measure the injections were no longer given directly into the veins but intramuscularly. The boy received a total of about seven injections with what Cantell called 'semi-purified material' – containing about 400,000 'international units' of interferon.[41] This time only a few minor flu-like side-effects were monitored. However encouraging, the injections did not make a difference to the already bad clinical condition of the boy. After treating another young patient without achieving any favourable clinical results, the paediatrician lost interest in Cantell's trial plans.[42]

The failure of the Helsinki endeavour coincided with a request from Hans Strander – who after leaving Cantell's Helsinki laboratory had started residency in oncology at the Karolinska Hospital in Stockholm – to do a collaborative study with interferon in cancer patients at the cancer ward ('Radiumhemmet').[43] The culture of clinical experimentation was strongly present at the Radiumhemmet. There were historically grown links between 'cutting-edge' research and cancer therapy. It was this combination of therapy and research that attracted cancer patients from all over Sweden in the hope of benefiting from the latest technological advances: experimental surgery, x-ray machines and experimental drugs.[44]

For residents like Strander it was normal practice to combine cancer therapy with clinical research. As a physician researcher he was supposed to straddle between the bedside and the laboratory bench in an attempt to bridge the gap between the laboratory and the clinic. In other words he had to question what experiments in the test-tube and in mice meant vis-à-vis human patients. Simultaneously, as a practising physician, he had to take care of patients suffering from more or less advanced cases of cancer, mostly with a bad prognosis. This made him familiar with the ambiguous dilemma of whether or not to treat patients with an experimental therapy that had not been tested sufficiently, in view of the general willingness of patients and doctors alike to try anything in search for a cure.[45]

However eager to collaborate with Strander and gear up for a test on cancer patients, Cantell realized that even without a lot of administrative hurdles setting up a proper trial would be far from easy. First of all there was only enough interferon to treat a few patients for a limited period of time, while Gresser's mouse studies indicated that it was necessary to continue treatment over a relatively long period. Furthermore the assessment of the clinical data would be difficult. Malignant diseases mostly had a variable course and owing to the fact that only in very desperate cases oncologists were prepared to omit other treatments, it was difficult to find proper controls.

Cantell was curious to hear from Strander which cases apart from leukaemia he considered suitable for treatment. 'What would you think about myeloma (or other lymphomas), fulminant melanomas or brain tumours?'.[46] Strander speculated on the basis of his clinical as well as research experience that interferon as a viral inhibitory substance would most likely have a beneficial effect in malignancies which were supposed to have a viral origin. Except for human leukaemia Strander identified at least four possibilities: multiple myeloma and another kind of lymphoma, Hodgkin's disease (malignant tumour of lymph nodes), osteosarcoma (a rare but extremely malignant bone cancer) and melanoma (highly malignant form of skin cancer). Since there already existed an experimental treatment programme for leukaemia patients at the Radiumhemmet, Strander opted for performing first a preliminary trial with the other problematic malignancies such as osteosarcoma and melanoma.[47]

In February 1971 Strander and Cantell were given the green light with the understanding that they would only treat patients with advanced cancer who had been unresponsive to conventional treatments. Since the study was primarily meant to see whether high-dose interferon therapy would produce adverse effects, a control group was not included. Their first patient was a 39-year-old man with malignant melanoma which originated from a birthmark in his skin and had spread all over the body. He appeared to be resistant to all existing drug therapies and had a rather poor prognosis. The patient consented to treatment with an 'experimental treatment'.[48] By injecting the semi-purified interferon preparation – containing about half a million units – directly into the bloodstream ('intravenously') Strander hoped to accomplish the most optimal exposure of the tumour cells.

Initially everything seemed to go well but after a couple of hours the patient's pulse suddenly quickened and his temperature rose sharply. Subsequently he developed chills, began vomiting and was shivering severely. The patient was diagnosed as suffering from a potentially life-threatening anaphylactic shock and Strander had to intervene with an emergency cortisone injection, whereafter the side-effects gradually subsided. The disappointment was great. Cantell and Strander did not know what to make of the patient's severe reaction to interferon. Was the patient allergic to interferon or was something different involved? They decided to investigate the case thoroughly before continuing with further experiments in patients.[49]

Cantell launched extensive series of animal experiments in his laboratory in Helsinki in order to learn more about the toxicology – supposedly due to the impurities in the preparation – and pharmacology – next to nothing was known about the fate of interferon in the organism – of his interferon preparations. It became a puzzling affair, tests with the same interferon preparation, even when given in massive doses, produced only a few side-effects – a slight fever or contraction of vessels – in the animals studied. Neither did allergic skin tests on the patient produce a reaction. In the end Cantell decided to see whether or not changing the mode of administration would make a difference. At least in rabbits, intramuscular and subcutaneous injections resulted in long-lasting and stable blood interferon levels without even the slightest contraction of the ear vessels.[50]

Three months after the first effort Strander and Cantell decided to give it another try. This time Strander administered the interferon intramuscularly, starting with a relatively small dose. To his great relief the patient tolerated the injection 'pretty well', with only a slight and transient rise in body temperature – a 'feeling of warmth' and malaise manifested itself a few hours after the injection.[51] Even much higher doses up to 1.5 million units per injection did not seem to harm the patient, except for the fact that the transient malaise and fever reaction grew more severe. The flu-like symptoms lasted for about half a day. Strander thought it most likely that the dose-dependent side-effects were due to contaminants in the preparations – which were known to be still far from pure – rather than interferon. As long as he could minimize the discomfort to the patient by giving Aspirin after each interferon injection he was prepared to try even higher dosages.[52]

By August 1971 two other patients had been enlisted in the interferon treatment programme: a 46-year-old woman with far advanced Hodgkin's disease and a 19-year-old female with advanced osteosarcoma who had just recovered from the surgical removal of a very large tumour in her frontal bone. All received three weekly injections with interferon without further complications being recorded. Everything seemed to go well, although Strander thought it too early yet to say anything about the effect on the malignancies.[53] The untimely death of the two older patients at the end of September only briefly affected morale and did not prevent Strander from including five more advanced cases of cancer in his treatment group. Strander spent a lot of time monitoring the condition of his patients at the bedside. Initially the clinical data accumulated without showing remarkable differences, but after a couple of months Strander had the feeling that he was seeing responses to interferon therapy in a few patients.

However questionable the nature and the endurance of the responses, he did not have difficulty in convincing Cantell to sent additional batches of interferon. On the contrary, Cantell assured him that with the NIH contract money he had been able to scale-up the production of interferon considerably. This would enable him to supply Strander with enough interferon to keep 10 patients under treatment.[54] Cantell also indicated that he had succeeded in improving his purification process. He would ship by air freight a container with both the conventional interferon material, which he had marked as 'c(rude)-IF', and a few tubes labelled 'p(urified)-IF' containing what he called 'superinterferon'.[55]

With the more highly purified and concentrated interferon preparations they hoped to see more definite clinical responses, but also a reduction in the dose-dependent side-effects.[56] Their hopes did not materialize in the sense that the patients tolerated the p-IF only marginally better, although the local reaction at the injection site almost disappeared. In none of the patients was treatment with interferon followed by changes in the clinical condition that were dramatic enough to support the claim that interferon exhibited anti-tumour or antiviral effects. Nonetheless, Strander, relying mainly on his 'clinical touch' as a physician, became increasingly convinced that he was getting some response: the three young osteosarcoma patients were all faring better than expected from the 'average' osteosarcoma patient at the Radiumhemmet and he observed how one of lymphoma patients became well, at least temporarily, from his tumour disease. Furthermore, he was fascinated by the quick clinical recoveries from shingles in a couple of the cancer patients treated with interferon, although he was aware that spontaneous recoveries in these cases were not uncommon.[57]

At least the clinical data looked promising enough to justify the decision that all future osteosarcoma patients admitted to the Karolinska Hospital would be treated with interferon and to extend his studies with myeloma patients.[58] Strander also became involved in laboratory studies at the nearby Institute of Tumor Biology, testing the effect of interferon on the potentially virus-related human tumour cell-lines that were in stock.[59] In doing so he organized his studies along two coexisting lines: laboratory investigations and bedside practices. Like Cantell, Strander seemed to adhere to the principle that establishing continuous links between laboratory data and clinical data would facilitate the systematic development of interferon therapy.

If we recall the series of experimental events that helped to shape the notion of interferon as a potential cancer therapy, two issues appear to figure prominently. First, I would like to point out the central role the question of purity had played so far in evaluating knowledge relating to interferon. Depending on the situation researchers used the purity or impurity of interferon preparations as a means to justify and criticize experimental results and claims. This is nicely illustrated by the controversy which arose over Gresser's anti-tumour claim and the way in which Cantell and Strander tried to come to terms with unexpected clinical effects in patients. By providing interferon researchers with a flexible arbiter in the context of justification the 'purity' parameter would remain invariably popular in the production of evidential arguments.

Second, the coming of age of the field of interferon research began to make itself felt through the dogmatic resistance to a change of concept of interferon as part of the host defence against viral infection and the pervasiveness of the non-toxicity dogma in evaluating clinical data. As we will see in following Mathilde Krim's efforts to mobilize interest, similar kinds of conflict and competition over what questions were important, what phenomena were interesting and what answers were acceptable continued to influence the development of interferon.

Mathilde Krim's interferon lobby

Despite strenuous efforts, Mathilde Krim did not fare well in mobilizing support for research into the anti-tumour properties of interferon in the USA in the early 1970s. Her applications for funding were rejected by the National Cancer Institute (NCI) and the National Institute of Allergy and Infectious Diseases (NIAID), which are both part of the National Institutes of Health

(NIH). During a visit to the NCI's office in Bethesda, MD, Krim learned that NCI officers were not willing to support the project because that would mean having 'to step on to NIAID's turf'.[60] The NIAID already had a formal interferon programme (as part of the Antiviral Substances Program) which coordinated NIH in-house research with regard to interferon and was responsible for all interferon extramural research – spending roughly 900,000 dollars or 1.5 per cent of the NIAID's annual research budget in 1970.[61] Requests for funding of extramural interferon research should therefore be addressed to the NIAID and not the NCI.

Apart from the territorial issue, the anti-tumour claims of the seemingly non-toxic (in marked contrast to most other available modes of cancer therapy) virus inhibitory substance, 'interferon' had little appeal, despite the high priority of NCI's special Virus Cancer Program.[62] The problematic and often conflicting reports regarding the tumour-inhibitory effects of both interferon inducers and interferon were hardly an inducement for support. Moreover, there were strong feelings within NCI quarters against funding research with a poorly defined and impure substance or, even worse, some sort of family of biologically active proteins. Published reports suggested a marked heterogeneity of human interferons with regard to molecular weight, electrical charge and differences in stability to heat and acid.[63] However, none of these interferons had been characterized in chemical terms and the production of interferon was known to be beset with difficulties. In many ways interferon seemed to resemble the kind of ambiguous complex biological activity the NCI had experienced in the past with unproven cancer therapies like 'Coley's toxins'.[64]

The NIAID, in turn, was not receptive to Krim's applications either. George Galasso, who headed the Antiviral Substances Program, was interested in interferon as a potential antiviral drug but was sceptical about the possibilities of interferon as an anti-tumour agent.[65] His sceptical attitude was based on a lack of evidence that inhibition of tumour growth was mediated by interferon and not by some other molecule, given the impurity of interferon preparations. Moreover, all interferon researchers on his Program Advisory Committee resisted Gresser's idea that the action of interferon might be pleiotropic or polypractic – inhibiting the growth of viruses as well as tumour cells. They held on to the principle that cells produced interferon in response to viruses and that, after release from the cell, interferon in turn specifically reacted with cells to induce the formation of another cellular protein which mediated the actual antiviral activity. This so-called 'interferon system' was considered to be an important part of the non-immunological host defence against viral infections. As such interferon fitted in rather well with the 'biomedical model' of disease and treatment – based on the theory of specific aetiology and the notion of specific therapy – on which the Antiviral Substances Program was largely based (see Figure 27).[66]

Despite the rather disappointing reception of her project proposals at the NIH, further discouraging reactions from private foundations such as the Rockefeller Foundation, Krim did not give up her fundraising efforts. She

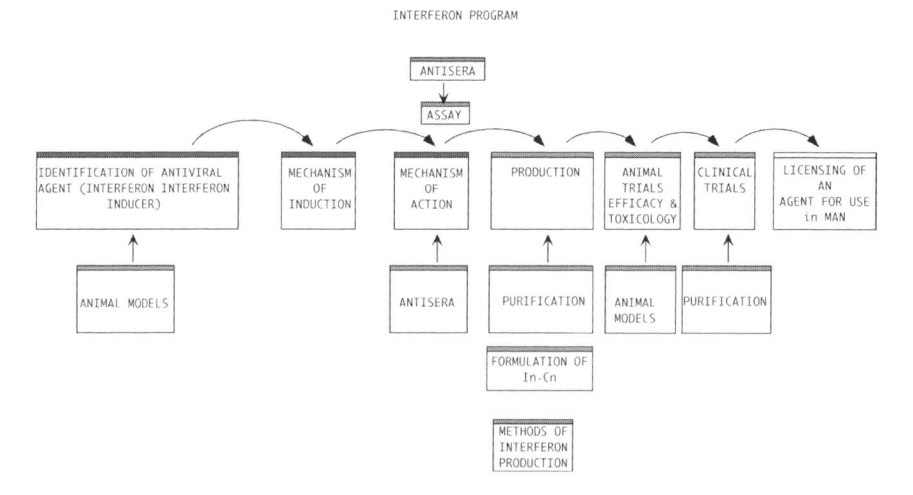

122 *Interferon: The science and selling of a miracle drug*

Figure 27 Interferon drug development programme as part of the NIAID Antiviral Substances Program (October 1969). Courtesy of the NIAID.

felt that hurdles were there to be taken and that any agent like interferon, which had shown inhibitory activity against several tumours in mice without any toxic side-effects, deserved further intensive investigation.[67] However, more than two years elapsed before, in early 1974, Krim's applications finally met with some positive response. Representatives of the German Behringwerke Company, which had become interested in the possibilities of interferon therapy at the peak of the interferon inducer 'fashion' around about 1969, approached the Memorial Sloan Kettering Cancer Center with an official request for a clinical trial with human interferon in cancer patients.[68] The German drug company offered to supply the human interferon that was needed for the trial. The newly appointed research director of Memorial Sloan Kettering, Robert Good, was rather taken by the German interest and decided that the trial should be carried out. Since he knew about Krim's vivid interest in interferon, he asked her to act as coordinator. Krim was encouraged to involve outside experts and thus she got in touch with the most outspoken believer in interferon's anti-tumour activity, Ion Gresser.[69]

In the spring of 1974, Krim visited Gresser in his laboratory. Gresser enthusiastically informed her of his latest investigations and publications concerning interferon's anti-tumour effects. Krim was intrigued by Gresser's speculations about interferon as a broad-spectrum cellular hormone serving to modulate cellular behaviour and enhance the host defence mechanism – acting locally or at a distance, stimulating when stimulation was needed and inhibiting when inhibition was needed.[70] This sounded similar to Robert Good's new approach to the treatment of cancer dubbed 'immunotherapy' aimed at the stimulation of the non-specific immune response by biological ('natural') means – involving non-specific factors that amplified immune reactions, as opposed to the specific antibody-related, classical immune mechanisms.[71]

As a result of extensive media coverage immunotherapy had a high visibility in the USA in the early 1970s.[72] Treatment with immunostimulating agents such as BCG (bacille Calmette-Guérin, a bacterium related to the tuberculosis bacillus) was presented in the media as a promising 'fourth modality of cancer treatment' – in addition to surgery, radiotherapy and chemotherapy. The clinical oncologists who were involved in conducting the immunotherapy trials viewed the new approach as potentially supplementing conventional cancer therapies, with a preference for a combination with cancer chemotherapy.[73]

Hearing Gresser argue that interferon's anti-tumour effects were most likely based on enhancing a variety of mechanisms of defence available to the host, Krim immediately saw possibilities to align interferon with the increasingly fashionable immunological approach to cancer therapy. Gresser's advise to contact Hans Strander, a physician at the Karolinska Hospital in Stockholm who was performing experiments with interferon in cancer patients, did not fall on deaf ears. Krim immediately booked a flight to Stockholm.[74]

In his room at the Radiumhemmet, Strander told Krim that over the years he had treated more than 30 cancer patients with interferon without noticing serious toxic effects. He considered the provisional results encouraging. In particular the osteosarcoma trial looked promising.[75] As a routine way to deal with trials concerning patients with advanced cancer – involving high and predictable mortality – he worked with historical controls: that is, Strander compared his interferon-treated patients with the medical histories of osteosarcoma patients treated by conventional means at the Karolinska Hospital. The historical record showed that about 80 per cent of the osteosarcoma patients developed metastatic tumours in the lungs within 12 months that were invariably fatal, following conventional surgery and radiotherapy.

Strander believed that a doctor should always choose the treatment, based on the doctor's best judgement and knowledge of the particular patient. He argued that a retrospective study – study that looks back into case histories – could supply the data needed to judge the effectiveness of a given therapy. Furthermore he did not think it ethically acceptable to tell a fatally ill patient, 'I have several treatments, but I am not sure which one is best. To find out I'll pick your treatment by a randomized procedure similar to flipping a coin. Is that okay with you?'.[76]

Strander's attitude is illustrative of the physician's conflict between loyalty to the perceived needs of the patients and to the objectives of research. Moreover, his course of action was in line with the culture of therapeutic evaluation in clinical oncology in the early 1970s. The spread of randomization was slow compared with other medical specialties and the presentation of research results based on the accurate observation of single cases and on comparison with historical cases was still widely practised. According to the historian of medicine, Ilana Löwy, the quasi-monopoly of oncologists over their patients – with little interest from other medical specialties in the treatment of the mostly hopeless cancer cases – might have contributed to the persistence of historical controls in clinical oncology.[77] As we will see in the next sections, what thus began as an unproblematic decision later became a matter of controversy and policy.

Staging a workshop on interferon in the treatment of cancer

Krim learned that up until the time of her visit to Strander of the ten osteosarcoma patients who had been enlisted in the treatment group, only two had shown signs of tumour progression after more than a year of being free of symptoms.[78] In addition, the data collected on lymphoma patients indicated that in single cases Strander was getting a response, at least temporarily. Strander emphasized the preliminary status of his studies and thought it premature to draw any firm conclusions. Yet the preliminary results of the Swedish trial sufficed to strengthen Krim's belief in interferon as a potential non-toxic therapeutic agent.[79] Krim decided to renew her efforts to interest the NCI in supporting research into the possible anti-tumour effects of interferon. With Strander's osteosarcoma trial as a vehicle to build an agenda for her cause, she was hopeful that NCI's Virus Cancer Program would make some funds available to interferon research.[80]

Krim's immediate aim was raising consciousness. As a 'woman of action' she wanted to show people in senior positions within the American cancer establishment how much potential there was. The best way she could think of was to stage an international conference on interferon and cancer. In developing funding for such a meeting she met with quite some resistance, in particular from NIH quarters. As I mentioned before, there was a basic mistrust of reports concerning interferon's anti-tumour properties. In addition there were strong feelings about the possibility that Mathilde Krim, an ambitious and politically well-connected outsider and woman, would interfere with the current distribution of federal funds and might gain too much control over the future course of interferon research. Moreover, the established interferon researcher Tom Merigan – who played a central role in the clinical studies part of the Antiviral Substances Program with his trials involving hepatitis patients and cancer patients suffering from viral infections – had already approached the NIH with plans for the organization of an international interferon symposium on the clinical potential of interferon as an antiviral agent.[81]

Despite the fact that Krim's meeting had a rather different objective, interferon and cancer, there were concerns that her workshop might upstage the NIH meeting.[82] Krim, however, persisted and finally in the autumn of 1974 she received an invitation from the head of NCI's brand new Molecular Control Program, Timothy O'Connor, who was to advise the Director of the NCI, the tumour virologist Frank Rauscher, on new directions in molecular biology with great potential in the fight against cancer.[83]

On 12 December 1974, Krim in the company of Strander – who had been asked by Krim to work-up his trial data into a presentable form and come over for the occasion – paid a visit to O'Connor at the NCI. O'Connor was impressed with Strander's latest data based on a group of 11 interferon-treated osteosarcoma patients. Strander told him that all these patients had been followed up for at least six months, during which time none had developed tumour progression. The data were judged as being increasingly exciting as they approached a statistical significance between the interferon-treated group and the historical control group consisting of 33 osteosarcoma patients treated between 1952 and 1972 at the same hospital by conventional means. Strander suggested that the preliminary observations were worthy of further expanded study. O'Connor agreed that interferon deserved further study and he was prepared to advocate funding within NCI quarters. Shortly after this meeting, Krim received an invitation to submit an application for a Planning Grant to support an international workshop that would aim at evaluating the state of the art in interferon research and formulate recommendations for further research concerning interferon and cancer. The application would be reviewed during a scheduled meeting of the Molecular Control Working Group on 7 January 1975.[84]

O'Connor's working group voted unanimously to approve Krim's subsequent grant proposal on behalf of the Memorial Sloan Kettering. However, another month was needed to overcome resistance within other parts of the NCI and Krim's application had to be resubmitted to go through a second review. Finally at the end of February 1975 Krim was informed that the Division of Cancer treatment of the NCI would co-sponsor the meeting. In the meantime Krim had succeeded in getting additional funds from the Rockefeller Foundation, the Sloan Kettering Institute for Cancer Research itself and a number of American pharmaceutical companies.[85] There was enough money available to do the unusual and offer each participant a free trip to New York City, plus accommodation at the Westbury Hotel and even a small honorarium.[86]

On Sunday 30 March about half of the participants gathered for an informal 'open house' reception in Krim's grand home at East 69th Street in New York. Krim felt that this would provide a good opportunity for getting acquainted and meeting old friends in a relaxed atmosphere before the start of the 'International Workshop on Interferon in the Treatment of Cancer', which was held at the Rockefeller Institute nearby.[87] Krim had succeeded in bringing together for the first time virologists, oncologists, immunologists, and representatives from the NIH, NCI and pharmaceutical industry, who submitted, discussed, evaluated and simply listened to a host of papers mainly concerning the anti-tumour effects of interferon. The size of the meeting, with about 200 invitees, was not unusually large. However, for quite a number of those present, the meeting represented their first educational experience with a field of research they had either never paid serious attention to or never heard of.

In spite of Krim's warm welcome, and the careful staging of the workshop, she could not prevent that at times the atmosphere became tense and the meeting threatened to turn into a battleground with believers and non-believers, Krim herself and NIH-interest groups fighting each other in their common drive for power and access to federal funds. At the reception, for example, rumours were spread that Strander's osteosarcoma study, which would be presented the next day, was flawed. Detailed information about the trial and the clinical pictures of each of the 14 osteosarcoma patients, which Strander had submitted to Alan Rabson, director of the Division of Cancer Biology and Diagnosis at the NCI shortly before the meeting to independently review, was said to disclose that most of the tumours were not osteosarcomas and that the Swedish pathologist's diagnosis had been wrong. Furthermore the trial was not a formally randomized comparison but relied on comparison with a historical group treated conventionally. Moreover, the results of Strander's trial did not seem to differ significantly from the recently published clinical results of a US-tested new chemotherapy regime of high-dose methotrexate, a cytotoxic chemical compound produced by the American Lederle company.[88]

Cantell, among others, became worried and rather annoyed about the premature judgement of research data, which had yet to be presented by Strander. 'A surprisingly malicious, suspicious and negative attitude seemed to breathe towards the research of Hans and mine . . .'.[89] Cantell had the gut feeling that the trial might become the focus of a political struggle. He therefore contacted his friend Merigan, whom he knew was part of the NIH establishment and had a reserved attitude towards Krim's meeting too. 'I told my worries to Tom and flashed the threat that I would terminate my collaboration with US scientists, if Hans's study would be publicly decried on unjustified grounds'.[90] Merigan, whose clinical studies depended largely on the continued supply of Cantell's interferon, took the threat seriously. He phoned Rabson in Washington, who would arrive the next day to review Strander's patient records. Rabson reassured both Merigan and Cantell that there were some differences in interpretation between the Swedish pathologists and his research group, but the majority of the cases indeed represented classic osteosarcoma. Cantell then let things go at that.[91]

The actual session termed 'treatment of human osteosarcoma with interferon' satisfied all parties. On the one hand Strander was given credit for having started a trial with interferon in cancer patients, which had produced some potentially interesting results. But on the other hand most agreed that the experimental sample size was rather small and that the diagnosis of osteosarcoma was questionable in a number of cases. Hence, it was too early to draw any conclusions. Further extended study should be awaited.[92]

Far from being a novel aspect of interferon-related research it is remarkable to see that a complete session was dedicated to interferon's interactions with immune defence functions and cellular regulatory mechanisms. In the past there had been occasional paper presentations pointing at interactions between

Figure 28 Lowel Glasgow used this picture in 1970 to support his controversial claim that host resistance against viruses was multifactorial with multiple interrelationships of the interferon system and the immune system. Courtesy of Alfons Billeau.

the interferon and immune system. These claims did not meet with general approval and were ascribed to impurities in the interferon preparations. Most interferon researchers supported the view of the interferon system as part of the non-immunological host defence against viruses (see Figure 28).[93]

Now for the first time the possible pleiotropic action of interferon as a kind of immune defence modifier was on the agenda (see Figures 29 and 30). The paper presentations at hand were far from univocal regarding the nature of the interactions with the immune system. On the one hand there were claims regarding an immunosuppressive activity of interferon preparations as witnessed by a slowing down of the rejection of transplanted grafts in mice. This would contradict the claim that interferon had an anti-tumour effect. On the other hand there were observations which could be interpreted as saying that interferon had an immune enhancing effect that might be related to its tumour inhibitory activity. Despite the conflicting and ambiguous nature of the experimental data, this session played an important role in mobilizing the interest of medical researchers working in the field of immunotherapy of cancer. In pioneering various ways of stimulating the cells of the immune system to inhibit tumour growth, with the aim of developing immunotherapies of cancer, they had become accustomed to the apparent paradox that non-specific biological agents with immunoregulating properties could have both immuno-enhancing and immunosuppressive effects.[94]

Figure 29 Around 1976 William Stewart II used the same allegory of the seven blind men and the elephant, to produce a caricature of several interferon workers who all argue in favour of their own favourite hypothesis. The drawing served as a means to convince the audience that the action of interferon is pleiotropic. Courtesy of Alfons Billeau.

Attempts to put interferon on the cancer map

The 1975 symposium was choreographed by a hopeful Krim as a watershed event. The conference was indeed instrumental in bridging the gap between the test-tube and the patient. Basically, it was the first major public attempt at translating interferon's anti-tumour effects into scientific terms that fitted both fundamental laboratory research and clinical research. However, Krim's grand objectives were to gain approval and support from the scientific community, funding and cooperation from the drug industry and to obtain enough funding from the NCI to expand interferon research – a house of cards based almost entirely on Strander's 14 cancer patients. But, neither scientists nor the representatives of government, private funding agencies and pharmaceutical companies who attended the meeting seemed impressed with the experimental data presented over the three-day period. The press all but ignored the event. Despite the fact that Krim succeeded in raising consciousness, the workshop as such did not suffice to gain interferon official recognition as a promising anti-tumour agent.[95]

The Blind Men and the Elephant

It was six men of Indostan
To Learning much inclined
Who went to see the Elephant
(Though all of them were blind),
that each by observation
might satisfy his mind.

The First Approached the Elephant,
And happening to fall
Against his broad and sturdy side,
At once began to bawl:
'God bless me! But the Elephant
Is very like a wall!

The Second, feeling of the tusk,
Cried, 'Ho! what have we here
So very round and smooth and Sharp?
To me 'tis mighty clear
This wonder of an Elephant
I very like a spear!'

★ ★ ★

The Sixth, no sooner had begun
About the beast to grope,
Than, seizing on the swinging tail
That fell within his scope,
'I see' quoth he 'the Elephant
I very like a rope!'

And so these men of Indostan
Disputed loud and long
Each in his own opinion
Exceeding stiff and strong,
Though each was partly right,
And all were in the wrong!

Moral
So oft in scientific wars
the disputants, I ween,
Rail on in utter ignorance
Of what each other mean.
And prate about an elephant
Not one of them has seen!

Figure 30 The drawings in Figures 28 and 29 were based on this poem by John Godfrey Saxe. Courtesy of Joseph Sonnabend.

As a member of the panel of consultants, Krim had experienced the political importance of producing an accessible scientific report with recommendations for further research. After the workshop she asked the interferon researchers Tom Merigan and Jan Vilcek and the NCI tumour virologist, Arthur Levine, to help her produce a report of the workshop.[96] While the complete report was being prepared, Krim was asked by *Nature* to write a summary for their News and Views column. Krim immediately put together a short report and sent copies to members of the organizing committee and the Division of Cancer Treatment of the NCI.

Krim's version of the workshop caused a little row at the NCI. In her drive to put interferon on the research map, Krim had left out accounts of dissenting voices, such as the statement by Richard Adamson of the NCI that he had been unable to demonstrate an anti-tumour effect of interferon in his mouse leukaemia system.[97] This policy was criticized severely within NCI quarters. Krim was told that unless she retracted the paper immediately and included what amounted almost to a disclaimer, detailing the shortcomings of Strander's study, an official dissenting opinion would be published in *Nature*. Krim decided to take the threat seriously. She agreed to add Adamson's comments and to share authorship with the other members of the editing committee of the official workshop report.[98] The outcome was a most carefully worded communication, which pointed at the possibility that interferon might exert its apparent anti-tumour effects both directly on the tumour cells and through a modification of components of the immune system as a kind of immune defence modifier.[99]

In the meantime, Krim exerted further pressure on the NCI to support interferon research. Shortly after the workshop she sent Rauscher, who to her great indignation had not attended the last day of the conference, a personal and confidential letter summarizing the final recommendations he had missed and an indication of the level of support she expected the NCI to provide.[100] Reluctantly, Rauscher submitted to Krim's pressure and invited her for a meeting with senior NCI officials on 15 April 1975. At this meeting the edited conclusions and recommendations of the workshop were presented by Krim. Following her presentation it was pointed out that the NCI was not prepared to take any action until the formal report was made available and the newly established NCI Interferon Working Group under the chairmanship of Alan Rabson had reviewed the report.[101]

In the autumn of 1975, Rabson sent the recommendations to Rauscher and the Director of the Division of Cancer Treatment, Vincent DeVita for further consideration. The Interferon Working Group recommended that 4 million dollars be allocated for interferon research to support clinical trials, basic research and the production of interferon. In addition clinical trials should be conducted under the supervision of the NCI's Division of Cancer Treatment.[102] However, within the Division of Cancer Treatment the recommendations were received with mixed feelings, and DeVita decided to bring the matter up for discussion at the November meeting of the advisory committee to his division, the Board of Scientific Counselors.[103]

The meeting of the Board on 11 November 1975 began with a presentation by Mathilde Krim on the use of interferon in the treatment of cancer. Right from the start Krim felt that she would have the odds stacked against her. Her report, its conclusions and recommendations were indeed severely criticized, especially Strander's clinical trial. The tumour data were questioned, the number of cases was thought to be too small and it was suggested that it was unlikely that biological activity could be obtained in a mixture containing one in a thousand active molecules. Others stated that the clinical trial was difficult to interpret, even if expanded, because of the use of a historical control group. One should at least include a concurrent control group in order to be able to evaluate the present data.[104]

The critics within the NCI Board were, at least, biased in their accusation of Strander for non-conforming or dissenting behaviour regarding the design of his trial. In 1975 of the 47 cancer trials published in the *New England Journal of Medicine* 34 per cent were uncontrolled, 13 historically controlled, 13 per cent had concurrent controls and 40 per cent were randomized.[105] Clinical oncologists who claimed that randomization of cancer trials was neither ethical nor necessary prevailed and at least half of the NCI-sponsored clinical trials were still non-randomized. In addition, the widely praised American chemotherapy studies in osteosarcoma patients which interfered with Strander's interferon claims were non-randomized. Criticizing the design of Strander's trial must have served the interests of the Randomized Clinical Trial lobby within the NCI, who were pushing the randomized clinical trial as the ultimate means of applying the scientific method to the practice of oncology. However, they could not have acted without the support of the powerful chemotherapy lobby within the NCI. Only after the NCI began to scale down its chemotherapy programme in the late 1970s did NCI statisticians begin to question the usefulness of chemotherapy in osteosarcoma.[106]

An air of hostility developed toward Krim during the meeting. Her strenuous efforts to bring interferon to the attention of NCI officials and the recommendations made by the Interferon Working Group were heavily criticized.[107] At the end of the meeting Krim was astonished to find that the board recommended that there was no compelling reason for establishing a special NCI research programme, to study interferon in humans. The majority position was that before large-scale clinical testing of interferon was to begin, it would be more fruitful first to concentrate on basic *in vitro* and *in vivo* studies with interferon. The work could not be funded simply on the basis of Krim's say-so. The NCI was advised to provide adequate funding by contract and grant mechanisms only.[108]

The same month, Krim's report was also presented to the National Cancer Advisory Board (NCAB). According to Richard Rettig 'it is appropriate to think of the NCAB as the "board of directors" of the National Cancer Program' with rather far-reaching powers regarding NCI's research agenda.[109] Part of the NCAB were former members of the panel of consultants, like Mary Lasker herself. Krim might have known that despite the negative recommendations of the Board of Scientific Counselors of the Division of Cancer Treatment with the support of the 'Laskerites' something positive would emerge.

Eventually, during the NCAB meeting, the differences in opinion between on the one hand Rabson's Interferon Working Group and Krim, and, on the other hand DeVita's Division of Cancer Treatment and Rauscher, over the kind of support that should be dedicated to interferon research, were settled in a compromise.[110] The NCI would make 1 million dollars available – which amounted to about 0.15 per cent of NCI's annual budget – through NIAID's Interferon Program to purchase human interferon for investigation of interferon as a possible anti-tumour agent.[111] At the same time DeVita was granted approval to organize a site visit by a blue-ribbon panel of experts – consisting of two oncologists, two pathologists and a statistician – chaired by Arthur Levine, to evaluate Strander's data at the Karolinska Hospital in Stockholm.[112]

Apparently Krim had succeeded in overcoming part of the resistance within the NCI against supporting interferon research. It was not the massive funding Krim had asked for and the NIAID had succeeded in maintaining its key position as interferon's funding authority. But interferon for the first time had gained official recognition as a potential anti-tumour agent.[113]

In May 1976 the panel visited the Radiumhemmet. It was at Levine's insistence that Strander added to his interferon-treated patients and the historical controls, a group of contemporary ('concurrent') controls consisting of osteosarcoma patients who had been treated simultaneously by conventional means in other Swedish Hospitals. In Levine's view by using historical controls Strander ran the risk of 'comparing apples and oranges'.[114] Levine did not appreciate Strander's defence that it was customary all over the world to employ historical controls in studies on osteosarcoma.[115] The eventual site visit report was a great blow to Strander.[116] The NCI panel concluded that by using a contemporary instead of a historical control group to evaluate the trial data, no evidence had as yet been found that interferon treatment 'significantly altered the present natural history of osteosarcoma of the long bones'.[117]

On behalf of the tax-paying public Rauscher congratulated Levine for his potentially money-saving investigative efforts.[118] In practice, however, the report hardly affected NCI's policy regarding interferon – the decision to earmark 1 million dollars for interferon research was carried through.[119] As it turns out, Rauscher overestimated the influence of the essentially negative site visit report and underestimated the impact of Krim's interferon lobby in combination with anecdotal media reports of the treatment's effectiveness in cancer therapy.[120] But why did the interferon promoters meet so much opposition in the first place and how did Krim in particular succeed in overcoming the resistance and secure priority status for interferon on the American cancer agenda?

Sandra Panem has already identified some of the factors that worked against Krim: apart from being a *woman* in a 'man's world', Panem noted that Krim was regarded as an outsider to the field of interferon research, neither knowledgeable enough nor sufficiently critical about interferon.[121] Among interferon researchers and the National Institutes of Health bureaucracy there was anxiety that the outsider Krim might gain too much control over the future course of interferon research.

Krim nevertheless succeeded in managing the opposition effectively by involving herself in a hybrid activity that combined elements of scientific evidence and reasoning with large doses of social and political judgement. Without trying to reach agreement on the controversial issues, she tried to address the common interests and characteristics of the various parties involved. On the one hand she emphasized the mysterious elegance of using a naturally occurring non-toxic substance, and on the other hand she underlined the fact that an impressive body of laboratory studies legitimated the potential value of interferon in the treatment of cancer. She properly valued the importance of linking basic research to clinical problems: convincing scientists and policy-makers alike that correlating laboratory findings with clinical findings seemed a perfectly straightforward goal in the case of interferon and cancer. However her attempts at linking laboratory with clinical data were beset with difficulties. In particular, the dispute arising over the exemplary status of Strander's trial data and the way it was managed deserve further attention.

The controversy over Strander's trial data illustrates that acceptance of experimental therapies in American clinical oncology in the 1970s hinged mainly on their evaluation by NCI-endorsed evaluators and methods. Moreover, it unequivocally shows that treatment evaluations shift and change in response to fluctuations in the balances of power in the context of evaluation. For instance, whereas initially Strander's trial data were well received within NCI quarters, gradually the mood changed with growing criticism of Strander for non-conforming behaviour regarding the design of the trial. I indicated with reference to the NIH trial-design statistics that the critics were biased in their accusations. In my view, criticizing the Strander trial served the interests of the increasingly powerful randomized controlled trial lobby within the NCI. The outsider, Strander made a good target to plead their case and push the randomized clinical trial as the ultimate means of applying the scientific method and the objective kind of evaluation it supposedly entails to the practice of oncology.

However, as I will show next, Krim's opponents underestimated the impact of her consciousness-raising efforts. As an inquisitive follower of Lasker she understood the importance of persistent political pressure and of the transfer of negotiations about the value of interferon in the treatment of cancer increasingly to public arenas outside the established biomedical order in America.

6 Interferon, audiences and cancer[1]

In the 1970s cancer therapy became the focus of the increasingly politically powerful, critical health movement in America, highlighting the failures and severe side-effects of conventional cancer treatments: surgery, radiation and chemotherapy. Faced with unprecedented public criticism of the low success rate of the costly 'battle against cancer', the American cancer establishment was seriously under attack. It needed a scientific promise that suited the growing public demand for effective and less toxic cancer remedies. The politically well informed supporters of interferon capitalized on this need and the popular desire for more natural and organic remedies. In publicly emphasizing the presumed non-toxic and natural qualities of interferon as both an unorthodox organic and science-based promise in the fight against cancer, they succeeded in getting interferon absorbed in the accelerating politics and economics of the American cancer scene. The media gave enthusiastic coverage to the claims for a promising cancer cure. A boom in expectations was fuelled, resulting in interferon acquiring 'miracle drug' status.

The focus in this chapter is on reconstructing and analysing the mangle of practice and culture through which the transient status of wonder drug was achieved. Special attention will be paid to the channels of communication and mediation, primarily in the American and British public arena.[2] Major actors who influenced the shaping and circulation of interferon both as individuals and as a collective ranged from doctors, journalists, laboratory researchers, patients, politicians and regulators, to drug company executives, each using their own specific arguments, images, facts and figures. I show how the press blossomed into the most important agenda-setting forum, resulting in a continuous process of mediating meaning and status as part of what has become known as the 'interferon crusade'.[3]

Capitalizing on a growing demand for unorthodox cancer remedies

Following further political groundwork by Mathilde Krim, a growing number of scientists and influential laymen in America lent their support to the interferon cause. Apart from the intrinsic lure to scientists and laymen alike of a natural substance that was reportedly devoid of the severe side-effects associated with most other anti-cancer drugs, each social group or individual had private motives to jump on the 'interferon bandwagon'.

The group of American 'interferonologists' came to realize that Krim's lobbying efforts proved effective in bringing in additional research money. Owing to the growing disappointment within NIAID quarters about interferon's therapeutic potential as an antiviral agent and the seemingly permanent inability to purify interferon, the field of interferon research had been facing a serious cutback in recent funding.[4] Together with the increasing costs of doing research this made them shift attention to interferon-related problems for which funds seemed more readily available – interferon and cancer.[5] They were swiftly followed by advocates of immunotherapy like the oncologist Jordan Gutterman, who worked at the Department of Developmental Therapeutics of the M.D. Anderson Hospital and Tumor Institute in Houston, Texas. Gutterman would become one of interferon's most avid promoters.

Immunotherapy of cancer received considerable attention in the early 1970s with claims of immunostimulating biological agents such as BCG. It was the promise of a cancer therapy which claimed to exert its anti-tumour effects naturally through a stimulation of the body's own immune system as opposed to the use of the highly toxic chemical drugs employed in conventional cancer therapy. As such it was a therapy well suited to the culturally unsettled climate of the 1970s which was characterized by the malaise associated with Vietnam, Watergate and the oil crisis, and by the growing number of Americans who began to question the nation's institutional policies regarding environment, medicine, and science and technology. The flourishing social reform movements, in particular the environmentalists and health-care reform groups were united both in their belief that man-made chemicals were harming the environment and responsible for the perceived cancer epidemic, and in their growing doubts about the Cancer Establishment and the effectiveness of the war on cancer. The members of the diverse but increasingly powerful cancer counter-culture favoured unorthodox 'natural' cancer remedies. Immunotherapy seemed to fulfil their needs.[6]

Immunotherapy was simultaneously legitimized scientifically through both an array of laboratory research and fundamental breakthroughs in immunological knowledge showing the importance of the cellular (non-specific) mechanisms underlying the immune responses. With the help of new immunological laboratory tools (e.g. lymphocytes) studies were performed resulting in the claim that the immune responses of cancer patients were suppressed and should therefore be stimulated. However, by the mid-1970s, as negative clinical trial data with immunostimulating agents like BCG accumulated – immunotherapy and chemotherapy were not found to be superior to chemotherapy alone – interest in immunotherapy dwindled.[7] As a promising anti-tumour substance which was claimed to be involved in one way or another in the non-specific immune response of the body, interferon was readily assimilable into a research programme that was in urgent need of a new impetus.[8]

Introduced to interferon at Krim's workshop, Gutterman became intrigued by the possibility of using the supposedly non-toxic biological substance in cancer patients.[9] Confronted with BCG's repeated failure to show a significant prolongation of remission and survival Gutterman decided to give interferon a try, although he continued to apply BCG as a last experimental resort for patients unable to benefit from conventional therapy.[10] Given his own expertise with the experimental combination of immunotherapy and chemotherapy of advanced breast cancer, he sent a request for interferon to the NCI to be used in a breast cancer study.[11] The NCI supply, however, appeared to be limited and had to be spread among several clinical centres. Like all other applicants Gutterman was granted only a small amount of what he had asked for – hardly enough to treat a mere handful of patients for a short period.

In complaining about the frustrating state of affairs to Krim he learned that she was increasingly successful in getting influential lay persons actively interested in interferon. Three prominent public figures: Mary Lasker, the 'patron saint' of American cancer research, Laurence Rockefeller, the conservationist and member of the NCAB, and Congressman Claude Pepper, Democrat of Florida, lent their support to the interferon effort. Krim told Gutterman to inform Lasker personally of his problems.[12]

The commitment of these patrons of American cancer research to interferon was influenced by the growing public criticism of the National Cancer Program. Despite the additional investment of more than one billion dollars of tax-payers's money in cancer research since 1972, the American cancer establishment could only claim credit for minor progress in their self-proclaimed 'war on cancer'. This poor research performance did not mesh with the public announcements by politicians, nor with the claims of senior NCI and American Cancer Society (ACS) officials that they were about to discover a cure against cancer. More and more critics emphasized the gaps in knowledge between laboratory and bedside and aired doubts about the prospects for any sudden breakthroughs in the field of cancer therapy.[13] Organized elements of the cancer counter-culture like the International Association of Cancer Victims disseminated information about alternative 'suppressed' cancer therapies and provided a platform for dissident cancer voices. They openly accused the NCI, the ACS, and virtually the entire American medical community of systematic opposition to alternative therapies, and emphasized the failures of the orthodox biomedical model and conventional cancer treatments in particular.[14]

The American cancer establishment was thus seriously under attack and needed a boost in the fight against cancer to satisfy public demand. As a product of scientific medicine and with its claimed naturalistic basis interferon seemed to offer the opportunity to silence the growing opposition. Moreover, individual experiences with the limitations of conventional cancer therapy played a role in their eagerness to encourage new approaches. Pepper's wife and close associates of both Rockefeller and Lasker suffered from the 'dread disease' and were not responding to conventional cancer chemotherapy.[15]

With the idea of fuelling extra pressure for a major interferon campaign, Mary Lasker provided 1 million dollars through her Foundation for the purchase of additional interferon for Gutterman's clinical programme.[16] Without further delay, in October 1977, Gutterman, who had planned to attend an immunotherapy meeting in Rome, changed his travel plans and flew to Finland to make arrangements for buying interferon from the Finnish Red Cross through Cantell. Gutterman knew that the relatively scarce and precious interferon material was not for sale simply to the highest bidder.[17]

In order to gain the maximum scientific and public impact for his human leucocyte interferon, Cantell cautiously controlled its distribution. Since the Finnish Red Cross was the only major producer of human interferon (see Figure 31), Gutterman had no alternative.[18] Not until Gutterman had given a full account of his research plans was Cantell prepared to start negotiations on the terms of a contract for the supply of interferon. They agreed on a price of 25,000 dollars by the billion units (daily dose between 1 and 10 million units a day) of partially purified interferon (p-IF).[19]

In early 1978 Gutterman was able to begin interferon treatment of patients with inoperable and recurrent malignancies. Like most American oncologists, Gutterman felt that it was vital to be aggressive in order to get a response – cancer chemotherapy was synonymous with the use of maximum-tolerated and often highly toxic doses of drugs.[20] He would usually push his patients

Figure 31 The large-scale production of interferon at the Finnish Red Cross Transfusion Service in the late 1970s. Courtesy of Dr K. Cantell.

to the limit of what they could bear. However, he had only just started to treat his first breast cancer patients with the primary aim of yielding information on the highest tolerated dose when he and his colleagues began to witness spectacular, partial regressions of mammary tumors in five of nine treated women for whom all treatment possibilities had been exhausted. Gutterman was elated at the response, although he realized that his study was small, lacking proper controls, and too short in duration to determine if the therapy really made a difference in prolonging life. Sudden and seemingly miraculous improvements in cancer patients did occasionally occur, sometimes simply as a result of their doctors paying extra attention to them when testing a brand-new experimental remedy.[21] This could not prevent Gutterman from presenting an anecdotal report of his promising preliminary results at the March 1978 Workshop on Clinical Trials with Interferon organized by the NIH.[22]

At the NIH meeting anecdotal reports presenting interferon as a lead in cancer therapy came from a couple of other researchers as well. Tom Merigan from Stanford University reported encouraging shrinkage of tumors in two out of five patients with advanced non-Hodgkin lymphoma and David Habif from Columbia University claimed marked tumor regression in six out of seven women with inoperable or recurrent breast carcinoma. In accordance with Gutterman's clinical experiment both studies were short in duration and lacking a control.[23] However anecdotal, the simultaneous presentation of three different American studies claiming that interferon produced encouraging clinical responses had an impact on the small international audience of invited interferon researchers and NIAID, NCI, and Food and Drug Administration (FDA) officials. Reduction in the size of malignant growths per se was not an unusual therapeutic effect, but interferon's ability to induce a shrinkage of advanced solid human tumors that resisted all conventional therapies was a rather uncommon phenomenon in clinical oncology.

The optimism about interferon's therapeutic potential was also shared by those involved in studying the possible efficacy of interferon as an antiviral therapy. In particular, Merigan's placebo-controlled, randomized double-blind study showing that daily treatment with interferon diminished both the spread and the pain of shingles in patients with cancer was reason for optimism. This indicated that interferon might at least turn out to be useful in controlling serious viral infections in immunosuppressed cancer patients. There were also anecdotal reports about beneficial effects in children suffering from recurrent laryngeal papilloma – virus-induced warts that fill up the whole larynx so that the child can barely speak or breathe and can be life-threatening – for whom the only treatment available was repeated surgery. At the same time there were also less promising reports.

The clinical data presented by Pieter de Somer from the Rega Institute in Belgium, and Huub Schellekens from the Erasmus Hospital in the Netherlands, called into question the efficacy of interferon in chronic hepatitis as claimed by Merigan's research group in 1976.[24] Because it had been performed under controlled conditions, the Dutch–Belgian study would, in formal terms, have an advantage over Merigan's uncontrolled observations. However, unlike Merigan, they had used human fibroblast interferon – produced from large-scale virus-infected cultures of human cells obtained from fetal tissue or the foreskin of circumcised baby boys. Since it had been established that fibroblast interferon differed from leucocyte interferon in its physicochemical properties, it might be possible that the two types of human interferon differed in their clinical activity too.[25] To Merigan and most other researchers present, giving up on what was considered one of the most promising clinical antiviral effects observed so far was not considered as being an option – hepatitis was a rather common virus infection that was held responsible for severe liver damage and for which no treatment was available.[26] They argued that apparently leucocyte interferon held more promise for treatment of hepatitis than did fibroblast interferon.[27]

When De Somer next brought up the question of side-effects, the atmosphere threatened to become rather tense. The year before at the international interferon meeting in Israel De Somer had caused a wave of indignation among his fellow interferonologists by openly criticizing the non-toxicity dogma.[28] Although one of the principal virtues of interferon was claimed to be its lack of toxicity, all interferon preparations tested so far were known to produce side-effects. Most patients receiving interferon treatment developed transient flu-like symptoms such as fever, chills, fatigue and loss of appetite, and in certain cases after prolonged administration, some hair loss. The reversible side-effects were never taken very seriously as most of the clinical researchers involved were used to the much more severe toxic side-effects arising from most other forms of viral and cancer chemotherapy. Moreover, since it appeared that the effects diminished when using more purified preparations, the consensus was that the side-effects were not due to the interferon molecule but rather to one or more impurities.

Few dissenters could be heard until 1977 when De Somer reported the observation of serious side-effects following administration of interferon in a Belgian patient. In particular, the fact that De Somer indicated that the side-effects may well be related to the interferon molecule itself, was difficult to swallow for his audience. His remarks did more than violate the non-toxicity dogma in the field of interferon research. The response to this news 'by some was that such negative results should not be made generally public because they might dampen popular enthusiasm for interferon and stall the interferon crusade just as it was gaining momentum'.[29]

To the relief of the 1978 NIH workshop invitees, De Somer did not show up with yet another dramatic patient history.[30] The side-effect issue hardly affected the final assessment of interferon as a promising therapeutic substance that justified testing on larger numbers of patients under controlled conditions. However, obtaining interferon for a major study required millions of dollars and, with the risk of ending up with just another bogus cancer treatment, NCI officials remained hesitant to commit themselves to interferon. Gutterman did not want to wait for a change in NCI policy to come about and turned to the wealthy American Cancer Society (ACS) for help. The ACS was known to administer a special fund of 5 million dollars for cancer research aimed at financing 'best bets' that might produce benefits to the cancer patient as quickly as possible.[31] Encouraged by Lasker, Gutterman wrote up a daring proposal for a large test programme on more than 100 patients with a variety of tumors, including metastatic breast cancer.

Once again Frank Rauscher, who had resigned as director of the NCI to become the ACS's senior vice-president for research, was confronted with a request for massive funding of interferon research. In his previous post at the NCI Rauscher had opposed a similar kind of request, but that was in 1975 when Krim's interferon crusade still had to gain momentum. Since then political support for a major interferon campaign had been building up rapidly. Furthermore, the managing staff of the American Cancer Society believed they were not making enough progress in the treatment of cancer to satisfy public demand. This played a role in Rauscher reconsidering the matter.[32]

Late in July 1978 a scientific advisory panel, consisting for the greater part of interferon supporters, reviewed the application. On the panel's recommendation Rauscher and the ACS Board of Directors quickly approved a 2 million dollar grant for clinical testing of interferon. Rauscher immediately appointed an ad hoc committee of clinical researchers under the co-chairmanship of Gutterman and Merigan to work out protocol details, such as the selection of research institutions, patients and types of cancer. Scarcity and costs of the interferon material was believed to limit the trials to no more than 150 advanced cancer patients, who for reasons of validity would be randomly selected. The committee sent the experimental protocols for approval to the FDA.[33]

On 30 August 1978, the American Cancer Society (ACS) publicly announced its plans to spend for the clinical testing of interferon the largest sum ever committed by the Society to a single project. The fact was obviously not lost on the media as the press release triggered a wave of interferon-related publicity with headlines such as 'Interferon: The cancer drug we have ignored', 'Test planned on substance used in cancer treatment', 'New cancer weapon?', or 'Natural body substance: $2 million test on cancer retardant'.[34] Together with additional interest shown by radio (e.g. by National Public Radio's news programme *All Things Considered*) and television networks (e.g. by ABC, CBS and NBC news) this brought interferon, as we shall see, out of the relative seclusion of the laboratory into the limelight of public attention.[35]

Interferon, scientists and the media

The picture conveyed most often and most vividly in the mass media showed interferon as a somewhat mysterious, clinically unharnessed, non-toxic natural body substance that was claimed to be the hottest, though long ignored – by a conservative cancer establishment – line of biomedical research currently being followed. The idea generated by the stories in the media was that if only enough of this extremely scarce and expensive naturally occurring protein could be made available by scientists, some kind of miracle cure was at hand for everything from cancer to the common cold.

The double framing of interferon as a natural solution to a dreaded disease and as the product of 'cutting-edge' biomedical research was reinforced by the illustrations employed in the media. A case in point was a photograph in *Newsweek* with the subscript 'searching for the natural key to interferon' showing scientists in laboratory coats staring hopefully at a sophisticated laboratory set-up composed of a tangle of wires, tubes, retorts and graduated cylinders filled with fluids, seeing things only they were supposed to recognize (see Figure 32).[36]

Except for the excessive elements in the public media, the public image of interferon as a promising product of a laboratory-supported scientific medicine was not dissimilar to the associations and legitimations presented by interferon researchers to their scientific audience – continually linking state-of-the-art basic research with achieving future cures at the bedside.[37] Moreover, the growing interaction between the field of interferon research and the fields of immunology (with the conceptualization of the body as a complex system) and molecular biology (with the conceptualization of the body as a system of networks that process information) helped to establish and elaborate on the image of interferon as part of a complex system held together by communication and

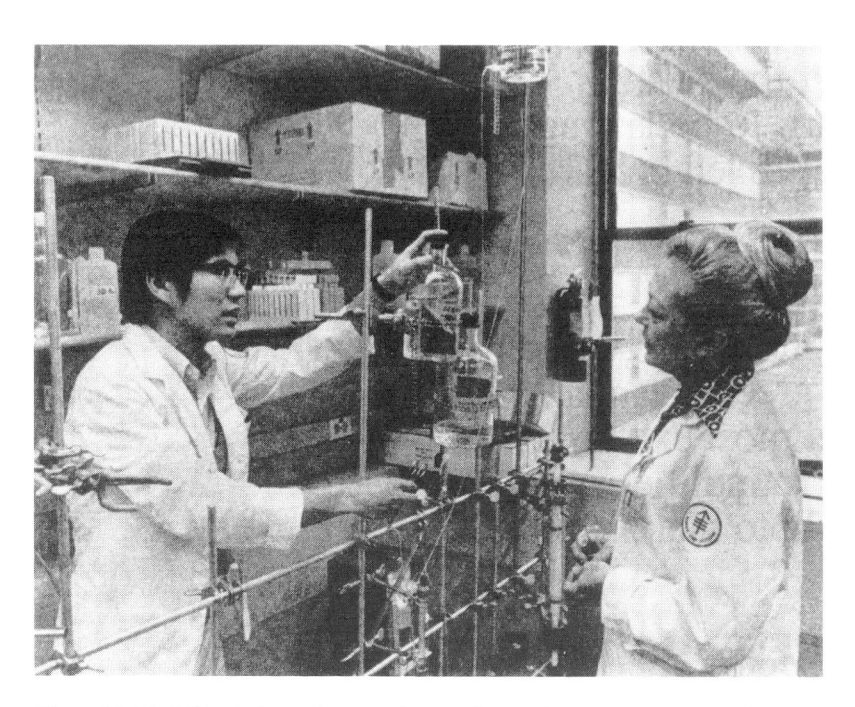

Figure 32 Mathilde Krim and co-worker in their Laboratory at New York's Memorial Sloan Kettering Cancer Center. Reproduced with kind permission from the publisher.

142 *Interferon: The science and selling of a miracle drug*

feedback.[38] This is nicely illustrated in the epilogue of the textbook *The Interferon System* where the interferon researcher William Stewart II described the action of interferon in the following terms: 'Interferon is only one of a large number of products by which cells communicate. It is as though we have learned only one of their words and pretend to speak their language'.[39]

Similar types of information metaphor were also increasingly employed to picture interferon in the media, mostly in combination with the usual military metaphors (like bullet, hunt, war and weapon). As such the media reinforced the interferon as-part-of-a-network image. For instance, the journalist Joseph Hixson depicted in the *New York Magazine* the body as a regulatory communications network that is continuously in the process of distinguishing between what 'is "self" and what is "foreign" and thus to be destroyed' such as viruses and tumors.[40] Interferon was presented as one of the intercellular messengers (using a superman visual image) which emerge whenever what is foreign manifests itself and triggers the bodily defence forces (see Figures 33 and 34). About the same time, the reporter of the *Medical Tribune* noted that interferon was exemplary for a new generation of therapeutic drugs which speak to the body

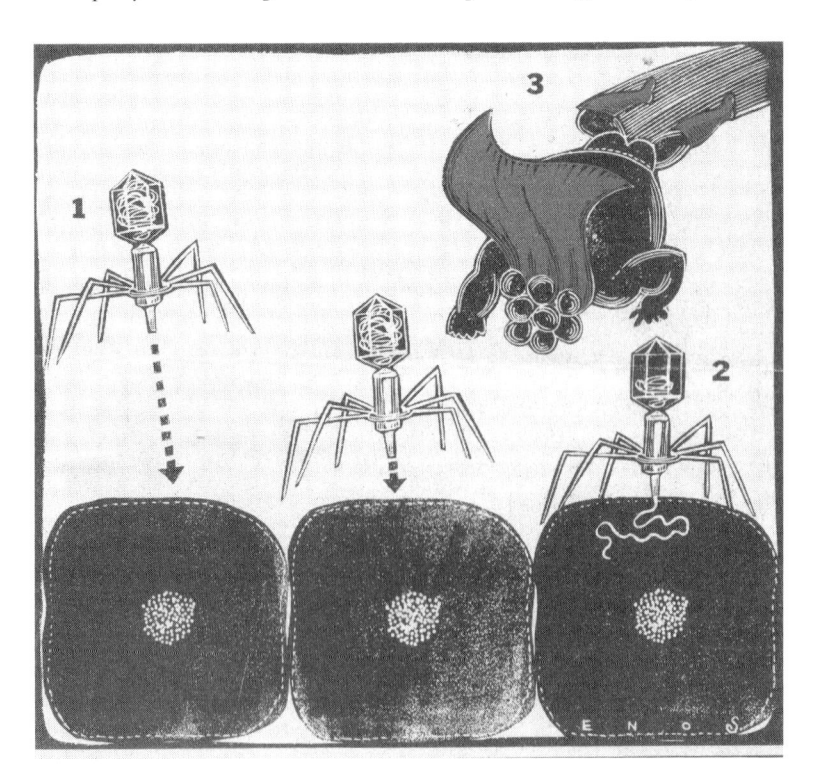

Figure 33 'The alert: Invading viruses (1) approach and release nucleic acid – a viral replicative blueprint – into the cells (2). The body responds by producing intercellular messengers, giant protein molecules called 'interferon' (3).' Reproduced with kind permission of the *New York Magazine*.

Figure 34 'All systems go: Interferon triggers a cellular anti-viral mechanism'. Reproduced with kind permission of the *New York Magazine*.

in their 'native biochemical language'. She pictured interferon as a sophisticated product of modern science: 'The difference between interferon and less refined medical approaches is the difference between the demanding precision of Obi-Wan Kenobe's high-energy laser sword, and the clumsy but often effective brute force of a tank'.[41]

Frequently linking interferon with advanced science and technology directly shaped arguments concerning the question of interferon's exploitation. The deployment of the new genetic engineering technology as a means to deal with the problems of interferon production and purification became a vital rhetorical strategy in the interferon crusade. The high public visibility and the scarcity of interferon as a potential new miracle drug in turn whetted the appetite of the newly established genetic engineering firms which were looking for feasible demonstration projects to lure public investments.

The genetic engineers used techniques (known as 'recombinant DNA' or 'rDNA' technology) developed in the late 1960s and early 1970s which allowed them to copy or 'clone' human genes by slipping them into bacteria or other microorganisms. Around 1977 – as soon as the first signs were there that the heyday of public DNA furore was over and the regulatory climate was slowly improving – the cloning and expression of genes became the object of intense research activity in the United States and Europe. The perceived allure of the new technology was that medically important, but mostly scarce and costly, human and animal proteins could be produced by the genetically engineered bacteria in unlimited quantities at modest cost.[42]

Confirmation of the feasibility of the commercial applications of rDNA techniques came in the autumn of 1977 when American scientists achieved the first complete bacterial synthesis of an animal protein, the small peptide hormone, somatostatin, produced in the human brain. Following the somatostatin experiment a growing number of genetic boutiques were established by venture capitalists anticipating new industrial applications of rDNA techniques with the active participation of a new generation of entrepreneurial scientists in academia who were at the forefront of molecular biology. By 1978, several private companies had been formed in the field of genetic engineering. The new firms, including Cetus, Genentech (in California) and Genex (in Maryland), and the Swiss-based Biogen, were eager to prove the commercial worth of the gene-splicing technology.

There were several reasons why the cloning of interferon was considered as an attractive demonstration project that could serve this purpose.[43] First, with public expectations running high, Wallstreet's capital investors would be eagerly waiting to pour money into interferon related commercial projects. Second, the production of interferon by conventional means was still beset with enormous problems after twenty years of hard labour: the scant amounts of interferon that were produced worldwide were impure and costly. It was the most expensive experimental drug ever used in clinical trials in the USA. Once you had succeeded in isolating the human interferon gene and splicing it into bacteria, the microbes were claimed to do two jobs for the price of one: they produced human interferon in large quantities and in a form (as a single protein product) that could be purified with far greater ease than for instance Cantell's natural interferon 'soup'.[44] Third, there was the technical challenge of isolating and cloning a gene coding for a protein that still awaited chemical identification.[45]

The claims by genetic engineering firms and leading molecular biologists about the imminent possibility of making available, in large quantities, at modest cost, substances that were expensive or difficult to make such as interferon, and the resulting imaginative race to clone the first human interferon gene, intensified public interest.[46] Krim seized the opportunity and worked to use the Second International Workshop on Interferons, which like the first, was held at the Sloan Kettering Memorial Cancer Center in New York, as a podium to the outside world: she made it a media event.[47]

Beside the 'old' group of interferonologists and representatives of the ACS, NCI, FDA and the press, the by-invitation-only gathering was attended by about 300 molecular geneticists, immunologists, biochemists and clinicians, ranging from infectious disease specialists and clinical oncologists to transplant surgeons. In addition more than fifty representatives of the genetic boutiques and the large established pharmaceutical companies were present. Apart from Burroughs Wellcome and Merck Sharpe & Dohme, also relative newcomers showed interest, such as Hoffmann-La Roche ('Roche'), the Schering-Plough Corporation ('Schering-Plough') and Bristol-Myers Company. The latter had entered into commercial contracts with, respectively, Genentech, Biogen and Genex not only as a means to develop interferon as a therapeutic drug but also to acquire new technical expertise that could pay off even if in the end interferon itself might never find a big commercial market.[48]

The growing involvement of the drug industry and the major molecular biology laboratories that became manifest at this meeting marked the move of interferon research to 'big science', involving large-scale research programmes. The 'small fish' (interferonologists) who had taken on this line of investigation were increasingly eclipsed by the team-led efforts of the leading industrial and academic research centres. Furthermore, the intense industrial interest in interferon began to affect the relatively free exchange of research information and materials in the field of interferon research. Certain previously accepted norms of openness and exchange were no longer matter-of-course with the development of commercially motivated forms of secrecy within the research community.[49]

In contrast to the 1975 Krim meeting, her second workshop went largely without dissenting voices being heard and had the effect desired by Krim. It amplified the wave of public enthusiasm regarding interferon. Without going into particulars, it is worth looking at some of the latest data and concepts which were being presented, discussed or rumoured over the three-day period in New York and were particularly helpful in stirring the imagination of the media and public at large.[50]

After more than twenty years of scientific 'go slow' regarding the purification and characterization of interferon, spectacular advances were being reported by biochemists employing new sophisticated laboratory techniques like high-performance liquid chromatography (HPLC) and the so-called 'SDS polyacrylamide' gel electrophoresis. In fierce competition a number of research groups not only claimed to have purified several different human interferons to homogeneity but also to have determined their amino acid composition. However, the chemical identity (or what many scientists still used to call 'lack of identity') of these proteins in terms of amino acid sequence was still the subject of considerable speculation.[51] At least the biochemical data strongly suggested the possibility that they were investigating a family of proteins. More than three different molecular forms of human interferon had reportedly been characterized. This, in combination with the fear of the 'old-timers' in interferon research of losing control of a field of research that was overwhelmed by an invasion of immunologists, molecular biologists, tumor biologists and clinicians unfamiliar with their means of communicating interferon research, provided an important incentive to devise a new system for their nomenclature.[52]

As of March 1980 the interferon research community would decide to abolish the old terminology of Type I ('leucocyte' and 'fibroblast') and Type II ('immune') interferons in favour of the type designations alpha, beta and gamma. Alpha and beta corresponded to what always had been called Type I interferons, but helped to make explicit the knowledge that leucocyte and fibroblast cells could each produce different sorts of interferons. The new naming also enabled them to react swiftly to future claims of new interferon-like factors that were different from current types by adding sequentially a delta, epsilon etc.[53]

Optimism was also expressed with regard to meeting the challenge of producing more and cheaper interferon. Rumours abounded that several 'gene splicing' groups were on the verge of cloning and expressing the human interferon gene in the bacterium *Escherichia coli* ('*E. coli*').[55] If successful, this was thought to open the way for producing unlimited quantities of relatively inexpensive human interferon. At the same time, researchers from Wellcome Research Laboratories had reportedly succeeded in piloting an industrial process based on conventional fermenter technology for the mass production of human interferon from continuous cultures of a specific line of lymphoblastoid cells, so-called Namalwa cells.[56] After initial concerns about product safety of this interferon prepared from a human cancer cell line, regulatory authorities in Britain and America were expected to grant permission to start limited clinical trials with this kind of interferon.[57] Apparently the Wellcome interferon closely resembled the common leucocyte interferon in chemical and immunological properties.

Notwithstanding the fact that the ACS trial had only just started there were further anecdotal reports about promising results with interferon in viral and malignant disease. Warnings that none of the trials was large enough or suitably designed, and that the interferon preparations had more side-effects than initially expected, could not prevent most participants from becoming excited about the long-range possibilities of interferon both as an antiviral and anti-tumour agent. According to one of the clinical researchers present, it might turn out to be one of the first natural products for use in human medicine, with an effect potentially as great as that of the corticosteroids. The whole thing might still be dubious – 'iffy'– but it was a promising topic, if only one looked at the impressive number of laboratory data that had been amassed regarding the molecular biology of the interferon system.

It is interesting to note that the question of whether or not the bewildering array of effects (cell growth inhibition, induction of a resistance to virus infections and regulation of immune responses) could be attributed to one molecule, no longer aroused fierce debate. The once popular argument that effects different from the antiviral activity were most likely due to contaminants in the preparation was no longer taken seriously after several 'pure' interferons had been tested *in vitro* and *in vivo*. Instead of the 'antiviral versus anti-tumour' type of argument, the idea was gaining ground that the many diverse effects of the interferons were related through common initial sites of action. No longer were interferons regarded as specific viral inhibitory substances, but as kinds of cell mediators or regulators, thought to modify and mediate cell functions in various ways through mechanisms similar to those described for hormones, probably as part of the non-specific host defence system.

It was becoming fashionable to speak about interferon in terms of a prototype of a 'new' class of hormone-like cell regulatory proteins, biological response modifiers, which were known to immunologists under the name 'cytokines'. In providing a rationale both for interferon's versatile action in the test-tube and the patient, this concept of interferon had an irresistible appeal to laboratory and clinical researchers alike. On the one hand, this concept fitted the regular biomedical framework of drug/receptor interaction, despite the stated difference between the polypractic interferon therapy and the 'magic bullet'-oriented conventional cancer chemotherapy. On the other hand, it was considered a breakthrough that promised to open up new fields of research and therapy.

Hailing a miracle drug

The implied claims of a potential medical breakthrough regarding both virus and cancer therapy at the New York meeting signalled yet another wave of public enthusiasm regarding interferon. The popular media gave enthusiastic coverage of the potential of the 'natural drug' interferon with dramatic lead paragraphs such as 'If IF works it could . . . be a vital weapon in the battle against cancer, protect against viral and bacterial disease, provide a cure for shingles, rabies, chicken pox, eye infections and prevent the common cold' or 'Interferon: the promising new drug against cancer'.[58] The interferon story had the drama of life and death, the horror of dying patients, the awesome picture of decades of obscure and difficult research, the dedication and persistence of a handful of scientists to produce minute amounts of a potential life-saver. It had all the ingredients that journalists love to write and talk about.

Encouraged by Lasker, who more than ever before was putting all her political weight behind interferon, Krim gave numerous interviews to newspapers and magazines, appeared on radio and TV programmes, and she testified at the hearings on new developments in cancer treatment before the House Select Committee on Aging in June 1979.[59]

An all star line up of witnesses who had lost a loved one to cancer, like Mrs Nat King Cole and the former actress Helen Cahagan Douglas, set the stage for the three days of hearings before the House Select Committee on Aging. The parade of famous witnesses asking for more support and better treatments (read, more effective and less 'debilitatingly' toxic) was immediately followed by testimonies by NCI Director Arthur Upton and NCI's Division of Cancer Treatment Director Vincent De Vita. De Vita seized the opportunity to draw attention to new NCI research initiatives, most notably the establishment of a Biological Response Modifier (BRM) programme. The BRM programme aimed at developing a new generation of therapeutic drugs for the treatment of cancer which would augment the capacity of the patient to react against his or her own tumor. It resulted from a compromise within NCI quarters between the 'laskerites' who lobbied hard for a special interferon programme and opponents like De Vita who resisted against funding a short-term interferon-only programme.[60] De Vita emphasized that the BRM programme focused on biologicals, natural substances that might be less toxic than synthetic anti-cancer drugs by virtue of the fact that their action was mediated through normal defence mechanisms of the patient. Despite his assurance that interferon was first on the list of substances to be explored, he was openly criticized for ignoring interferon.[61]

When asked why the NCI had not taken the lead in sponsoring clinical trials De Vita responded that this was certainly not the case as the NCI had purchased more than 1 million dollars of the agent and supplied some for clinical trials long before the ACS announcement. However, in order to be able to proceed with work on interferon and other promising biologicals within a structured and comprehensive new drug development programme the NCI would need extra funding. De Vita indicated that it would cost an additional 15 million a year to develop three biologicals for the treatment of cancer. Sensing intense congressional interest in interferon he clearly wanted to have his cake and eat it. He was openly encouraged in his demands for additional funding by the chairman of the Congress Committee, Claude Pepper who bombarded the audience with militaristic metaphors, in the following way: 'If there were an enemy at our borders killing a thousand of our citizens a day, any member of Congress who would say that we couldn't afford to spend more to defeat that enemy would be hooted down'.[62]

His peptalk-like performance was swiftly followed by the testimonies of Krim, Strander and Gutterman. Krim's statement in the American Congress reflected her indisputable political talents:

> I want to comment on how we foresee, today, interferon's place in the medicine of tomorrow. We believe that as important as it may become, it is going to be one among many other biologicals used. . . .We do not think that, in the treatment of cancer, interferon will be a panacea, the cure all. Rather, it is likely to be most useful in conjunction with other treatments, such as surgery, chemotherapy and radiation therapy . . .[63]

Instead of committing herself completely to interferon Krim presented interferon as a prototype of a new form of cancer therapy that should be viewed as supplementing conventional treatments. This was in line with her newly acquired status as member of NCI's Subcommittee on Biological Response Modifiers. There was a risk that temporary setbacks in work on interferon might put a premature end to research support. Obviously, Krim wanted to reduce that risk by presenting work on interferon not as a purpose in itself but as part of a new therapeutic approach that would help to constitute a medicine of tomorrow. Having said this I have to add that at the same time Krim was not scrupulous about promoting interferon on its own merits in discounting undesirable effects produced by the administration of interferon. She presented them as temporary shortcomings, though she did not go as far as Gutterman who boldly stated: 'Interferon opens up a new form of cancer treatment which is non-toxic. There is no nausea, no vomiting, no diarrhoea, or the other side-effects of chemotherapy . . . It is the most promising cancer lead we have'.[64]

The ends seemed to justify the means. By withholding information on interferon's side-effects – responses (in general reversible) as observed during interferon treatment at the M.D. Anderson Hospital and Tumor Institute included fever, nausea, vomiting, lassitude, hair loss and bone marrow suppression (which, though less severe in nature, resembled those experienced during conventional chemotherapy) – Krim and Gutterman succeeded in further fuelling the enthusiasm for interferon and mobilizing public and private funds.[65]

Following the hearings, the American Congress awarded additional funds to the NCI of 13.5 million dollars (about 3 per cent of the annual NCI budget), specifically earmarked for the BRM programme but only on the condition that it would focus attention on interferon.[66] The news that the NCI was going to pour millions of dollars into interferon research and production, and simultaneous public claims by Gutterman that 50 per cent of his cancer patients showed remissions after interferon treatment, snowballed the flow of interferon-related publicity: 'A wonder drug in the making', 'Potent protein; Medical researchers say the drug interferon holds great promise', 'Interferon: The body's own wonder drug' and 'Race is on for miracle drug'.[67]

Despite the sobering facts about interferon's questionable therapeutic exploitability which underlaid these bold headlines, the frequent associations of interferon with terms like 'cancer', 'weapon', 'natural body substance', 'miracle' and 'scarce and costly' fuelled public expectations. The fact that this potential cancer therapy was extremely hard to come by and costly, and the knowledge that no other biological substance could match interferon's extraordinary biological activity, all contributed to the popular belief that interferon must be highly effective. Most of these descriptions in the media were directly or indirectly inspired by enthusiastic accounts of research events by biomedical researchers. According to the *Washington Post* columnist Nicholas von Hoffman, the uncertainty about interferon's abilities as an anti-cancer cure did not discourage 'any number of persons in white smocks from making their debuts before the cameras to conjure up rose-hued dreams of therapeutic miracles'. Of course they did their best to pepper their public declarations with 'if', 'could' and 'might', but already the mere fact that interferon was taken seriously by a scientific community highly critical of other unorthodox cancer remedies like vitamin C and laetrile was sufficient to let the promise and hope part of the message stick for the audiences.[68]

The public message of hope was corroborated by accompanying speculative story appetizers in the media. Reporters included famous and rich cancer patients like the film star John Wayne, Ted Kennedy's son and the exiled Shah of Iran to blow up their interferon stories.[69] In doing so they stimulated further public speculation about the therapeutic potential of this natural substance that was extremely hard to come by. The fact that the doctors of those famed cancer patients seriously considered treatment with interferon and the knowledge that no other biological substance could match interferon's extraordinary biological activity as well as cost price, all contributed to the inflation of expectations. In combination with references to interferon in the media like 'interfering with cancer' and 'the body's own wonder drug' this created a breeding ground for the popular belief that interferon not only must be highly effective but that this was more like a magic potion that came close to God's own elixir.[70]

Gene dreams and the inflation of expectations

The claims by genetic engineering firms and leading molecular biologists about the imminent possibility of making available, in large quantities and at modest cost, this potential breakthrough in the fight against cancer intensified public interest. According to *The Observer* the race was 'on for miracle drug'.[71] The fierce scientific competition involved in the efforts to clone the interferon gene was fascinating in its own right; not because it was one of the first medically significant human genes to be cloned nor because it proved that genetic engineering had passed, as the *Wall Street Journal* put it, 'from science-fiction fantasy to fact'.[72] Rather, it was the metaphor of a race in combination with both the promises of a new and wondrous production technology and of a billion-dollar miracle molecule that made interferon so fascinating for the general public.[73]

The euphoria reached a peak after Biogen and Schering-Plough announced at a joint press conference at the Park Plaza Hotel in Boston on 19 January 1980 – skillfully orchestrated to arouse media attention (see Figure 35) – that through the use of genetic engineering the molecular biologist Charles Weissmann and his team were the first to succeed in getting bacteria to produce human interferon in biologically active form. While admitting that there were still a lot of questions to be answered, Nobel Prize winner, Walter Gilbert, a Harvard professor and chairman of the board of Biogen, predicted that within one or two years the mass production of interferon for use in clinical trials would be feasible.[74]

But did this announcement really deserve the connotation 'breakthrough'? For one thing, the cloning of interferon had already been accomplished by a Japanese research team and published in the *Proceedings of the Japanese Academy* a few months earlier.[75] In addition, no one knew whether the engineered protein would behave in the same way as the natural one did. It was known that the bacterial cells were not programmed to add sugar molecules to the interferon product as the human cells would naturally do. However, hardly any knowledge was available about the biological function and effects of these missing sugar residues. Moreover, the commercial significance of the news, with several competitors in the race to produce human interferon, was far from obvious. Yet, the linkage between genetic engineering and the possibility of manufacturing a promising anti-cancer drug sufficed to make journalists report it as a 'breakthrough'. With a few exceptions they failed to ask critical questions that might have helped to place the work, which had yet to be published in the scientific literature, in a critical perspective.[76]

During the next days and months the story appeared on the front pages of most American newspapers and magazines and others around the world – in Europe but also in Asian, African and South American countries – ran major

Figure 35 Charles Weissman delivers the news of the successful cloning of the human leucocyte interferon gene and its expression in biologically active form at a press conference held at the Boston Park Plaza Hotel. Reproduced with kind permission of the publisher.

pieces on it.[77] The major networks led their prime-time news broadcasts with the latest promising news on interferon. The readers and viewers were bombarded with headlines such as 'Medical breakthrough reported: "Glamour stock" could help cancer patients', 'Cancer treatment available soon', or 'The big IF in cancer: will the natural drug interferon fulfill its early promise?', and 'At only $100 million a gram, this "miracle" has a future'.[78] These kinds of leads were supported by familiar but still powerful images of scientists and doctors in white coats busy working with tubes and syringes in their sophisticated, high-technology laboratories and clinics. Exemplary in this respect is the frequent use of the 1960 Flash Gordon comic strip depicting the first successful clinical trial with interferon (see Figure 13, chapter 2).

Worldwide interferon became a hype. The event was described by Nicholas Wade in *Science* in terms of a 'cloning gold rush' that turned molecular biology into big business – with biotechnology stocks rising to record levels at the international stock markets. Popular magazines talked about genetic engineering as the solution to the problem of producing a 'priceless miracle drug'.[79] All segments of society had a part in the hype. As Sandra Panem aptly portrayed the situation:

> . . . scientists who genuinely believed that they were on the right track and that money solicited at the expense of candour would be wisely used; investors and the public who wanted interferon to be a wonder drug and

did not choose to ask whether the claims might be overstated; and those representatives of the media who reported anecdotes with unbridled enthusiasm.[80]

Until the Biogen announcement, the dramatic portrayal of interferon seemed to work to the advantage of all, with enormous amounts of energy and money pouring into efforts related to interferon in times of otherwise sharp financial cutbacks.

However, there was another side to the hype. Cancer centres in America and Europe were inundated with requests for interferon. Countless desperate patients and their families were begging hospitals, doctors, research centres and drug companies to provide them with the new wonder drug. Cantell, who like Strander was widely portrayed in the media as one of the divine heroes of interferon research, was literally besieged with requests for interferon from Europe and the US to Japan, Malaysia and India – confronting him with the horror of hundreds of fatally ill husbands, wives, relatives and children seeking a cure, but also more benign cases like the following man: 'I am having the biggest health problem of my life. I am having cold sores in my penis. . . . I beg you help me . . .'[81] Even those familiar with the rage surrounding the development and testing of the 'miracle' drug cortisone in the early 1950s had not seen anything like this happen on this scale and with such vehemence.[82]

The interferon mania only added to the media interest: 'Wonder drug hope for Miss Anneli: Amazing case book of the wonder drug doctor' (see Figure 36), 'Dad's wonder drug plea' and 'Drug brings hope for tumor boy Daniel'.[83] The interferon story had the drama of life and death, the horror of dying patients, the awesome picture of decades of obscure and difficult research, the dedication and persistence of a handful of scientists to produce minute amounts of a potential life-saver. Personification of scientific medicine and disease – projecting the inherent benevolence of medical science on individual scientists, doctors and patients – was a powerful rhetorical tool that journalists routinely employed. Imagery got inextricably bound up with content and this helped to intensify the interferon furore.

This can be exemplified by looking at bits and pieces of the British ITV programme *TV Eye* entitled 'Cancer: The new Weapon' that was broadcast on 19 June 1980.[84] The documentary started with a close-up of an ampoule of interferon which the voice-over 'modestly' introduced as follows:

> This phial of medicine contains one of the rarest and most expensive substances in the world. It's also one of the most exciting. It's already been used successfully to treat some cancer patients . . . which could offer hope to millions of cancer patients throughout the world. We report on its successes, its failures and the controversy surrounding its use.

The programme proceeded with a portrait of four seemingly successfully treated cancer patients. The very same 16-year-old Swedish girl Anneli, who

Figure 36 Sunday Mirror front page on 1 June 1980. Reproduced with kind permission from MSI/ABC Press.

figured prominently in the British tabloid the *Sunday Mirror* shortly before as the happy dance-loving and horse-riding interferon 'miracle', went first (see Figure 36). While the camera zoomed in on an exultant looking Anneli the voice-over informed the viewer that this teenager with a highly malignant form of cancer owed her life to interferon. 'Six years ago her chances of surviving a cancer operation were one in five. Today, she is free of cancer'. By subsequently emphasizing that the doctor involved, the world's leading expert on interferon Hans Strander, remained cautious despite interferon's most startling abilities to heal (showing a young intently looking but self-effacing medical scientist), the question 'Does interferon work against cancer?' became more and more rhetorical in nature. Add this to the eager reproduction of the powerful penicillin analogy and the viewer seemed to have no other choice than to believe that interferon must work – if not yet then certainly in the near future – in spite of all the question marks still circulating among medical professionals.

> If it lives up to its promise and as its name implies, that's still a very big IF, its discovery will rank with Salk's development of the Polio vaccine and the discovery of penicillin by Alexander Fleming [accompanied by familiar

library pictures of these medical heroes]. In fact there are many similarities between the story of penicillin and that of interferon. Both were discovered in Britain but developed elsewhere. Fleming's discovery was also expensive at first, now it's cheaper than the bottle it's put in. Fleming's use of penicillin was surrounded by controversy. So too is interferon.

The reiterative nature of the public comparison between interferon's journey from laboratory to clinic, and the long, obstacle-filled road that penicillin had travelled from Fleming's laboratory to the pharmacist's shelf lent not only additional credibility to interferon's workings but also reinforced interferon's status as a miracle drug.[85] In the British context these inflationary forces were stronger than anywhere else due to a specific cultural condition, the 'penicillin trauma', which also played an important role in the public reception of interferon as 'the antiviral penicillin' in the early 1960s. Inadvertently or not, the programme capitalized on the slumbering public sentiment that once again Britain lost out by failing to exploit its own discoveries with the following statement:

> So, throughout Europe and America the pace of interferon research, development and testing is moving ahead very quickly. But what of Britain? Where after all interferon was originally discovered. Well sadly, as so often happens with British inventions, Britain is now lagging behind the rest of the world . . . The Imperial Cancer Research Fund is still negotiating for supplies and can not start clinical trials until autumn. Last year they collected more than twelve million pounds in donations from the British public. At this stage they refused to discuss their plans.
> (*TV Eye*, 'Cancer: The new weapon', 19 June 1980)

Add to this the postscript 'please do not ask your doctor about interferon because supplies just do not exist' and the viewer is left with the impression that the only barriers to interferon's application were financial support from a sluggish Government and opposition from a conservative medical establishment.

Coping with an imminent black market

In the light of media messages and imagery prone to speculation of this kind, it is unsurprising to find that there was massive demand from patients suffering from cancer or a severe viral disease wanting access to this 'miracle drug' despite it being in short supply and as yet untested. Much distress was involved. In their face-to-face contact with patients, doctors found themselves besieged by demands for a drug that they could not supply, and frustrated by hopes of a cure they could not deliver. At centres where interferon was being tested everyone wanted to be in the trial and doctors had a hard time explaining that nobody could or would be favoured in their selection of trial candidates. In order to determine whether interferon had any 'real' efficacy medical researchers had to work in accordance with stringent testing protocols which implied that only patients meeting the highly specific trial requirements would be allowed to participate. And given the test subjects's own enthusiasm for interferon in combination with their special social status as lucky dog, this was already asking for the moon. The doctors involved in testing interferon feared that the high hopes of 'the chosen' might work against what was defined as 'objective benefit' and as such might undermine the scientific validity of the interferon.[86]

Confronted with the disturbing side-effects of the hype, the biomedical establishment increasingly regarded the media frenzy as problematic. Medical research organizations, like the British Medical Research Council and the Imperial Cancer Research Fund (ICRF), and health administrators saw the public demands for interferon as a challenge to their authority of drug evaluation, registration and supply procedures.[87] Media stories about distressed families of terminally ill cancer patients willing to sell their homes and business to pay for their loved ones to be treated with interferon led to worries that manufacturers would yield to public pressure to provide interferon outside the formal testing channels and, even worse, that a black market for interferon might develop.[88] Medical authorities realized that it was hard to explain to a patient suffering from cancer that there might be a more effective treatment in the pipeline but that it could not be generally available for some years, until licence procedures were completed, by which time he or she might be dead.[89]

Distribution of the unlicensed drug interferon outside the approved medical trials was not only believed to undermine the ongoing drug evaluation process but also the whole state-regulated drug testing practice – established in the 1960s to maintain certain quality and safety standards.[90] With the prospect of improving supplies it seemed even more difficult to resist the public's disinclination to wait and to ensure that all available material would be channelled into the official trajectory of controlled clinical studies, required for the proper evaluation and licensing of new drugs.[91] Confronted with the desperate efforts of countless families and friends of cancer sufferers to obtain supplies of interferon at any price, the British Government and the major funding bodies for cancer research in the UK decided to issue a joint press notice in May 1980, cautioning against over-optimism and attempts at by-passing the formal drug evaluation route.[92]

Despite the fact that the excitement and motives of its own members had played an obvious part in the interferon mania, the scientific community swiftly left the media to carry the blame. As long as the public enthusiasm interferon elicited was not associated with the term 'uncontrollable', the dramatic portrayal of interferon was widely accepted as necessary for the advancement of a most promising area of biomedical research. But once defined as a threat to the practice of medicine and professional autonomy, physicians on both sides of the Atlantic started making public appeals for a moratorium on publicity about interferon. The interferon community openly supported the physicians in their efforts to blame the journalists for raising patients' hopes through irresponsible reporting.[93] The primary concern of the scientists was that the continued media frenzy, which had initially worked to the professional advantage of interferon research, might in the end rebound to the discredit of interferon research itself.[94]

It can be no accident that, while doctors and authorities worldwide were struggling to cope with what they regarded as mass hysteria about interferon, preliminary, unexciting results of the ACS sponsored interferon clinical trials were announced at the annual meeting of the American Association for Cancer Research.[95] Whereas in February 1980 the consensus of the chief investigators associated with the ACS study was still optimistic and Gutterman's 50 per cent response rate claim spurred on the ACS to allocate another 3.4 million dollars for further study, by the end of May things had changed.[96]

The information made available cooled expectations by suggesting that interferon was no more active than other available chemotherapeutic agents in treating breast cancer and multiple myeloma. Moreover, contrary to what was hoped for (in particular by outsiders who were sold on interferon as a non-toxic agent), side-effects similar to those of other cancer medications were reported. The patient who responded best was said to suffer the most serious side-effects such as abnormal liver function and even cardiac toxicity, and therapy had to be interrupted.[97] These results placed interferon in perspective, just like any of the other many substances being tested for anti-tumour activity.

Immediately following the premature release of the first ACS trial results the American Society of Clinical Oncology, with the prominent oncologist Charles Moertel of the Mayo Clinic as its immediate past president and spokesperson, took the unprecedented step of issuing a strong warning against over-expectations for interferon.[98] In a personal statement to journalists Moertel emphasized that, while the jury was still out on interferon, cancer patients should realize that cancer treatment was a complicated matter and the best thing a patient could do was 'get in the hands of knowledgeable cancer specialists and let them call the shots'.[99] In saying so Moertel articulated the worries of doctors about the eventual damaging effects of public enthusiasm on professional autonomy.

The message basically was that interferon as a non-toxic wonder drug for cancer was a myth created by the media, unsustainable in view of the available scientific test results. Whereas before the side-effects associated with interferon treatment had always been played down, now the very same side-effects were emphasized to show that interferon was hardly any different in its toxic effects from other available cancer treatments.

The redefinition of interferon as a potentially harmful experimental treatment with uncertain therapeutic benefits proved effective in undermining its public image as a wonder drug and put a lid on what Rauscher described as his continuing nightmare: 'the unquenchable desire for interferon by cancer victims'.[100] Almost overnight the media picked up the message with headlines such as 'Interferon: Studies put cancer use in doubt', 'Is it a wonder cure for cancer or the most expensive flop in history?'.[101] Just as the interferon alliance had been overly successful in seeking a favourable press, the biomedical establishment managed to reshape the nature of the media coverage of interferon – from forceful promotion of interferon's benefits to disappointment about its failed promise. As I will show in the next chapter, however similar the course of events in the case of interferon to those of other so-called cancer cures, the most common dead-end scenario of potential cancer remedies would not materialize.

Examining interferon stories in the American and British media, there is essentially one major script centring around the theme of 'promise and hope'. Initially the media highlighted the dramatic curative potential of interferon against the dreaded disease cancer. They framed interferon as a somewhat mysterious and scarce non-toxic body substance which exerted its anti-tumour effects in a natural way, and at the same time as the latest promising product of a laboratory-supported scientific medicine. Subsequently, attention shifted to interferon as the most promising demonstration project of a revolutionary production technology, genetic engineering. I described how, by blowing up the imaginative combination of what were widely perceived as two wondrous products of modern science and technology, the mass media helped to trigger an international interferon hype, which is unparalleled in medical history in terms of speed of circulation, intensity and extension. Given the media's impressive capacity to amplify the message of promise and hope, I would like to stress that the media coverage of interferon was for the greater part a mirror of the expectations, legitimations and opinions circulating within the biomedical and public realm.

7 Yet another twist
Marketing interferon as a helpful neighbour[1]

'When it comes to clinical trials, few issues are simple. And many are controversial' wrote the *Science* correspondent Gary Taubes in 1995.[2] Taubes's dictum seems to be at odds with the public model of controlled clinical trials as a very helpful tool to relieve medical practice of that most feared element known to scientists and regulators: subjectivity. Are most doctors and regulators who firmly believe in randomized controlled trials as the key to an 'evidence-based medicine' mistaken? Given an ideal world without social, professional and economic interests affecting judgements of the efficacy and risks of medical therapies one might have answered 'no'. It is impossible, however, to conceive of such a trial taking place in a human vacuum. Conducting clinical trials involves establishing links and commitments between many different individuals and organizations, including clinicians, laboratory researchers, patients and their families, regulators and last but not least the drug companies.

In being shaped by the specific context of medical practice, clinical trials, even the most sophisticated randomized controlled trials, as I already have shown in chapters 3 and 5, are not value-free measuring devices that objectively evaluate the efficacy of new therapies. Like any other medical device associated with our daily lives randomized clinical trials incorporate the beliefs and ideas of the persons and organizations who developed them and then are moulded by those implementing the methodology.[3] The final part of the book's narrative provides an exemplary case to illustrate the complexities surrounding the process of therapeutic evaluation by means of clinical trials.

Given intense disappointment with what by 1983 became dubbed 'the miracle drug looking for a disease', it might come as a surprise to hear that interferon would ultimately succeed in finding a promising niche in clinical practice. This then brings us to the following question: how did interferon become naturalized as part of medical practice in the 1990s?

Beyond interferon

In response to the drastic turn in the nature of reporting on interferon in the course of 1980, Mathilde Krim and her staff wrote a letter to the *New York Times*. They warned against the negative effects such qualified reporting might have on the public support of a field of research that sooner or later might provide society with a new immunotherapeutic approach to human cancer. The core argument in this letter – emphasizing the unique property of interferons to enhance the activity of the body's own defence mechanisms – can be understood as a first attempt at developing alternative strategies to legitimate work on interferon: picturing interferon as part of a promising new but experimental therapeutic approach in the future treatment of cancer and viral infection.[4] A similar message of promise and hope was conveyed at the first annual International Congress for Interferon Research in Washington, DC, in December of the same year.[5]

At the Washington meeting, which was organized by an American advertising and marketing company on behalf of the publishers of the newly established *Journal of Interferon Research*, there were numerous displays with sequences of interferon DNA and derived amino acid structures. Together with the presentation by the molecular biologist Charles Weissmann of a slide with the first photograph ever shown of – what had always been considered a vain hope – highly purified crystals of alpha interferon, this promoted interferon to the tangible world of chemical facts (see Figure 37).

The prospect of the mass circulation of unproblematic samples of crystalline interferons was expected to incite a considerably larger scientific public to use these molecules as a fresh starting point for their research. However appealing, the excitement was to a greater extent related to the claims of Biogen and its competitor Genentech that they both had succeeded in creating new, hybrid interferons by cleaving and rejoining ('recombining') pieces of two different interferon genes. According to Weissmann this was the first time that proteins had been engineered in this way, resulting in a new kind of biosynthetic

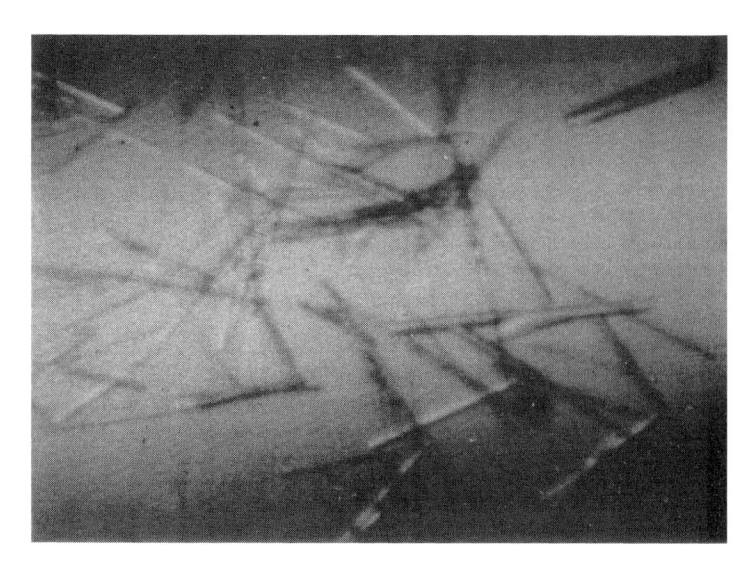

Figure 37 Photomicrograph of interferon $\alpha 2$ crystals. Courtesy of Dr C. Weissmann.

interferons with unique biological properties. Genetic engineering, so he argued, could in this way be used to 'create an unending array of "unnatural" interferon molecules'.[6]

While several teams of biochemists had been working day and night to purify interferon to homogeneity, produce its chemical formula and structure and try synthesizing the molecular entity from pure amino acids on the basis of available structural data, they were, as the American biochemist, Pete Knight, expressed it 'run over from behind by the cloners'.[7] What the competing teams of molecular biologists basically had done was to beat the biochemists by genetically engineering microorganisms that did the synthesizing job for them.[8] In addition, the newly developed laboratory tools to engineer and decode DNA molecules provided the genetic engineers with the technical means to 'read and write in the language' of the genes.[9] It was a laboratory exercise that helped to reconstruct the protein structure from the DNA sequence of what was considered the gene coding for human interferon.

The claims of the molecular biologists can be understood as an attempt at showing the drug makers and the public that there was more to interferon than just a demonstration project of a promising new technology. It should be viewed not only as a test case of a new production technology but also as a pilot project for a new approach to drug development. In a similar way to the chemists being routinely involved in engineering an endless array of pharmacologically active chemicals, the molecular biologists would offer their expertise to drug makers to synthesize, modify and test new biosynthetic molecules. While the representatives of biotechnology firms began 'advertising' interferon as a means to establish an additional method of drug development, other interferon advocates increasingly legitimated work on interferon as part of a promising new multi-modality approach in the treatment of cancer.

Most interferon researchers emphasized that cancer was too complex a disease to allow for a comparison between the search for a cancer cure and the development of wonder cures against polio or bacterial disease. In the case of a highly differentiated illness such as cancer it would be better not to think in terms of overall cures, but rather of treatments. The consensus view was that though as a single agent interferon might turn out to be medically more useful in treating viral infection, ultimately it might prove most valuable as part of the increasingly popular multi-modality approach in cancer treatment. They claimed that attaining this end and learning how to use interferon, would require a further expansion of basic and preclinical investigations as well as clinical studies.[10]

Clinicians and laboratory researchers alike conveyed the impression that with more questions than answers they were just beginning to explore the potential of interferon, not as a single substance but as a large family of biologicals. So far more than ten different interferon molecules were claimed to circulate worldwide. The diversity of interferons with distinct and complementary biological activities seemed to grow every day.[11] According to the interferon researcher Frances Balkwill in a 1982 progress report in the popular British science magazine *The New Scientist*: 'It is clearly going to be difficult to know which interferon to use for any particular treatment or what combinations of "cloned" alpha interferons will be most effective'.[12]

Interferons were said to make up a new modality of cancer treatment in acting not as chemical toxins, directly on the tumour cells, but as biologicals through activation or modulation of key elements of the body's immunology system. They could be used as biological enhancers – helping to increase the host's own response against the tumour – in conjunction with the three main cancer therapies: surgery, chemotherapy and radiation. The NCI publicly singled out its programme on interferon and other biological response modifiers (BRM) as 'most exciting' in terms of benefits to be expected for future cancer treatment.[13]

In the opinion of the immunologist Louis Epstein of the University of California, San Francisco, the success of interferon therapy ultimately depended on the ability to forge further links between the laboratory and the clinic. She claimed to have developed a methodology in her laboratory which, if further elaborated on, could separate out those people whose cancers had the best chance of responding to a particular interferon. Her approach might make it possible in the future to individualize cancer treatment. She envisioned clinical oncologists sending tumour samples to her laboratory in order for her to figure out the susceptibility of a specific tumour to interferons and other compounds. This would enable the doctors to account for the problematic individual variations among their cancer patients.[14]

The overall message was as a science reporter of the *Washington Post* aptly expressed it in his headline, 'Beyond interferon'.[15] By creating an image of interferon as a prototype of a promising, new, but still poorly understood, area of cancer therapy known as immunotherapy – one that was going to have an important role in future cancer practices – the promoters of interferon had succeeded in establishing a more permanent base for support.

Despite the growing public scepticism in the light of emerging stories about the iatrogenic consequences of interferon treatment and the rather disappointing response rates in the ACS-sponsored trials using Cantell interferon, the interest in interferon as part of a fourth modality of cancer therapy, known as 'immunotherapy', was on the rise. American cancer treatment centres that aspired to maintain an image of being at the cutting edge of the field of clinical oncology could not afford *not* to study an experimental therapy that was closely linked with the latest developments in tumour biology and molecular biology.[16] In these centres great pressures were often exerted on practitioners to develop close relationships with the laboratory, to conduct preclinical and clinical research, and to enrol as many patients as possible in clinical trials.[17] Clinical oncologists were used to observing side-effects and overall low response rates in their day-to-day therapeutic practices. The mere knowledge that interferon as a scientifically endorsed experimental therapy reportedly produced clinically significant anti-tumour activity in a limited number of patients suffering from a variety of highly problematic tumours sufficed to give it the benefit of the doubt. Stephen Sherwin, an American clinical oncologist, put his colleagues feelings aptly by stating: 'Interferon may not be the wonder drug it was expected to be, but it may reduce tumours in specific cases. We have to give it a try'.[18]

Only the relative scarcity and impurity of interferon preparations withheld them from pursuing interferon research more actively. However, as soon as the more pure and homogeneous rDNA-produced interferons became more widely available in the course of 1982, a growing number of American and European clinical oncologists became involved in preclinical and clinical testing.[19] Their testing 'zealotry' was hardly affected by the news in November 1982 that 4 cancer patients out of 11 in a Paris Hospital had died from a heart attack while being treated with French alpha-interferon.[20] Following the French in their decision to stop only the trial in question, Robert Oldham, head of NCI's BRM Program, played down the significance of the tragic event. Thus far in American studies there was no evidence that interferon had ill effects on the heart. Most likely the fatal incidents were due to a problem with the manufacture of the French interferon or it might also be possible that the patients involved already had serious heart conditions before the trials began.[21]

At the January 1983 meeting of the NCI's Board of Scientific Counselors Oldham did issue a warning that although biologicals like interferon were natural products this did not necessarily mean that they were non-toxic. In the same breath he called interferon one of the more active new agents tested in the last several years, perhaps just below the level of the most successful chemotherapeutic agents and recommended further clinical testing. Even if interferon treatment had been officially associated with cardiac toxicity it is unlikely that the clinical testing would have stopped. Going by the reaction to cardiotoxicity problems involving experimental cancer drugs in the early 1980s, the NCI would have confined itself to sending letters to doctors across the country warning of the heart toxicities of interferon and advising limited use in patients with a history of heart problems.[22]

High risk:benefit ratios were and still are part and parcel of the culture of cancer treatment. Severe side-effects from experimental drugs had become an accepted part of life in most American and European cancer treatment centres. The treatment-related mortality rates ran from as high as 10 per cent for experimental regimens combining, for instance, radiation and aggressive chemotherapy to between 2 and 5 per cent in established, non-experimental chemotherapy regimens.[23] To fully understand the situation, we should take a glimpse at the world of cancer patients and doctors. According to the 1979 ACS' statistics only 33 per cent of Americans who developed cancer would be alive five years after diagnosis and treatment.[24] Even by 1989 with five-year survival rates rising to an estimated 40 per cent this still meant that for the majority of cancer patients death was inevitable.[25] In this world, doctors, trained in healing and frustrated by the frequent experience of being unable to deliver a cure, were prepared to go to any extreme if patients, for whom no other hope seemed to exist, allowed them to. At the same time most patients, in a desperate effort to avert their 'death sentence', were willing to try almost anything in the hope of a cure or at least a postponement of their death. In this context it is hardly surprising that institutes for cancer research like the NCI or cancer treatment centres placed treatment side-effects second to potential benefits.

Potential for beneficial effect was considered as sufficient basis by NCI's Board of Scientific Counselors to agree to an expansion of clinical studies and to pursuing the combined modality therapeutic approach in September 1983.[26] The apparently unremitting optimism within NCI quarters contrasted with the growing scepticism about interferon's potential among senior executives in pharmaceutical companies. Of the twenty or more firms which had announced their intention of investing in interferon development back in 1979, a growing number ceased work on interferon over the period of one year, 1983. They had either lost out in the 'cloning race' or management had negatively assessed interferon's potential both as a therapeutic drug and as a means to attract capital investment from the stock markets.[27]

Apart from the disappointing cancer trial data there were also growing doubts about interferon's potential as an antiviral agent. The expectations – fuelled by the latest reports from the Common Cold Unit at Salisbury that nasal sprays of alpha interferons markedly reduced the occurrence of colds in volunteers – about opening up the commercially interesting niche market of common cold prevention were dashed with the latest news about its side-effects.[28] Even at low-dosage levels interferon appeared to induce side-effects similar to the symptoms responsible for most of the miseries of the very same cold that it was supposed to prevent. 'A single high dose of zinc may do just as good a job at substantially lower cost', was the scathing judgement of the science correspondent, Joseph Alper, in *The New York Times*.[29]

In the boardrooms of the three 'interferon champions' that had most heavily invested in the therapeutic substance – Wellcome and, most notably, Roche and Schering-Plough, interferon's lacklustre clinical performance in treating tumours as well as virus infections was reason for growing concern.[30] The company officials responsible for investment in interferon had a hard time defending the costly interferon research and development (R&D) programmes. However, they succeeded in convincing the sceptics in management that although interferon might not be a 'magic bullet' in itself, it had enormous potential as a prototype of a new generation of custom-designed biosynthetic drugs that was expected to generate a therapeutic revolution. They presented it in terms of the unique possibility of learning to tap into an entire medicine chest of new drugs from proteins within the human body. Unlike most chemical drugs these genetically engineered biologicals could be viewed in their 'natural' (physiological) role as regulators or modifiers of a variety of pharmacologically interesting biological mechanisms. They argued that in order to develop these biosynthetic molecules as therapeutics, new testing methods and procedures would be needed to bring them to the clinic effectively. For instance, since human interferons were only active in humans and monkeys, researchers were forced to do the unusual and skip most of the preclinical animal testing and go straight into humans to determine a toxicological and pharmacological profile.

However different, work on this new generation of biologicals at the same time was argued to be easily integrated into the current system for drug development based on tinkering with molecules as a means to create variants with the most optimal pharmacological characteristics in terms of risk:benefit ratios. The promise of scientific and therapeutic innovation appealed most to senior drug company executives who were less worried about today's profits than about tomorrow's prospects. With patents on most of the top prescription pills expiring by 1990 and with an unprecedented small range of chemical compounds in various stages of development in their laboratories, ready to take over as money-making ('blockbuster') products, drug companies had come to recognize that a new wave of innovation was needed to position themselves for long-term survival.[31]

However, those responsible for the interferon-related R&D programmes were not able to get round the iron business rule of profit generation. The pressure for return on investments in interferon research and in production facilities for the first genetically engineered interferon products (Schering-Plough's homogeneous interferon alpha-2 and differing by only one single amino acid, Roche's homogeneous interferon alpha-A preparation) and Wellcome's 'natural' interferon product – containing the full spectrum of alpha interferons produced by cultured human cells – did force the interferon advocates within the companies to adapt their R&D strategies. In line with government-supported research programmes, the pharmaceutical industry focused on interferon as part of a new kind of disease management: immunotherapy within a multi-modality treatment framework. Apparently the drug companies recognized the strategic and commercial importance of taking advantage of the more general move across medicine towards combination therapy.[32]

A drug looking for a disease

The drug regulatory authorities found the combined-modality approach difficult to assess. Their evaluative practice and standards were still governed by a single-agent therapeutic philosophy. For interferon to be considered legally as a new therapeutic drug, it had to be officially evaluated as a single agent. This implied that before licensing procedures could be taken into consideration, the companies had to look for a disease, rare though it might be, that justified a need for interferon. Moreover, under FDA regulations each interferon variant was considered a new therapeutic drug and should be independently evaluated.[33]

In search of suitable diseases as candidates for interferon as a treatment, the drug companies actively supported clinical trials to evaluate the effects of interferon on as wide a variety of diseases as possible (see Figure 38). They offered clinical investigators worldwide – free of charge – large quantities of their

Figure 38 Record from clinical trial with Hoffmann-La Roche's recombinant alpha-A interferon, showing a patient's blood levels of interferon. Courtesy of Dr L. Gauci.

interferon products. Interferons were tested on hepatitis B, various lymphomas, colds, breast cancer, prostatic cancer, multiple sclerosis, herpes keratitis, malaria, AIDS (as soon as the epidemic began to manifest itself around 1982) and many other diseases related to cancer and viruses. The drug companies mounted one of the most intensive clinical trial programmes ever set up to evaluate a new pharmaceutical agent. In the first three years (1982–1984) the recombinant interferons alone were tested in over 4000 patients.[34]

By the end of 1983 the large-scale testing efforts finally showed signs of paying off. Spurred by Strander's chance observations at the Karolinska Hospital in Stockholm, alpha interferons were also tested on patients suffering from papilloma virus-associated non-cancerous tumours ('warts'). The warts ranged from being a cosmetic problem or undergoing spontaneous regression to being life-threatening. The greatest interest in the use of interferon had been in connection with genital warts ('condyloma acuminatum') and a less common but potentially life-threatening form of wart named 'laryngeal papillomatosis', located in the respiratory tract, particularly of children. This represented a serious clinical problem due to the tendency of the wart to recur and obstruct breathing. Some patients required surgery as often as every 2 months. Responses to interferon treatment were claimed to be of long duration in at least 50 per cent of the patients suffering from these warts and so far the risk:benefit ratio looked promising. None of the studies was controlled yet but large controlled trials were underway.[35] However encouraging for the patients and the doctors involved, the drug companies paid initially more attention to the research news coming from the field of clinical oncology. As this was more likely to open a window into a commercially interesting family of disorders and medical specialty.[36]

Drug company officials swung into action as soon as the news broke in the autumn of 1983 that Gutterman's research group had achieved a higher than

80 per cent trial response to 'natural' alpha interferon therapy (with Finnish Red Cross material produced by Cantell) in patients with a rare form of chronic leukaemia known as 'hairy-cell leukaemia' (so-called because of the 'hairy' appearance of the malignant cells).[37] The knowledge that there was hardly any viable treatment for this form of cancer, with estimated mortality rates of about 15 per cent per year, sufficed for those looking for a suitable disease indication.[38] They seemed to take for granted the fact that among clinical oncologists scepticism prevailed about the overwhelming efficacy claim by Gutterman and his team, who had earned themselves a name for over-optimistic interpretation of their interferon trial data.[39] Since the successful hairy cell leukaemia trial was performed with the relatively impure (less than 1% pure) and therefore pharmaceutically problematic Finnish alpha interferon product, the company's first priority was to demonstrate that the dramatic clinical effect could be reproduced with their own 'pure' interferon products. Following extensive clinical testing they ultimately succeeded in their efforts. Consequently licence applications for what both Schering-Plough and Roche dubbed 'the world's first anticancer interferon' were sent for evaluation to, among other drug regulatory agencies, the FDA.[40]

In having closely monitored and guided the proceeding of interferon trials throughout the years the FDA's Office of Biologics Research and Review (OBRR) had built up considerable expertise in evaluating both 'natural' and biosynthetic interferons. The questions of identity, purity, potency and safety featured very prominently in the frequent interactions and negotiations between the OBRR and all those interested in obtaining the licence 'Notice of Claimed Investigational Exemption for a New Drug' (IND) that allowed for clinical trials with a particular interferon. First, the question of identity was related to the taken-for-granted concept of interferon as a family of proteins and corresponding genes. Interferon was made up of at least three different types, each being made up of multiple subtypes, each of which might have different effects on the body. Every interferon product made by a different method, manufacturer, organism or interferon clone could in principle be different from the others. Each biosynthetic variant, even if it differed from another interferon by only one amino acid, was considered a distinct protein and likewise had to be evaluated independently under FDA regulations. There were more questions than answers to this complicated issue.

Second, the question of purity was hardly less problematic. For instance, genetically engineered interferon products which were claimed to come close to 100 per cent purity presented additional purification concerns such as the presence of unwanted bacterial contaminants or of altered or aggregated inter-feron species (due to gene mutations) which may be either non-functional or antigenic. Compared to this, the third question of interferon potency was a relatively unproblematic matter which was dealt with on a consensual basis during the successive standardization conferences in the late 1970s. Finally, the safety question was coped with in terms of a benefit:risk ratio assessment on a case-by-case basis. Since, as OBRR senior staff member Petricciani argued, there was nothing absolute to say about safety of experimental drugs in the clinical research setting.[41]

The test procedures required by the OBRR before any experimental clinical use had evolved in parallel with the development of new interferons and new production methods. Changes in the procedures represented the emergence of a new consensus on what was acceptable and counted as 'state-of-the-art testing'.[42] To facilitate the learning process and secure the interferon researchers' cooperation Kathy Zoon, who had an important part in the work that led to purification of the human interferons, was assigned as interferon expert to the OBRR and subsequently to the interferon licensing committees. She played a key role in evaluating the preclinical and clinical volumes of data on the various alpha interferon preparations presented by the respective companies and in reaching a judgement on whether each individual product reached an acceptable level of safety, quality and efficacy. The FDA was able to secure approval for the company's product licence applications for hairy cell leukaemia within only one year of their initial submissions in the spring of 1985. According to FDA Commissioner Frank Young the direct involvement of 'bench scientists' like Zoon had proven to shorten the official review and approval process by at least half a year.[43]

However, there is more to say to the relatively short review time than only crediting the close involvement of expert scientists for this achievement. By the time the interferon applications were filed the unconditional support of FDA's mandate on drug regulation, legislatively set by the Kefauver–Harris Amendments of 1962 following the thalidomide tragedy, was eroding. The FDA was confronted with an increasingly intense lobby for relaxation and speeding up of its drug approval procedures by a novel coalition of the pharmaceutical industry and consumer groups. Historically the organized consumer movement in America had always played an important role in the further tightening of regulatory requirements and product liability laws as a means to safeguard the public from injudicious trial and marketing of new pharmaceutical products. However, in the context of the American AIDS epidemic, consumers showed that their support of stringent regulatory arrangements was not unconditional. In particular, if they perceived the regulations as slowing down the availability of promising experimental treatments for urgent disease conditions such as AIDS.

The industry in turn had always sought to protect its position by politically highlighting the problem of stifling innovation through what they considered as an over-rigorous bureaucratic system of controls.[44] In an effort to silence the critics, reduction of the average review time – which in 1977 still averaged more than two years – was given priority status within FDA quarters.[45] Thus, interferon came to serve as a high visibility showcase of the FDA's increased efficiency and its decisiveness to act in the case of medical 'urgency'. Moreover, as I will show, in blurring the distinctions between experimental and established treatments, FDA's IND regulation and priority review policy unintentionally helped the marketing of interferon through the 'industry' of clinical trials.

On 4 June 1986, the FDA and the US Department of Health and Human Services staged a joint press-conference to announce the swift licensing of biologically engineered alpha interferons to treat a rare leukaemia for which there had been no effective therapy. The event showed that the FDA too had learned to capitalize on the power of the media. The FDA's spokesperson Young hailed the development and approval of interferon as 'a milestone, a promise of what will be coming in the future'.[46] Thus Young showed that interferon had also become a priority test-case of what he subsequently referred to as 'a second revolution of pharmaceutical innovation, akin to the discovery of antibiotics in the 1940s'.[47] By investing heavily in establishing state-of-the-art testing procedures for biosynthetic biologicals like interferon, FDA officials tried to anticipate an expected explosion of biosynthetic therapeutic agents emanating from the rapidly expanding bioengineering R&D programmes in the international drug industry.[48]

It is interesting to see that Wellcome with its non-recombinant 'natural' interferon product fell by the wayside. In most Western countries like the USA, Wellcome's natural alpha interferon 'cocktail' (with the brand name Wellferon) met with difficulties from the regulatory authorities except for Britain, the home market for Wellcome products. Asiatic countries – which historically do not oppose heterogeneous pharmaceutical preparations because of a different therapeutic drug culture – had hardly any problem accepting the interferon cocktail. By the time the FDA put aside its doubts about Wellferon, Wellcome decided that with the market already flooded with Roche's and Schering-Plough's rDNA-produced interferons, it was no longer commercially interesting to rush for a product licence application.[49]

Naturalizing interferons as biological response modifiers and cytokines

In the meantime the therapeutic development of interferon had helped to spawn a spectacular investigative enterprise. This included numerous academic, federal and industrial research programmes, the *Journal of Interferon Research* and the formation of the International Society for Interferon Research. Going by the growth in the number of papers published each year with the word 'interferon' in their titles from about 800 in 1980 to more than 3000 in the year 1987, the field experienced an information explosion – in terms of new knowledge of cellular mechanisms of immunity and of new immunological explanations of the biological effects of interferon.[50]

The fact that the accumulation of knowledge based on laboratory data was of relatively little practical help to clinicians using interferon treatment at the bedside was not regarded as a failure, but instead was used to advertise interferon's unwieldy and complex nature as part of an intricate system of checks and balances. This so-called cytokine network was said to constitute a key part of the body's natural defences. The complexity of the cytokine interactions was argued to complicate interferon's study in the test-tube as well as its clinical application. Researchers maintained that additional knowledge of these interactions was needed to improve or optimize the use of biological response modifiers or cytokines like interferon. The questions piled up but there were few satisfactory answers. Did all interferons have the same or different functions within this network? Which interferons should be tested first clinically, on what diseases and how? Was it justified to start from maximum tolerated dosage regimens knowing that more not always appeared to be better with these substances? Would combination treatment indeed add up to more than the sum of its parts? Interferon advocates called them difficult substances to work with, far more complex than traditional therapeutic drugs. As a justification for their research efforts they kept emphasizing that interferon generated intriguing research horizons. It was claimed that work on interferon provided a window both onto biological processes at the molecular level and new approaches to the treatment of disease.[51]

Laboratory and clinical researchers alike stressed that the clinical potential of interferons would be enhanced by further investigation and understanding of the molecular mechanisms underlying the effects of interferon in patients. The solidarity implicit in both their emphasis on the importance of linking biomedical research at the cellular and molecular levels to clinical interventions as part of what had become known as 'molecular medicine' stemmed from shared interests in the maintenance of credibility and funding.[52] Disappointing as well as successful clinical results could be used by bench scientists to claim that further basic research was mandated. At the same time clinical researchers were able to defend their poor results by referring to the slow progress in the understanding of the underlying mechanisms by which interferons exerted their biological effects both *in vitro* and *in vivo*. Occasionally achieving high response rates at the bedside, however rare the disease condition, seemed to justify the further expansion of both preclinical and clinical research. Nonetheless the actual expansion of interferon-related activities was largely dependent on the support and coordination of the pharmaceutical industry.

During the transition to the large-scale production of genetically engineered interferons, laboratory scientists and clinicians had become increasingly dependent on the industrial infrastructure. By possessing the key to the material resources and research materials the companies began to play a central role in structuring the development of interferon.[53] Laboratory scientists were encouraged to deploy the new techniques of molecular biology, cellular physiology and biochemistry in the search for additional naturally occurring molecules that might provide genetic blueprints for the development of these future therapeutic drugs.[54] Moreover, by providing a new powerful research tool, unlimited quantities of pure crystalline interferon, the research community was stimulated in efforts to investigate the underlying molecular mechanisms of interferon's actions.[55] At the same time clinical investigators were encouraged to start experimenting in humans with this first representative of a rapidly growing family of clinically active biologicals, the so-called 'biological response modifiers' or 'cytokines'. The impression was conveyed that participating clinicians would stand out as pioneers of a new era of disease treatment.[56]

In 1986, once the FDA, swiftly followed by other regulatory agencies, licensed Roche's and Schering-Plough's recombinant alpha interferon products for the treatment of hairy cell leukaemia, the marketing branches of the companies worked hard to create a need for interferon.[57] In order to interest health care professionals in the new area of interferon therapy and to convince them of its scientific character, Hoffmann-La Roche and Schering-Plough established freely accessible electronic data bases (e.g. Schering-Plough's *ICON*), specific journals (e.g. Roche's *Progress in Oncology*, *Progress in Virology* and *Interferons Today and Tomorrow*), and financed medical book publications (e.g. Schering-Plough sponsored the publication of *Interferons in Cancer Treatment*).[58] The way in which the companies presented interferon by means of a combination of visuals and narrative to doctors, decision-makers and journalists in the specialist press drew upon the same range of images that had influenced the public perception of interferon through the mass electronic and print media; framing interferon both as a naturally occurring substance in the body and as a state-of-the-art product of modern science and technology.

Optimizing response rates seemed to be the explicit aim of virtually any clinical research project dealing with interferon.[59] Except for hairy cell leukaemia (response rate higher than 80 per cent), clinical researchers involved in testing interferons claimed response rates anywhere between 10 and 50 per cent. The problem and advantage of using percentages was that success appeared to be a highly ambiguous term. Overall response rates ('efficacy') resulting from clinical studies under controlled circumstances might look promising, even when it remained unclear what this actually meant for individual chances of success and how well a treatment might perform in everyday clinical practice ('effectiveness').[60] Regardless of interpretation, however, response rates remained invariably low in most diseases, suggesting that it could help only some of the patients some of the time.

Under normal disease conditions this kind of negative scientific assessment would dissuade doctors from applying a therapeutic drug. But, as I mentioned before, in circumstances where there is no hope for a cure, the rules of the game appear to be different for doctors and patients alike. In diseases in which successful treatment is rare, seeking treatment through medical intervention is a gamble which can have a few winners. Gambling is an alluring analogy for all parties as it turns poor clinical results into acceptable chances, allotting responsibility for failure to bad luck rather than medical or other capacities (see Figure 39).[61]

In 1986 the Dutch interferon researcher Huub Schellekens predicted that interferon would be used more widely in ten years time than might be expected on the basis of the approved list of indications because of the psychology of interferon as a potential 'multi-drug' from common colds to cancer.[62] Schering-Plough and Roche of course fully endorsed this perceived mechanism and were keen to turn this into a self-fulfilling prophesy. They hoped that if they

Figure 39 The following subscript was added to this picture in the *Scientific American* (May 1994): 'Lela Rector of Marion, was a winner in an unusual lottery last year. After a form of interferon-beta gained approval for treating multiple sclerosis in July 1993, demand far exceeded supply. So the distributor, Berlex Laboratories, held a lottery for some 67,000 qualified patients. Rector, who drew number 109, was among some 30,000 people reportedly in treatment by March 1, 1994. The other registrants are expected to have access to the drug by the end of this year'. Reproduced with kind permission from the publisher.

were able to show that interferon as a cytokine enhanced – even if marginally – the efficacy of routine and semi-routine therapies for problematic cancers and virus diseases and chronic crippling diseases like multiple sclerosis, the molecule could become a commercial success as an auxiliary immunotherapeutic drug. They therefore actively promoted trials and investigator's meetings that would evaluate the possible synergism between interferons and various conventional therapies.

The drug industry was well aware that it was highly dependent on the co-operation of clinicians, foremost the clinical oncologists – who were highly skilled in the art of bringing laboratory compounds to the clinic and shaping drug profiles – to define additional clinical situations in which the interferons might be applied. In line with the culture of experimentation in clinical oncology the message was optimizing response rates through modulating the maximum tolerated dosage regimens, varying modes of administration and trying different combinations of therapies. At the same time they began marketing interferon as part of the comprehensive though open-ended (in terms of number of factors involved and nature of interactions) cytokine network; positioning interferon as a *helpful neighbour*, compatible with and supportive of existing treatment practices. In the process the Roche-sponsored journal *Interferons*

Today and Tomorrow changed its name to *Interferons and Cytokines*, while *The International Society for Interferon Research* became *The International Society for Interferon and Cytokine Research*, and the official journal of this society became the *Journal of Interferon and Cytokine Research* by 1995.[63]

With the relentless support of the drug industry and patients in desperate need of a cure, and through the combination of scientific drive and professional ambition, most notably clinical oncologists, infectious disease specialists and neurologists continued to work on the trial design. They tried different combinations (e.g. in combination with other forms of chemotherapy or a combination of various biological response modifiers) and different routes and durations of administration.[64] In doing so, they ultimately tinkered toward success in terms of establishing new therapeutic drug practices for interferon and actively working on the treatment's effectiveness. Although superior treatments for the treatment of hairy cell leukaemia became available – and have largely replaced the use of interferon for this indication – interferon as an adjunct to other therapeutic modalities became part of the routine treatments of a growing number of major malignancies such as non-Hodgkin's lymphoma, multiple myeloma and kidney cancer. In the 1997 *Cancer Chemotherapy Handbook* interferon can be found under the separate heading 'Biological Response Modifiers' (though in actual clinical oncology practice the term 'cytokine' prevails) with emphasis on the differences (in administration, dosage regimens and mechanisms of action as immunomodulating agents) from other cytotoxic therapeutic drugs in use in clinical oncology and in combination treatment as having the greatest promise for successful therapeutic application.[65]

Interferon, almost given up on in terms of having a role to play in medicine by business analysts, the press, the public, scientists and doctors alike in the early 1980s, has become exemplary not only for a new generation of biosynthetic drugs but also for the health benefits that result from advances in molecular medicine. Fuelled by a potent combination of scientific drive, professional ambition, marketing efforts and a lot of hard work between 'clone and clinic', interferon was transformed from a so-called 'orphan drug' for a rare terminal disease into a most successful family of drugs looking for diseases.

By the year 2005 interferons as therapeutics have been approved in the US for use against more than ten different diseases, including genital warts, Kaposi's sarcoma, several forms of hepatitis, malignant melanoma and multiple sclerosis. This makes for a rather diverse therapeutic profile with interferons being classified as part of the therapeutic class of anti-neoplastic drugs, of biological response modifiers, of multiple sclerosis agents and of antivirals.[66] The approval in 1997 for use of interferon in conjunction with chemotherapy in patients suffering from a specific form of malignant tumour of the lymph nodes, so-called follicular lymphoma, shows that over the years the FDA changed its attitude toward combination therapy, from an unmentionable registration issue towards a negotiable case.[67]

The world market for interferons – the whole spectrum of alpha, beta and gamma type interferon products – was valued at $13.6 million in 1986, rose to $751 million in 1992, $1.57 billion in 1993, down to $928 million in 1995, up (resulting from the global market introduction of interferon beta for application in treating multiple sclerosis and the licensing of interferon alpha for the treatment of hepatitis B and C) to $2.4 billion in 1999, $3.8 billion in 2001 and continues to rise with estimated global sales of more than $5 billion in 2005.[68]

However alluring the commercial perspectives of combination therapy, interferon's public barometer has always been most visibly affected by monotherapeutic promises. A case in point is the international wave of publicity on interferon following promising scientific reports in 2003 and 2004 about the potential benefits of interferon as a treatment against the notorious viral disease SARS (severe acute respiratory syndrome).[69] This latest episode in the eventful life of a 'wonderous' molecule is reminiscent of the early days of interferon as an 'antiviral penicillin' in the British public sphere.

The widespread application of interferon as an effective medical commodity is almost beyond dispute; the need for it appears to be a premise, rather than a contestable assumption. As a consequence, opposition to the rather costly forms of interferon therapy revolves less around questions of need than around questions of cost or economic feasibility, which increasingly dominate the political agenda of 'marketplace' medicine.[70] The extensive use by government regulatory bodies of cost utility studies of interferon beta in multiple sclerosis shows that economic considerations are easily linked with medical discourse and therefore are difficult to separate from medical arguments showing the necessity for interferon treatment.[71]

The last paragraph on economics nicely illustrates that shaping therapeutic drug practices is not exclusively a medical or scientific affair. To think so would deny the multiple links between science, medicine and society which have been made manifest in my description of the development and dissemination of interferon as part of a new therapeutic approach in the treatment of cancer, viral disease and chronic degenerative autoimmune diseases like multiple sclerosis. Of course, in order to become part of the therapeutic armament of doctors, interferon had to meet scientific standards of what constituted legitimate evidence for efficacy. But does meeting those standards suffice to explain the exponential growth in the clinical use of a wide variety of interferons?

As far as the spokespersons of the pharmaceutical companies are concerned it does. The growing market for interferon treatments is claimed to be a 'natural consequence' of a growing need of doctors and patients alike. They apparently regard the new drugs as valuable medical assets. The growing range of therapeutic effects, so they argue, reflect the need for interferons. However, if there is one thing the interferon crusade can teach us, it is that the equation of need and the widening of the pharmacological spectrum is a logical fallacy. The need for interferons was never 'simply' there. Likewise, the development and dissemination of interferons was never a self-evident, inevitable process.

This brings us to the question, how did the interferons become naturalized and legitimized as part of medical practice?

By the early 1980s strong links had formed between interferon research, the booming field of molecular biology and private industry. Powerful commercial, institutional and professional interests had become aligned with what was considered a major showcase of a revolutionary new approach to drug development and drug therapy: genetic medicine. The international pharmaceutical industry, which was seeking to position itself for long-term survival through scientific and therapeutic innovation, became the most dynamic and strategic actor of all those involved in work on interferon. In achieving a key position in the distribution of research resources and materials, the pharmaceutical industry increasingly dictated the development of interferon. However dominant their role, for the profiling and marketing of interferon the drug companies were largely dependent on their ability to enter into cross-fertilizing relationships with laboratory scientists, clinicians and regulatory bodies.

I have shown that in order to become accepted, interferon needed a therapeutic profile and treatment concept that could be integrated or combined with existing therapeutic practices. This required that the drug makers in collaboration with laboratory researchers and clinicians actively created and made available a sort of rough-and-ready 'therapeutic rationale'. I dubbed the rhetorical strategy that proved effective in establishing a more permanent base for support 'beyond interferon': picturing interferon both as a new turn in drug development and as a first step in the development of immunotherapy – a new therapeutic modality that was going to constitute a medicine of tomorrow. In positioning interferon as a promising product of molecular biology, a cytokine, which fitted in well with the 'molecularization' of medicine and was supportive of existing research and treatment practices, the pharmaceutical industry anticipated the successful integration of interferon into medical practice. However promising in scientific and therapeutic terms, the combined-modality therapeutic approach of interferon met with resistance from drug regulatory authorities.

As I pointed out, the official international drug regulation practice was still largely governed by a single substance-dominated therapeutic philosophy. In order to legally consider a pharmacologically active compound as a new therapeutic drug it had to be officially evaluated as a single agent. This implied that before licensing procedures could be taken into consideration the companies had to look for a disease that justified a need for interferon. With most trial responses to single-agent interferon therapy comparing unfavourably with drugs already available in the clinic, the industrial interferon programme managers faced the seemingly Herculean task of establishing an unambiguous justification for clinical use.

The search for suitable disease candidates through clinical trials became a growth industry in its own right, which played a central role in transforming the interferons into the therapeutics as we know them today. By being simultaneously sites of scientific research and medical care, clinical trials do more than just evaluating the safety and benefit of new drugs. In serving as bridges between bench and bedside they connect a diverse assortment of actors: biomedical researchers, physicians, cancer foundations, pharmaceutical companies, patients and regulators. These diverse parties face the difficult task of participating in scientific research and at the same time making decisions in an environment where the most basic actions are intertwined with matters of sickness and health and where no one is able to draw sharp boundaries among scientific, clinical, political and economic questions.[72] What the interferon case underlines is that there is an important social and cultural dimension to the evaluation of a new remedy. Not only scientific, professional and economic interests affected judgements of interferon's efficacy and safety but also, as I will argue in the final chapter, the powerful symbolic significance which these natural body substances have acquired on their journey between bench, bedside and the public sphere.

8 Interferons in restrospect and prospect

Patients at work

Thursday, June 5, 1997. 'With all these uncertainties about the therapeutic effect of interferon beta I can not prevent from feeling like a potential guinea-pig.' These are the words of a multiple sclerosis patient attending the meeting 'Choosing for interferon?'. The scene of action: the 'Doctor Soer' auditorium of 'Sunny Home' somewhere in the Netherlands. At the right-hand side of the entrance is a provisional podium flanked by two big billboards with promotion posters of the Multiple Sclerosis (MS) Society. The slogan 'Diagnosis MS, what next?' immediately strikes the eye.[1]

The local meeting is organized by the Dutch MS Society as part of a national government-sponsored campaign to inform MS patients of the pro's and con's of treating MS with the newly registered therapeutic drug, interferon beta. Among those present are MS patients in various stages of the disease – from experienced wheelchair individuals, to less disabled 'stick' people, to the seemingly healthy group of 'greenies' who are still in the early stages of the disease. They are accompanied by their relatives and friends. In one way or another all are trying to cope with a chronic degenerative nervous disease for which there is no effective treatment available yet, other than a few drugs that provide symptomatic relief.

The speaker is a neurologist. In an effort to make contact with the audience he addresses the ambiguous feelings that might have risen after reading the information leaflet.

> *Our pamphlet might have dashed your high hopes but it is not our aim to sell promises today, but to present a realistic picture of what this new therapeutic drug has to offer to you as MS patients.*

The neurologist emphasizes the problem of judging the effectiveness of therapeutic drugs in a poorly defined disease like MS with a problematic correlation between laboratory and clinical parameters.

> *It is a therapy that is claimed to produce therapeutic effects with a less than 30% response rate; in order to achieve a therapeutic effect in one patient seven others have to be treated and to postpone wheelchair dependence by 9 months the number needed to treat almost triples. Those prepared to seize the opportunity*

have to accept a reduction of their quality of life, with no immediate benefits. Using interferon means coping with side-effects ranging from flu-like symptoms, painful swellings at the site of injection, to suicidal tendencies, and learning how to fit in the storage, preparation and injection procedures into one's daily life. Summarizing, it might be said that while in the short-term interferon users are worse off than fellow patients who decide not to use interferon, in the long term the user group as a whole will be better off. However we have to keep in mind that the actual benefits will differ from one person to another, with a relatively large number of users who will not experience any beneficial effects at all. It certainly is not the wonder drug that the media have tried to make of it. Are there any questions so far?

The audience seizes the opportunity to pose questions with both hands.

> What should I make of your rather negative image of interferon treatment with less than 10% positive and 90% negative news?
> There seem to exist two different types of beta interferon, beta-1a and beta-1b. Are we talking about one and the same therapeutic drug or two closely related but different drugs?
> Why are there differences in dosaging between the various hospitals?
> What is the predictive value of brain-imaging techniques with regard to interferon's therapeutic benefits?
> What are the chances that the disease will break through this new medication in the long term?
> If the history of MS is punctuated with false-dawn cures, what guarantee is there that interferon will not turn out to be the next false hope on the list?

In one of the last remarks before the end of the meeting the MS patient – quoted at the beginning – indicates that with more questions than answers about the application of interferon beta she feels subjected to medical experimentation. Interferon beta is not any different in this respect from any other therapeutic drug currently in use in clinical practice, claims the doctor in response.

> The problem is that certain side-effects will not show up in the limited clinical studies which are required to qualify for a therapeutic drug licence but will make themselves felt only after widespread and long-term use. For the sake of future users it is of utmost importance that clinicians as well as patients cooperate in closely monitoring the effects of therapeutic drugs, in particular the newly introduced ones.

His comments invite support from an MS patient who justifies her decision to take part in an interferon trial by pointing out the possible benefits that may result from these kind of studies for future generations of MS patients. She feels she can make a contribution, however small, to the progress of medicine.

At times the nature of the discussion between the patients and the doctor on the platform – both parties discussing similarities and differences in research results and questioning the credibility of and rationale behind claims as well as procedures – makes one feel like attending a scientific congress. As such this open-information event questions the self-evident certainty with which we think we can pinpoint who is a scientist and who is a layperson. Speaking the language of medical science clearly enables MS patients to redraw the boundaries between the entities 'science' and the 'public' and to reconstitute the expert/lay divide. In other words this capacity empowers patients in relation to medical and scientific professionals.[2]

Most important, the example of the MS-patient meeting shows us that there is more to the introduction of prospective consumers to a new therapeutic drug than the transfer of readymade knowledge. The questions of quality, safety, research, evaluation and professional control are raised by prospective consumers in such a manner as to suggest that the organization of the treatment setting is negotiable. The prescription and consumption of interferon beta appear to require the patient's belief and trust in its effectiveness. Achieving this requires hard work by all the parties involved – doctors, researchers, patients, company executives and regulators.

This interactive process involves not just moulding the laboratory and therapy routines of interferon beta, but also the therapeutic and social context in which it is used. The moulding, however, is constrained in major ways by the cumulative impact of years of work between bench and bedside, on which the therapeutic profile of interferon beta basically draws. This involved a continuous but creeping and retrospectively hidden process of (re-)evaluating and (re-)shaping interferon-related research problems, questions and practices. As I will show in this final chapter, the notion of moulding proves very helpful in accounting for the full array of interferon's multiple identities as a versatile research object, a plastic medical tool and commodity, but also as an icon of twentieth-century miracle medicine.

Crafting interferon(s)

The book's narrative has led us deep into the particulars of practice *and* culture in a particular field of biomedicine in the second half of the twentieth century. To conceive of interferon research as mere intervention and representation to create new and better evidence-based ways of 'knowing' and 'doing' appears to be rather one-sided.[3] The same applies to picturing work on interferon as the opportunistic management of interests and investments that has allowed researchers to pursue careers, biomedical research centres and hospitals to enhance prestige and industry to maximize profits. Neither description suffices to account for the complex dynamics of the process of research and development of interferon(-s).

Basically work between bench and bedside involved acts of:

1   learning through active interventions, engineering experiments and tests with the aim of interfering with health and disease, generating data of

somesort to ultimately bring under surveillance new phenomena and develop new tools of intervention;
2   accounting for the sheer endless production of laboratory and clinical data in a culturally meaningful way which makes them serve specific scientific, social and economic purposes – with control and exploitation of the molecular processes of life at the heart of the enterprise.

A major part of research on interferon went into fine-tuning experiments and tests so that these yielded results that bear directly on the problems and questions at hand or could be used to improve on and multiply the manipulative options at bench and bedside; hence generating new problems and questions. The nature, range and priority of research problems and questions underwent regular changes and was closely bound up with changes in the experimental context. On one occasion practical problems and questions resulting from the local organization of research materials and resources, and from the local preference of experimental techniques or testing methods (e.g. the specific choice of a bio-assay, or of a trial design) prevailed. On another, work on interferon was governed by institutional policy or what counted as front-line concepts (e.g. interferon as part of a reductive, mono-causal system of host resistance against viruses *versus* interferon as part of a complex, multi-factorial immune system) and state-of-the-art techniques and research approaches (e.g. virological, biochemical or genetic approaches). Quite some perseverance was required on the part of those researchers wanting to challenge specific paradigms or dogmas in the field of interferon research (e.g. the non-toxicity dogma).

An important window onto the nature of development and change in work on interferon is provided by cross-local and cross-national comparison. I showed that the various workplaces each had their own characteristics, infrastructure and resources. The local distribution of and access to, as well as the familiarity with, particular materials, techniques, treatments and testing methods did much to shape the pace, direction and hence content of the research work. Seemingly small differences in experimental design and set-up could make major differences to the outcome of experiments and tests. Some major sources of differences and contingencies were and still are related to the nature, scale, selection and treatment of experimental systems (*in vitro*, *in vivo* or in humans) or of test subjects. When different biological organisms (in terms of type of cell culture or animal species) or different selection criteria for human testing are used to investigate the very same question, one may expect to yield systematically different sets of data. The same, however, may occur when using the same organism or the same selection criteria for human volunteers in different places. Even minor differences in the features and treatment of biological materials employed in the same laboratory or in the selection and treatment of human volunteers tested in the same research unit can alter the outcome of experiments.

An additional problem of doing research with biological organisms is that at every level of biological organization it is an open question how to generalize the findings. I agree with the science historian Richard Burian that this makes application of the basic biological sciences to clinical practice such an extraordinarily difficult and unpredictable endeavour.[4] New knowledge of disease processes at the cellular and molecular level does not necessarily yield medical applications and simultaneously effective treatments are developed without appropriate understanding of the underlying pathological processes. In principle all laboratory systems for identifying and testing new therapeutic agents are faulty, as there is no way to mimic exactly the way the human body works – neither laboratory animals carrying human tumours nor human tumour cells in culture seem to be faring any better in this respect.[5] Yet as the interferon case shows, the acknowledged problem of biological contingencies and differences hardly seems to discourage biomedical scientists from extrapolating *in vitro* and animal experiments and producing bold therapeutic claims based on that work. The current and future efforts to genetically engineer human features into experimental organisms may succeed in narrowing the biological gap but still leaves us with the classic problem of translating laboratory studies with experimental disease models to the 'odd mice out' – whether it concerns multiple sclerosis patients or any other category of patient.[6]

To what extent and how did researchers in the field of interferon research cope with local differences in research? I showed how initially the researchers involved managed differences by informally agreeing on exchange mechanisms, without rigorously defining common units of measurement. Through circulating, comparing and combining research materials, techniques and skills in a relatively small and flexible network of personal contacts, researchers managed to reduce extreme variations of experimental results. This enabled them to narrow down the margins of interpretation to workable levels. Not until the informal exchange mechanism within this 'gift culture' began to show flaws – with expansion of the field of research, the shift away from qualitative to quantitative research and the intrusion of the pharmaceutical industry and national centres of standardization into the relatively private domain of 'laboratory life' – did a need for formal types of standardization and comparison begin to manifest itself.[7] However, as it turned out the actual use of formal standards not only helped to master differences in the context of 'big science' but was also instrumental in establishing new differences through reshaping and redefining the outcome of experiments.

By being dependent on experiments – which in principle are open-ended events – for its qualifications, interferon's identity was and has remained necessarily unstable. At each point in time interferon's properties and qualities were defined through the complete set of experiments and tests which interferon had undergone. In principle interferon changed with every new trial.

Over time the scientific notion of interferon underwent a gradual transformation. First, as a *viral interfering factor* it was provisionally considered a *deviant viral entity*. By the mid-1960s interferon came to refer to *active antiviral substances* in particular culture fluids – as a generic term of reductive molecular nature – which complied with a specific set of experimental criteria and which were considered *essential products of the cell* that played a part in resistance to viral infection. More than four decades later interferon is still regarded as a generic term, but of yet a different, far more complex, molecular nature; it refers to a specific part of the body's natural defences consisting of a group of closely related *immunomodulatory and regulatory proteins* that form the so-called 'cytokine network'. But there is far more to say about interferon's 'fantastic voyage' than commenting on the particular micro-dynamics of scientific intervention and representation over time.

Seeking 'magic bullets'

Throughout the history of research and development of interferon we have seen how claims and judgements regarding laboratory and clinical data were presented, contested, translated and negotiated by a variety of groups with competing interests and diverging backgrounds. Earlier on in the British context, and subsequently in America in the 1970s, interferon came to be perceived as a potential medical breakthrough. Yet, in both cases interferon faced quite different circumstances.

This book shows that there are no simple, one-dimensional explanations for how interferon garnered headlines and achieved popular notoriety successively in the 1960s and the 1970s. Neither technical arguments about interferon's innovative therapeutic qualities (real or apparent) nor the interest of a specific social group suffice to explain the successive cycles of public enthusiasm and disappointment. In both cases a convergence of different historical factors made interferon temporarily into an acclaimed medical breakthrough. Scientists, doctors, industrialists, patients, journalists and government bureaucrats operated in a complex social world with historical depth that cannot be assumed to be passive. Nevertheless they were experienced as important constraints or incentives and are therefore worth examining. To begin with I shall examine the initial success in mobilizing the British government and the drug industry for interferon.

The scientist fundamental to the early development of interferon Alick Isaacs exemplified a culture of and identity for biomedical research that is not conventionally associated with biomedical scientists in Britain in this epoch. As a champion of persuasive public demonstrations and a strong competitor for public credibility Isaacs played a pivotal role in bringing interferon, however unsubstantiated its therapeutic effects, to the public stage. Isaacs, unlike most of his British academic colleagues in biomedicine and despite his fragile state of health made his work into a resource for scientific, political and industrial action. As a scientist-entrepreneur who boldly risks everything on what he believes is a promising phenomenon, hypothesis or potential medicine, Isaacs exemplifies a scientific way of life that came into prominence most visibly in the United States in the 1950s.[8]

In analysing Isaacs's early success at connecting his private science with public issues of disease and cure in the British context, I discussed in chapters 2 and 3 a strong public engagement with spectacular medical advances. In the post-Second World War period, antibiotics, steroids and neuroleptics had provided physicians with treatments never before seen. The wide and intense acclaim for this seemingly endless flow of new wonders at the disposal of doctors helped to establish in mass culture a firm belief in major therapeutic achievements as a hallmark of scientific medicine.[9] However, this condition was far from culturally specific and cannot explain, therefore, the specific British enthusiasm for interferon as the new 'antiviral penicillin'.

In his most stimulating essay 'Penicillin and the New Elizabethans' Robert Bud points out the powerful and multiform symbolic role of penicillin in British post-war culture as an icon of medical progress, but also of Britain's erratic research potential and of America's ominous industrial dominance. In line with Bud, I argue that making public parallels with penicillin helped to link interferon rhetorically with major issues in the popular consciousness in post-war Britain: pride in scientific and technological abilities, resentment at the impotence to capitalize on the innovative potential and jealousy of American success.[10] As such the very formation and use of the penicillin-associated imagery in the British media served to give high visibility to a seemingly promising medical discovery, and boost public expectations about interferon's future as yet another wonder of modern medicine. In both tapping and stimulating the public sentiment of national interest, the propagation of the penicillin-oriented imagery played a pivotal role in both the first and second wave of British enthusiasm in the 1960s and 1970s.

In carrying the greater part of the public blame for the failure to develop penicillin in Britain, the Medical Research Council (MRC) took a particularly strong interest in interferon. The MRC chose interferon as a high-visibility showcase to convince the British public of their preparedness to invest in and pursue potential 'homegrown' medical innovations from the very first moment. Since the MRC lacked the expertise and resources to develop interferon as a therapeutic drug, they sought collaboration with the drug industry. Setting up a collaborative programme with British pharmaceutical companies provided an additional opportunity for the MRC to present itself as a public policy stronghold in the defence of national interests.

However promising as a potential 'antiviral penicillin', British drug companies initially showed restraint with regard to interferon. Within an industrial research and development environment mainly oriented toward medical chemistry and pharmacology, and with the overall product pipeline filled with chemical compounds, biological preparations were perceived as troublesome and high-risk commodities with uncertain manufacturing as well as uncertain commercial prospects. MRC's offer to join forces made a difference though. From industry's point of view, risk-sharing with the MRC, which offered their unique expertise in interferon research and the eventual rights in the developed product, made work on interferon a far more attractive option.

Moreover, they shared the determination of the MRC to prevent British research products with penicillin-like commercial potential from ending up with the label 'made in the USA'. The combination of these historically and culturally specific conditions led to the first successful post-war initiative to secure a collaboration between the British government and the drug industry.

In chapter 3, I showed how Ehrlich's 'magic bullet' concept of chemotherapy (which is intimately linked with scientific medicine's dominant theory of specific aetiology), as well as the closely related issue of chemical purity, played a central role in the subsequent collaborative efforts to develop interferon as a therapeutic drug. Initially the association of interferon with Paul Ehrlich's ideal of generating therapeutic molecules which like 'magic bullets' seek out and destroy the enemy and injure nothing else was helpful in keeping the pharmaceutical companies interested. The promise of a specific (that is truly curative) and innocuous agent that would knock out viral disease made senior drug company executives believe that interferon, despite its problematic status as an unruly biological, would fit into their unprecedentedly successful Ehrlichean chemotherapeutic drug development programme. Within this chemotherapeutic research and development framework, though, the purification and chemical identification of the pharmacologically active part – conceptualized as a single and pure chemical compound – was considered a prerequisite for successful drug development.

So when attempts at purifying and characterizing interferon as well as efforts to translate the experimental laboratory methods into practical technologies for large-scale industrial production continued to fail, the evaluation of interferon's product potential in the light of the Ehrlichean chemotherapeutic programme changed dramatically. In resisting all efforts to make it fit into conventional drug development practices and without additional demonstrations of dramatic therapeutic effects, interferon rapidly lost its status as a most promising lead toward antiviral drug therapy and was relegated to the motley and ever-increasing collection of research objects and tools under study in the field of biomedicine. But living in obscurity would turn out to be a temporary sidestep in a shifting biography.

'Doctoring' the media

The story of how interferon achieved popular notoriety as an anti-tumour agent in America in the 1970s reflects the preoccupations of a different society in a different epoch. The theme of the 1950s and 1960s had been the firm and unconditional belief in a scientific medicine that was synonymous with therapeutic success and progress. In the 1970s, however, it was precisely this mythologizing image of progress and cure, that had become inextricably bound up with modern medicine, that came under challenge.[11] The fast-growing American medical counter-culture gained a political hearing for its critique of the technological-interventionist and reductive orientation of a high technology medicine that did not seem to have lived up to its promises. Cancer medicine, in particular, was challenged for being far less clinically effective than was officially asserted and even for being counterproductive in doing more harm than good. This challenge went hand in hand with attacks on the medical establishment for wasting tax-payer's money on developing new, questionable forms of cancer therapy with a reductionist scientific basis while neglecting the apparently promising search for the natural and organic as part of a more holistic approach to health and disease.[12]

As I have shown, in achieving a high degree of compatibility between interferon as an innovative approach to cancer therapy and this specific 1970s culture of American medicine, the interferon promoters succeeded in building support for their case. To the most influential interferon champion, Mathilde Krim, interferon represented as much marketing as scientific testing. She was well aware of the ever-growing importance of the mass media in generating public support and was very skilful at 'doctoring' the media. Krim played a pivotal role in shaping the media coverage of the research with the goal of promoting interferon. The interferon lobby lent journalists a hand to frame interferon both as a natural remedy and the latest product of the most up-to-date scientific research long ignored by a conservative cancer establishment. Thus they succeeded in addressing both the American public's habitual fascination with medical novelties and its more fashionable appetite for unorthodox therapies.[13]

However, the role of the press went beyond journalists acting as mediators between the interferon lobby and the public: it often dramatized the accounts of interferon research in such a way as to captivate the public's attention. Notwithstanding the fact that the media coverage of interferon reflected in part the expectations, legitimations and opinions circulating within the biomedical and public realm, the very mode of selection, emphasis and presentation of the information in the media played a significant part in shaping the cycle of promise and disappointment. Judging by the particular examples in chapter 6 interferon research offered ample opportunity to produce meaningful and arresting imagery which appealed not only to popular notions about health and medicine but also easily invited the audience to hyperbole: such as the magic and mystique of a scarce and costly natural substance with healing potential, the fearful association with a life-threatening disease, the promise of a medical breakthrough, the neglect of the unorthodox in scientific medicine, scientific and technological competition and the familiar parallels with the icon of medical progress and cure, penicillin. The media involved in the interferon saga was of course a diverse and colourful enterprise which ranged from large circulation weekly news magazines, more specialized popular science magazines, to national and local newspapers, tabloids and, last but not least, commercial and non-commercial TV and radio networks. Yet a striking feature of the interferon reports is the homogeneity in their choice of imagery (metaphors as well as pictures) and terms. Apart from the fact that there was a tendency to mimic each other the imitative nature of the reporting can be related to a shared frame of reference which Anne Karpf has labelled the 'medical approach'.[14]

The reproductive nature of the reporting implied the use of familiar medical icons such as the white-coated male doctor, the hopeful patient who is desperate for a cure, test-tubes and syringes, the high technology laboratory and penicillin as the perfect example of the magic bullet. This kind of imagery gave to the interferon stories an aura of medical progress and the self-evidence of finding a cure, and however predictable in nature it made them understandable and appealing to readers and viewers. Most interferon news reports were actually 'olds' (that is, not topical) and preferred the human interest story and a few select glorious scientific results or dramatic therapeutic events, such as single acute disease cases, interferon shortages, the race to clone the interferon gene and seemingly miraculous cures. Wider social, political or economic issues were neglected. Journalists pursued the apparent biomedical opinion leaders and rarely questioned the political or economic motives behind their claims regarding interferon's therapeutic potential.

However predominant this rather uncritical medical angle it did not deter the same reporters from covering the unorthodox, natural healing part of the interferon story with the same wonder and elan. With a keen sense of contemporary ideas and beliefs about health and disease they chose to adopt Krim's dualistic projection of interferon as both a product of science *and* nature. The rhetorical use of the ambiguous term if or IF in 'mediating' interferon did everything but undermine the science-based promise of a natural cancer cure. On the contrary this only contributed to the intensity of border traffic between physic and faith; scientific evidence and exalted expectations of medicine's potential for healing.

But did the fact that interferon became the public embodiment of scientific prowess and hopes of Elysian health suffice to turn interferon into a miracle drug? Of course, as the promise of a cancer cure circulated between scientists, the media and the public and back again, it gathered potency. However strong this self-enhancing inflationary force it was not enough to amplify public expectations beyond measure. It was the conjunction of a hoped-for cure with the promise of a wonderful new technology, genetic engineering, which produced an unprecedented global tidal wave of publicity and catapulted interferon to the elite league of miracle drugs. Neither penicillin nor streptomycin nor cortisone could match the way interferon was rocketed to Elysian heights by a pounding cocktail of modern mass media.

This brings us to the question of what characteristics these four therapeutic substances had in common that helped them to achieve miracle drug status in the process of circulating between media and public and back again. All four were experimental, new, and initially in short supply, and opened up the possibility of cure for life-threatening or crippling disease conditions that had hitherto been incurable. In addition, each had been involved in wonderful Lazarus-like healings at the bedside; breathing new life into the critically ill body. In playing the leading part in these miracle stories they became both object and subject of experience; fathering upon them autonomous powers in the act of healing. Furthermore, as the latest science-promoted promising products of modern medicine they all had 'natural' audience appeal as breakthroughs and medical triumphs. Hailing them as wonder drugs required a religious faith in a scientific medicine that was equated with health and the quest for immortality. It is part of what Roy Porter used to call the 'rising expectations trap' of modern medicine.[15] But there is more to the label of miracle drug than a periodic religious hymning of medical progress and cure. The miracle drug emblem is Janus-faced and transitory for a number of reasons: the fact that interferon was made of natural components facilitated its aura as an elixir of life, which intimately linked to gullible beliefs in alternative medicine; there was also the association with the disappointing everyday experience of incurable disease and failing remedies.

There is yet another repetitious quality to the interferon story. In the case of other miracle drugs like penicillin, streptomycin and cortisone a similar kind of deliberate creation of disillusionment occurred by a government-led coalition as a means to dampen public enthusiasm and stem the tide of public requests which it could not meet. An element of optimism in advertising a new remedy is widely regarded and even promoted as necessary for the advance of medicine, as long as the public enthusiasm it elicits can not be associated with the term 'uncontrollable'.

My claim is that the limit of acceptable, controlled enthusiasm depended in each case on the extent to which the government and medical community feared a loss of authority over an increasingly evidence-based healing culture.[16] But, by conveying the warning message of the transient quality to therapeutic success stories and wonder drugs the patrons of science and medicine in neither case succeeded in separating physic and faith. The temporary hailing of interferon as a miracle drug left a public imprint of exceptional, compelling but elusive natural healing qualities that preserved interferon's ambivalent but most vivid role as object and subject of experience. Bestowing on interferon autonomy in the act of 'interfering with nature' helped to create a niche for further work on interferon and to give this wonderous molecule a headstart on the highly competitive market of recombinant protein drugs in the 1990s.[17]

Toward genetic medicine

The growing interference between interferon research and the field of molecular biology played a major role in bringing about a redefinition of interferon's prospects as an object of research and development for the pharmaceutical industry in the late 1970s and early 1980s. In particular, implementing the newly developed genetic engineering technology made an important difference. Originally deployed with the aim of making highly purified interferon preparations available in large quantities, at modest cost, the molecular biologists, closely linked with the emerging genetic engineering industry, helped to redraw the interferon landscape beyond recognition.

In efforts to clone the interferon gene, produce the corresponding interferon protein product and reach agreement on interferon's chemical structure, the 'genetic engineers' succeeded in promoting interferon not only to the world of 'big science' and industry, but also to the tangible world of chemical facts. By making attainable the 'Holy Grail' of interferon research – pure interferon – cloning put an end to the important role of chemical purity as an arbiter in disputes concerning work on interferon both in industry and academia.

However, with the intention of achieving homogeneity and progress as well as unity in the understanding of the interferon system, the molecular biologists had paradoxically created heterogeneity and new complexity by reinforcing the idea that 'interferon' represented a dynamic family of genes and corresponding proteins. In an attempt to manage the unwieldy complexity, they fell back on the concepts of 'command', 'control' and 'communication networks' which had proved successful explanatory tools in molecular biology. In doing so the molecular biologists promoted the conceptualization of interferon as part of a complex host defence system of cellular messengers, so-called 'biological response modifiers', held together by communication and feedback. This was part of a much broader shift in the biomedical realm from 'substance thinking', looking at the biochemical 'nuts and bolts' in the machinery of the body, to 'information thinking', picturing the body as a complex regulatory–communications network.[18]

The cloning efforts not only strengthened the idea of interferon as one of the body's regulatory defence systems, but also opened the door to an alternative approach to drug development. The genetic engineers promised to create a new horizon in pharmaceutical research. In concentrating on molecular biology (e.g. DNA and proteins) rather than on traditional chemistry, they would provide the drug companies with an entire apothecary of new biosynthetic drugs from natural substances within the body. As promoters of this new genetic approach, they stressed novelty but at the same time they presented genetic engineering as fitting in well with the classic chemistry-based therapeutic drug development practice. With a steady decline in commercially viable and innovative therapeutic compounds from their chemistry-dependent 'pipelines', the promise of a new approach to drug research and development kept the pharmaceutical industry interested when the news about interferon's less-than-spectacular performance in clinical trials began to spread in the early 1980s.[19]

In chapter 7, I have shown that in order to become legitimized as part of medical practice against all odds, interferon needed a therapeutic profile and treatment concept that could be integrated or combined with existing therapeutic routines. This required that the drug makers in collaboration with laboratory researchers and clinicians actively created and made available a sort of rough-and-ready 'therapeutic rationale'. Clinicians and laboratory scientists alike increasingly perceived cancer as a family of diseases in which complex cellular and molecular regulatory functions have gone awry, instead of a unitary disease characterized by the uncontrolled growth of malignant cells that spread out aggressively and destroy the body. In line with this shift, the framework in which to consider patients, cancer and its treatments could no longer be effectively and acceptably linked with the therapeutic ideal of generating and using universal cancer medications which like 'magic bullets' seek out and destroy the enemy.[20]

Talking about the 'Big C' as a multi-factorial molecular disease was consistent with the emergence of immunotherapy as part of a combined-modality, laboratory-supported therapeutic framework in clinical oncology. This linked up perfectly with the efforts in the field of interferon research to redefine interferon as a means to develop a new complementary modality of cancer treatment that might be used in conjunction with conventional therapies. Redefining interferons as biological response modifiers or cytokines which could be used therapeutically to strengthen the body's self-defence as part of a laboratory-guided, multi-modality, therapy framework legitimated and reinforced the new molecular perception and treatable nature of cancer. And this allowed all parties to see interferon therapy *as* therapeutic. To put it differently, as advertised products of molecular biology the interferons fitted in well with the 'molecularization' of medicine and this helped to make them 'work' at the bedside. I pointed to the solidarity implicit in the emphasis of both laboratory researchers and clinicians on the importance of linking through interferon bench and bedside as part of a new kind of molecular medicine: genetic medicine. In always being able to justify that either further laboratory or clinical research was mandated, it made all parties involved relatively immune to disappointing research news. Moreover, in achieving a key position in the distribution of research materials and resources needed to set up preclinical and clinical studies, the pharmaceutical industry increasingly dictated the development and use of interferon.

So far we have seen a wide range of factors and pressures that shaped the differentiated profile and use of the interferons: the endless series of interventions and representations between bench and bedside, the consecutive cycles of promise and disappointment in the public sphere, the genetic turn in drug research and development and the political and social economics of evaluation and regulation. Given the complexity of the environment in which the safety and effectiveness of this family of protein drugs had to be assessed is it any wonder that changes within the context of the medical market can have unintended effects? A case in point is the FDA's priority review policy that was established in the 1980s. Introduced with the aim to make available more rapidly promising experimental remedies to desperately ill patients it appears to do far more than that. In blurring the distinctions between experimental and established treatments this new regulatory policy unintentionally helped market interferon through the fast growing industry of clinical trials.

Discussing drug regulatory affairs in the 1970s the pharmacologist Louis Lasagna claimed that 'flexibility may not appeal to either the pendant or the legalist in the Food and Drug Administration (FDA) or the academic world, but it seems eminently defensible as public policy'.[21] In his view, regulatory decisions should assess the impact on society of the various alternative courses of action possible in each given situation. However penetrating his remarks,

Lasagna still underestimated the flexible nature of the actual practice of evaluating and regulating therapeutic drugs. What the example of interferon underlines is the capability of science and industry to influence and redirect regulatory reforms to their own advantage. If we want to regulate therapeutic drugs in ways which will speed up the evaluation of new remedies while maintaining safety and efficacy standards, it is not sufficient to monitor closely new developments between bench and bedside and to keep an open mind about what will count as valid medical evidence in twenty-first century genetic medicine. In order to prevent us making a serious step backward with regard to trust in and identification with the institutions controlling medicines, drug regulators should finally start to take Lasagna's advice seriously, and develop more sophisticated tools for integrating medico-scientific developments and their socio-institutional as well as everyday life consequences into their assessments.

Patients at risk?

The story of how interferon managed to become part of the 'doctor's bag' suggests that, rather than the therapeutic evaluation of pharmaceutical substances being a straightforward and strictly evidence-based process, it is a thoroughly practice-based activity involving the conflicting therapeutic horizons of efficacy ('can it work'), effectiveness ('does it help') and efficiency ('is it worth it').[22] My report of 'patients at work' at the beginning of this final chapter is exemplary for the profound changes the dynamic landscape of therapeutic evaluation has undergone since the official licensing of interferon alpha for the treatment of a rare form of leukaemia in 1986. In the traditional, that is the pre-AIDS therapeutic framework, the testing and judgement of new therapies in terms of statistical significance, clinical relevance or economic yield hinged on their evaluation by professionally endorsed evaluators and professionally endorsed methods. In the late 1980s, however, the AIDS-crisis and the associated patient-activist interventions in biomedicine – which led to new liberalized drug regulatory and testing practices for the desperately ill – demonstrated compellingly the potential for change. Ever since, patient grass-root organizations have been increasingly successful in actively participating in the biomedical realm. In funding disease-related research and services, and establishing strategic alliances with academic research, industry and government, they have managed to exert a growing influence on the review and decision-making process of science and medicine.

At the beginning of the twenty-first century worldwide associations of patients like the American, British or Dutch Multiple Sclerosis Societies permanently monitor research and policy issues and try to convert these into information that is meaningful for individual consumers' health. Through magazines, newsletters and other publications and Internet websites they keep people living with the disease, their families and health-care providers informed about the latest disease-related news items. However, the consumer organizations in medicine are more than clearing houses of information. In establishing and supporting counselling groups, public meetings, training workshops for health-care professionals, scientific conferences and most lately Internet news groups, they have come to serve as national and international platforms for the exchange and critical appraisal of information on medical research, care and policy.[23]

This recent shift of health-care consumers towards 'shopping for health' seems to fit rather smoothly with a long-term process of reordering public and private responsibilities for health. It may also correspond with a shift in the balance of contexts in which issues of health and disease are articulated: from the workplace to the household. Within this changing context consumers increasingly take charge of their own health management. Armed with better and more readily available information about health and disease, they more consciously weigh the benefits, risks and costs of medical services on offer.

The rapid developments in Internet technology – with personal computers evolving into powerful communication tools – have played a most important role in changing the rules of the game in marketplace medicine. Through the Web, patients can freely exchange experiences with fellow sufferers, ask treatment advice from doctors who are not theirs and may obtain the latest biomedical updates before health-care professionals have been able to do so.[24] In principle, having unprecedented access to the world of science and medicine should put them in an equal position of knowledge with the other parties involved.

However promising this scenario of health-care consumers becoming masters of information and needs, it remains to be seen whether patients are indeed the ones to profit most from the combination of liberalized regulatory and drug testing practices, and a freely accessible medical 'cyber mall', but with increasingly costly new medical therapies on offer. The sociologist, Steven Epstein, rightly indicates that however vigilant the fast-growing consumer movement may seem, they underestimate the capability of science and industry to influence and redirect the consumer's interventions to their own advantage.[25] The rise, fall and subsequent rise of interferon certainly reflects the considerable power of the coalition between laboratory medicine and the drug industry to monopolize the information on the safety and efficacy of therapeutic drugs. This makes it hard for consumers and certainly the desperately ill, who are dedicated to try anything that enables them to feel less than totally impotent in the face of a wasting disease, to seriously challenge knowledge claims and critically weigh the benefits, risks and costs. Paradoxically, the efforts to open up the biomedical stronghold and push medical research in directions that serve patient interests more directly reinforce the dependence on the very same academic and industrial producers of promise and hope of new medical products and services.

In carefully avoiding labelling this chapter 'conclusion' I want to emphasize that interferon's 'fantastic voyage' has far from ended. Reminiscent of the 1980 interferon press coverage was the promising headline 'beyond interferon'. 25 years later, as an inseparable element of the culture of genetic medicine, the interferons help to shape the goals, organization and promise of what science, medicine and industry advertise as 'Cytokines and Beyond'.[26] To what extent there will be parallels between past and future events in the interferon 'saga' remains an open question. As a historian I am all too well aware of the fallacies and risks involved in trying to interrogate the past as a means to help direct current and future policies and developments. The unpredictable and dynamic nature of the mangle of practice and culture described makes it hazardous to predict the shape that this part of the 'marketplace' medicine will take in the future let alone suggesting new life forms. However, knowing that the application of historical knowledge to cope with current and future situations is a problematic endeavour should not withhold us from doing so. As the philosopher Thomas Nickles puts it succinctly: 'the only resources available to a generation, besides its own ingenuity, are those bequeathed by its past'.[27]

Appendix

All persons who have been interviewed in the course of the research reported on in this book are listed below. Apart from Dr, all other titles (professor, etc.) are omitted. Asterisks indicate that the interview was tape-recorded.

Pharmaceutical industry

Dr L. Adami★ (Roche)	Basel	June 19, 1990
Dr John Beale★ (Wellcome)	London	May 24, 1990
Dr M. Fernex★ (Roche)	Basel	June 18, 1990
Dr Karl Fantes★ (Wellcome)	Beckenham, UK	May 25, 1990
Mr. G. Farmer★ (Roche)	Basel	June 20, 1990
Dr Norman Finter★ (Wellcome)	Seven Oaks, UK	May 25, 1990
Dr Leon Gauci★ (Roche)	Basel	June 18, 1990
Dr G. Garotta★ (Roche)	Basel	June 19, 1990
Mrs. C. v. Helsen (Schering-Plough)	Brussels	June 23, 1988
Dr P. Herman (Schering-Plough)	Brussels	June 23, 1988
Dr Maurice Hilleman★ (MSD)	West Point, PA	Oct. 10, 1992
Dr P. Hunold★ (Roche)	Basel	June 18, 1990
Dr E. Hochuli★ (Roche)	Basel	June 21, 1990
Dr A. Keller★ (Roche)	Basel	June 20, 1990
Dr H. Leuenberger★ (Roche)	Basel	June 20, 1990
Dr M. Placchi★ (Roche)	Basel	June 19, 1990
Mr. P. de Pourq (Schering-Plough)	Brussels	June 23, 1988
Dr J. Ryff★ (Roche)	Basel	June 18, 1990
Dr David Secher★ (Cell-Tech)	London	May 23, 1990
Dr F. Schenker★ (Roche)	Basel	June 18, 1990
Dr T. Staehlin★ (Ciba-Geigy)	Basel	June 18, 1990
Dr J. P. Warbeer (Schering-Plough)	Brussels	June 23, 1988

Academic research and medicine

Dr Frances Balkwill★	London	May 31, 1990
Dr Samuel Baron★	Galveston, TX	Sep. 24, 1992
Dr Alfons Billiau★	Leuven, B.	Aug. 4, 1992

Dr Derek Burke★   Norwich   May 15, 1990
Dr Werner Boll★   Zurich   June 21, 1990
Dr Ernest Borden★   Milwaukee, WI   Oct. 12, 1992
Dr Derek Bangham★   London   Mar. 3, 1992
Dr Kari Cantell★   Helsinki   May 5,   1992
Dr Charles Chany★   Paris   Sep. 2, 1992
Dr Lois Epstein★   San Francisco   Oct. 8, 1992
Dr Ernesto Falcoff★   Paris   Sep. 3, 1992
Dr Emil J. Freireich★   Houston   Sep. 23, 1992
Dr Robert Friedman★   Bethesda, MD   Oct. 22, 1992
Dr Jordan Gutterman★   Houston   Sep. 23, 1992
Dr Ion Gresser★   Paris   Sep. 1, 1992
Dr Susanna Isaacs Elmhirst★ London   May 28, 1990
Dr Monto Ho★   Pittsburgh, PA   Oct. 29, 1992
Dr Pete Knight★   West Chester, PA   Nov. 4, 1992
Dr Mathilde Krim★   New York   Nov. 11, 1992
Dr Hilton Levy★   Frederick, MD   Oct. 10, 1992
Dr Jean Lindenmann★   Zurich   June 16, 1990
Dr Hilary Koprowski   Amsterdam   Mar. 19, 1990
Dr Edward de Maeyer★   Paris   Dec. 4, 1993
Dr Jacqueline de Maeyer★ Paris   Dec. 4, 1993
Dr Thomas Merigan★   Stanford, CA   Oct. 7, 1992
Dr Sidney Pestka★   Piscataway, NJ   Oct. 27, 1992
Dr M. Playfair★   London   May 29, 1990
Dr James Porterfield   (tel. interview)   Oct. 11, 1993
Dr Huub Schellekens★   Delft, NL   July 10, 1992
Dr Gerald Sonnenfeld★   Louisville, KY   Oct. 14, 1992
Dr Joseph Sonnabend★   New York   Nov. 12, 1992
Dr Hans Strander★   Stockholm   Apr. 30, 1992
Dr Michael Streuli★   Zurich   June 18, 1990
Dr Joyce Taylor★   London   May 30, 1990
Dr David Tyrrell★   Salisbury, UK   May 21, 1990
Dr Jan Vilcek★   New York   Nov. 10, 1992
Dr Charles Weissman★   Zurich   June 15, 1990
Dr Robert Wagner★   Charlottesville, VA Oct. 13, 1992
Dr Frederick Wheelock★ Philadelphia   Nov. 5, 1992

Government institutions and regulatory bodies

Dr George Galasso★ (NIAID)   Bethesda, MD   Oct. 19, 1992
Dr Kathy Laughlin★ (NIAID)   Bethesda, MD   Oct. 23, 1992
Dr Arthur Levine★ (NCI)   Bethesda, MD   Oct. 22, 1992
Dr John Petricciani★ (FDA)   Cambridge, MA   Nov. 6, 1992
Dr Frank Rauscher★ (ACS)   Stamford, CT   Oct. 16, 1992
Dr Kathryn Zoon★ (FDA)   Bethesda, MD   Oct. 23, 1992

Notes

* M. Holub, *Interferon, or on Theater*, Ohio: The Field Translation Series 7, Oberlin College, 1982, p. 1.

Preface

1 H. Schellekens, 'De weersomstuit van interferon', *Het Parool*, 13 November 1999.
2 S. S. Hall, *A Commotion in the Blood*, New York: Henry Holt and Company, 1997, 517.

Introduction

1 'The drug with a value beyond price', *The Weekly News*, Britain, 8 December 1979.
2 M. Edelhart, *Interferon: The New Hope for Cancer*, Reading: Addison-Wesley Publishing Company, 1981, 187.
3 Loutfy, M. R., Blatt, L. M. *et al.* 'Interferon Alfacon-1 plus cortocosteroids in severe acute respiratory syndrome', *JAMA* 290 (2003), 3222–8; Haagmans B. L. Pegylated interferon-α protects type 1 pneumocytes against SARS coronavirus infection in macaques *Nature Medicine*, published online doi: 10.1038/nm1001 (2004).
4 K. L. Hale, Interferon: the evolution of a biological therapy, M. D. Anderson Oncolog, 39, 4 (1994).
5 K. Cantell, *The Story of Interferon: The Ups and Downs in the Life of a Scientist*, Singapore: World Scientific, 1998.
6 Cantell, *The Story of Interferon*, p. viii. As far as *Nature's* honourable book reviewer and co-discoverer of interferon, Jean Lindenmann, was concerned, Cantell's personal diary offered a useful source to future historians and even modern sociologists of science might find in it 'some grist to their mill'. Interestingly Lindenmann's own positivistic judgement, calling the book's factual content entirely accurate, was put up for discussion a couple of weeks later in *Nature* by the family of the late Frank Rauscher, a former vice-president of research of the American Cancer Society (ACS). In a correction statement Rauscher was said to have been wrongly incriminated by Cantell for conflicts of interest in the purchase of interferon on behalf of the ACS in the early 1980s; J. Lindenmann, 'How to avoid making a fortune in medicine', *Nature* 394 (1998), 844–5; 'Correction', *Nature* 395 (1998), 342.
7 See, for studies confined to describing the careers of modern drugs as an autonomous developmental process governed by the logic of cumulative scientific and technological advance: J. T. Mahoney, *The Merchants of Life. An Account of the American Pharmaceutical Industry*, New York: Harper, 1959; D. Wilson, *Penicillin in Perspective*, London: Faber & Faber, 1976; D. Schwartzman, *Innovation in the Pharmaceutical Industry*, Baltimore: Johns Hopkins University Press, 1976; W. Sneader, *Drug*

Discovery: the Evolution of Modern Medicines, Chichester: John Wiley & Sons, 1985; M. Weatherall, *In Search of a Cure*, Oxford: Oxford University Press, 1990; R. P. T. DavenPort Hines and Judy Slinn, *Glaxo: A History to 1962*, Cambridge: Cambridge University Press, 1992; H. C. Peyer, *Roche: A Company History*, Basel: Editiones Roche, 1996; In the more culturally oriented studies with an emphasis on the social and political careers of therapeutic drugs, the neglect of the content and context of scientific and industrial research is striking; H. D. Walker, *Market Power and Price Levels in the Ethical Drug Industry*, Bloomington: Indiana University Press, 1971; M. Silvermann and P. R. Lee, *Pills Profits & Politics*, Berkely: California Press, 1974; and P. Temin, *Taking Your Medicine: Drug Regulation in the United States*, Cambridge: Harvard University Press, 1980. The notable exceptions to this separation between the social and scientific aspects of drug research, development and regulation are studies by John Swann, Jonathan Liebenau and John Abraham: J. Liebenau, *Medical Science and Medical Industry*, Baltimore: The Johns Hopkins University Press, 1987; J. P. Swann, *Academic Scientists and the Pharmaceutical Industry*, Baltimore: The Johns Hopkins University Press, 1988; J. Abraham, *Science, Politics and the Pharmaceutical Industry*, London: UCL Press, 1995.

8 Empirical studies carried out within this constructivist methodological perspective have succeeded in providing us with detailed and illuminating multi-layered and symmetric accounts of drug research and development. Taking cardiac drugs as an example Rein Vos has shown that the origin of modern medicines lies in the interface between the worlds of the laboratory and the clinic. The exchange of information between bench and bedside is argued to stimulate both therapeutic and diagnostic innovations. He argues that with the introduction of cardiac drugs like the beta blockers new subsets of heart disease came into being which in turn created new targets for therapeutic intervention. Nelly Oudshoorn, Louis Galambos, Vivien Walsh and Jordan Goodman followed suit with their case studies on twentieth-century drug trajectories, respectively concentrating on researching, testing, producing, marketing and disseminating sex hormones, vaccines, the anti-cancer drug Taxol and the male pill; R. Vos, *Drugs Looking for Diseases: Innovative Drug Research and the Development of the Beta Blockers and the Calcium Antagonists*, Amsterdam: Kluwer Academic Publishers, 1991; N. Oudshoorn. *Beyond the Natural Body: An Archaeology of Sex Hormones*, London: Routledge, 1994; L. Galambos and J. E. Sewell, *Networks of Innovation: Vaccine Development At Merck, Sharp & Dohme, and Mulford 1895–1995*, Cambridge: Cambridge University Press, 1995; J. Goodman and V. Walsh, *The Story of Taxol, Nature and Politics in the Pursuit of an Anti-cancer Drug*, Cambridge: Cambridge University Press, 2001; N. Oudshoorn, *The Male Pill*, Durham: Duke University Press, 2003.
9 P. Davis, *Managing Medicines*, Buckingham: Open University Press, 1997, 30–50.
10 J. Parascandola, 'Alkaloids to arsenicals: systematic drug discovery before the first world war', in G. J. Higby and E. C. Stroud (eds) *The Inside Story of Medicines: A symposium*, Madison (WI): The American Institute of the History of Pharmacy, 1997, pp. 77–91, 78.
11 F. Leuret and H. Bon, *Modern Miraculous Cures*, London: Peter Davies, 1957; C. Panati, *Breakthroughs*, London: MacMillan, 1980; A. Wycke, *21st-Century Miracle Medicine*, New York: Plenum Trade, 1997; L. Payer, *Disease Mongers*, New York: John Wiley, 1992; S. Fried, *Bitter Pills*, New York: Bantam Books, 1998; T. J. Moore, *Prescription for Disaster*, New York: Dell, 1999.
12 R. Freeman, *The Politics of Health in Europe*, Manchester: Manchester University Press, 2000; T. Pieters and S. Snelders, 'Seige's cycle: The careers of psychotropic drugs in mental health care since 1869 – A programmatic essay' (in preparation).
13 B. Spilker, *Multinational Drug Companies: Issues in Drug Discovery and Development*, New York: Raven Press, 1989, pp. 421–423; J. Drews, *In Quest of Tomorrow's*

196   *Interferon: The science and selling of a miracle drug*

    Medicines, Berlin: Springer, 1999; M. M. Weber, *Die Entwickelung der psychopharmakologie im Zeitalter der Naturwissenschaftlichen Medizin*, München, 1999.
14 P. L. Kahn, E. J. Yang, J. W. Egan, H. N. Higinbotham and J. F. Weston, *Economics of the Pharmaceutical Industry*, New York: Praeger Publishers, 1982.
15 Surprisingly, decades later thalidomide would start a new life as a drug. It made a comeback at the end of the 1990s as an indispensable medicine for those suffering from leprosy and from severe chronic autoimmune conditions like multiple sclerosis. The clinical use of thalidomide is restricted and carefully monitored by the international drug regulatory agencies. T. Stephens and R. Brynner, *Dark Remedy*, Cambridge (MA): Perseus Publishing, 2001.
16 In addition there is an interesting category of drugs that initially have been removed from the market by the pharmaceutical industry primarily for commercial reasons and have been reintroduced by their manufacturers in response to pressure from consumers and prescribers of medications; M. Weintraub and F. K. Northington, 'Drugs that wouldn't die' *JAMA* 255 (1986), 2327–8.
17 C. C. Mann and M. L. Plummer, *The Aspirin Wars*, New York: Knopf, 1991.
18 See, for the tenets of an ethnomethodological approach of science, which is based on the methodological principle of following scientists and studying the production of scientific facts and technical artefacts by using the notion of 'anthropological strangeness' – treating the making of natural knowledge as a strange and puzzling activity: B. Latour and S. Woolgar, *Laboratory Life; The Construction of Scientific Facts*, Princeton: Princeton University Press, 1986, 277–9; and B. Latour, *Science in Action*, Milton Keynes: Open University Press, 1987. See, for a comprehensive discussion of the use of this approach in the history of science: J. Golinski, *Making Natural Knowledge: Constructivism and the History of Science*, Cambridge: Cambridge University Press, 1998, 162–85.
19 See, for a detailed historical description and analysis of the rise of the molecular culture in biomedicine in the twentieth century: S. de Chadarevian and H. Kamminga (eds) *Molecularizing Biology and Medicine*, Amsterdam: Harwood Publishers, 1998.
20 See, A. Pickering, *The Mangle of Practice*, Chicago: The University of Chicago Press, 1995, 4.
21 See, for a similar though less media and public sphere centred use of the term culture: J. H. Fujimura, *Crafting Science: A Sociohistory of the Quest for the Genetics of Cancer*, Cambridge (MA): Harvard University Press, 1996.
22 See, M. Casper and M. Berg, 'Constructivist perspectives on medical work', *Science, Technology & Human Values* 20, 1995, 395–407; M. Berg and A. Mol (eds) *Differences in Medicine*, Durham: Duke University Press, 1998.

1   Interferon's birth

1 A preliminary version of chapter 2 was published in *Studies in History and Philosophy of Science*; T. Pieters, 'Shaping a New Biological Factor, "The Interferon", in Room 215 of the National Institute for Medical Research, 1956/57', *Stud. Hist. Phil. Sci.* 28 (1997), 27–73.
2 J. Austoker and L. Bryder, 'The National Institute for medical research and related activities of the MRC', in J. Austoker and L. Bryder (eds) *Historical Perspectives on the Role of the MRC*, Oxford: Oxford University Press, 1989, pp. 56–7.
3 This impressionistic account of NIMR's interior is based on: A. Landsborough Thomson, *Half a Century of Medical Research Vol I*, London: Her Majesty's Stationery Office, 1973, pp. 108–33; D. Tyrrell, 'Personal memories of the early days', *J. Interferon Res.* 7 (1987), 443–4; D. Burke, 'Early days with interferon', *J. Interferon Res.* 7 (1987), 441–3; David Tyrrell, interview, Salisbury, UK, 21 May, 1990; J. Lindenmann, 'The National Institute for Medical Research, Mill Hill: Personal recollec-

tions from 1956/57', *Arch. Virol.* 140 (1995), 1687–91; J. Lindenmann to Schweiz. Akademie der Mediz. Wissenschaften, erster Semesterbericht dated December 1956, Jean Lindenmann correspondence, personal archives; interviews with Derick Burke, Jean Lindenmann and Joseph Sonnabend; and, personal visits to the NIMR in September 1991.
4 See, J. Lindenmann, 'The National Institute for Medical Research', op. cit., pp. 1687–91.
5 T. van Helvoort, *Research Styles in Virus Studies in the Twentieth Century: Controversies and the Formation of Consensus*, Maastricht: PhD diss., Univ. Limburg (1993), p. 186; and interview with Jean Lindenmann.
6 This characterization of viruses is more or less comparable with that given today, although the notion that the nucleic acid of the virus encoded the hereditary information of the virus was rather vague. The mechanism of virus reproduction or replication had yet to be explored. At least there was cautious consensus that viruses went through an eclipse.
7 Interview with Jean Lindenmann.
8 Before starting his research work Lindenman had to sign an official document relinquishing in perpetuity all patent rights to whatever he might find during his stay at the NIMR to Britain's Medical Research Council; Interview with Jean Lindenmann.
9 The practice of tissue culture technique had been plagued for years by bacterial contamination. It was not until the arrival of antibiotics like penicillin in the late 1940s, that the *in vitro* culture of viruses in vessels with growing tissue embedded in suitable media evolved as a major tool in viral research. By the mid-1950s this technology was known to virologists as a relatively simple and reliable means to grow an endless number of viruses in their laboratories; see, F. C. Robbins and J. F. Enders, 'Tissue culture techniques in the study of animal viruses', *American Journal of the Medical Sciences* 220 (1950), 316–38.
10 Polio virus causes poliomyelitis, an infectious, pathogenic disease that induces paralysis, mostly in infants. Until mass vaccination against polio was made possible by the advent of the Salk vaccine in 1954, a number of devastating epidemics swept through Europe and America from the 1930s onwards, leaving a trail of paralysed people with some deaths. Among those affected was president Roosevelt. He was mainly responsible for the attention that was focused in the USA on polio as a major health threat, which after the Second World War sparkled off the 'war on polio'. During this publicly led search for a cure against poliomyelitis, large sums of money were poured into virus research, boosting the newly emerging field of virology; see A. J. Levine, *Viruses*, New York: Scientific American Library, W. H. Freeman and Company, 1992, pp. 62–5.
11 Myxomatosis virus was considered to be closely related to smallpox virus. In rabbits it had been shown to produce myxomatosis, a highly contagious and fatal disease, which was characterized by warts and skin tumours; see Levine, op. cit., pp. 209–10; and J. Lindenmann, 'The National Institute for Medical, op. cit., pp. 1687–91.
12 Interview with Jean Lindenmann.
13 J. Lindenmann, 'Erster Semesterbericht' dated December, 1956. Jean Lindenmann correspondence, personal archives.
14 By the term 'artefacts' I do not mean material objects which are related to conditions of human use, such as archaeological findings, machines, designs or technical processes. In laboratory research the word 'artefact' has a different connotation. Artefacts are described by most laboratory scientists as troublesome and unintentional events, which distort or confuse the identifiability of 'natural' phenomena in their experimental arrangements. Such artefacts are usually attributed to flaws or inefficacies in experimental procedures. For an analysis of artefact accounts in

laboratory research, see M. Lynch, *Art and Artefact in Laboratory Science*, London: Routledge & Kegan Paul, 1985, pp. 81–139.
15 C. H. Andrewes, 'Alick Isaacs', *Biographical Memoirs of Fellows of the Royal Society* 13 (1967) 205–21; J. Austoker and L. Bryder, op. cit., p. 43.
16 The experiments were to be submitted for publication in January 1957. See, H. Mooser und J. Lindenmann, 'Homologe interferenz durch hitzeinaktiviertes, an erythrozyten absorbiertes influenza-B-virus, *Experientia* XIII (1957), 147–8.
17 In retrospect one may find earlier observations concerning inhibition phenomena between viruses, both in animals and plants. These observations go back even to the pre-bacteriological era: see W. Henle, 'Interference phenomena between animal viruses: a review.' *Journal of Immunology* 64 (1950), 203–35, p. 205.
18 G. M. Findlay and F. O. MacCallum, 'An interference phenomenon in relation to yellow fever and other viruses.', *J. Path. Bact.* 44 (1937), 405–24, p. 420.
19 C. Andrewes, 'Virus diseases of man: A review of recent progress', *Br. Med. Bull.* 2 (1944), 265–9, p. 265.
20 D. Tyrrell, 'Personal memories of the early days', *J. Interferon Res.* 7 (1987), 443–4, p. 443.
21 Rickettsia were considered to be insect-borne infectious agents, which at least were thought to be different from both bacteria and viruses. See, S. Smith Hughes, *The Virus: a History of the Concept*, London: Heinemann Educational Books, Science History Publications, 1977, pp. 93–7; and H. Mooser und J. Lindenmann, 'Homologe Interferenz', op. cit., p. 148.
22 In this case the term 'heat-inactivated virus' refers to the process of heating the virus at 56°C for an hour in a buffer solution, whereupon it lost its infectious and cell destructive properties while retaining its capacity to induce viral interference in the host cell. Moreover, in the Australian paper heat-inactivated influenza virus had been shown to have superior interfering properties, while lacking enzymatic activity that would elute the virus particles from the surface of the red blood cells. For further details of the technical procedures see, A. Isaacs and M. Edney, 'I. Quantitative aspects of interference', *Austr. J. Exp. Biol.* 28 (1950), 219–30.
23 The chick embryo–influenza system was claimed to offer the following advantages. Employing inactivated virus instead of active virus as the interfering agent was thought to stimulate fewer complicating reactions in the host organism. In addition, the chick embryo–influenza system seemed to offer advantages such as the absence of antibody formation, the ready accessibility of the host cells and the availability of simple and sensitive methods for the detection of virus. Furthermore, throughout the years this particular experimental system had provided the virology community with a wealth of information on the nature of the interfering principle and the mechanism of interference.
24 Lindenmann to Mooser, letter dated 8 August 1958, Jean Lindenmann correspondence, personal archives.
25 S. Fazekas de St. Groth, A. Isaacs, and M. Edney, 'Multiplication of influenza virus under conditions of interference', *Nature* 170 (1952), 573–4; and interview with Jean Lindenmann.
26 A. Isaacs, 'Viral interference', *Symp. Soc. Gen. Microbiol.* 9 (1959), 102–21, p. 108; J. Lindenmann, 'Neuere Aspekte der Virus-Interferenz', *Z. Hyg. Infektionskrankh.* 146 (1960), 369–97, p. 383; J. Lindenmann, 'Induction of chick interferon: Procedures of the original experiments', *Methods in Enzymology* 78 (1981), 181–88, p. 182.
27 Interview with Jean Lindenmann.
28 Hughes, op. cit., p.101; A. Grafe, *A History of Experimental Virology*, Berlin: Springer-Verlag, 1991, p. 145; L. Kay, *The Molecular Vision of Life*, Oxford: Oxford University Press, 1993, p. 270; and interview with Jean Lindenmann.

29 T. van Helvoort, 'History of virus research in the twentieth century: The problem of conceptual continuity', *Hist. Sci.* xxii (1994), 185–235, pp. 213–18; and interview with Jean Lindenmann.
30 Interview with Lindenmann; Alick Isaacs to Prof. Mooser, letter dated 18 November, 1957, Jean Lindenmann correspondence, personal archives.
31 Isaacs and Valentine had just produced a series of electron microscopic images of virus preparations, which were said to show virus particles with ring-like structures inside, a kind of micro 'railway-lines' which disappeared after treatment with an RNA destroying enzyme (ribonuclease) and was believed to be RNA; see, C. H. Andrewes, 'Alick Isaacs', op. cit., 205–21, p. 215; and, A. Isaacs and R. Valentine, 'The structure of influenza virus filaments and spheres', *J. Gen. Microbiol.* 16 (1957), 195–204, p. 195.
32 The notion of co-labouring or collaborating in the sense of minimally working together toward a common product, was borrowed from Griesemer and Gerson; see, J. Griesemer and E. M. Gerson, 'Colloboration in the Museum of Vertebrate Zoology', *Journal of the History of Biology* 26 (1993), 185–203, pp. 196–203.
33 A. Isaacs, 'Interferon', *Scientific American* 204 (1961), 51–7, p. 51.
34 Thanks to Isaacs's carefully recorded laboratory notebooks, and Lindenmann's correspondence and retrospective descriptions of his work in Isaacs's laboratory, a day-to-day account of the laboratory work can be reconstructed. Holmes has convincingly shown that combining laboratory notebooks and other unpublished documents with published papers can yield fairly plausible accounts of investigative pathways through which researchers arrive at the knowledge which they ultimately report in published papers. On scientific discovery, investigative pathways and historical reconstructions, see F. Holmes, *Lavoisier and the Chemistry of Life*, Madison: Wisconsin Press, 1985; and F. Holmes, 'Scientific writing and scientific discovery', *Isis* 78 (1987), 220–35. This section is largely based on the following letters, articles, notebook and interview: Alick Isaacs to Herman Mooser, letter dated 18 November, 1957, Jean Lindenmann correspondence, personal archives; Jean Lindenmann to Herman Mooser, letter dated 7 August, 1958, Jean Lindenmann correspondence, personal archives, Zürich; Jean Lindenmann to Herman Mooser, letter dated 13 August, 1958, Jean Lindenmann correspondence, personal archives; J. Lindenmann, 'Induction of chick interferon: procedures of the original experiments', *Methods in Enzymology* 78 (1981), 181–8; J. Lindenmann, 'From interference to interferon: a brief historical introduction', *Phil. Trans. R. Soc. Lond.*, series B 299 (1982), 3–5; A. Isaacs, Laboratory Notebooks, 1946–65. National Library of Medicine (Bethesda), Film number: reel 91–28, Lab Note Book, S.O. Book 135, Code 28-72-0, 'Alick Isaacs Interference Expts. V'; and interview with Jean Lindenmann.
35 A. Isaacs, Laboratory Notebooks, 1946–65. National Library of Medicine (Bethesda), Film number: reel 91–28, Lab Note Book, S.O. Book 135, Code 28-72-0, 'Alick Isaacs Interference Expts. V', 4 September 1956.
36 For a detailed description of this *in vitro* technique see, F. Fulton and P. Armitrage, 'Surviving Tissue Suspensions for Influenza Virus Titration', *J. Hyg.* 49 (1951), 247–63, pp. 251–2.
37 An important incentive for the development of this *in vitro* technique had been the limited supply of fertilized eggs in Europe after the Second World War; F. Fulton and A. Isaacs, 'Influenza virus multiplication in the chick chorioallantoic membrane', *J. Gen. Microbiol.* 9 (1953), 119–31; W. Henle and G. Henle, 'The road to interferon: interference by inactivated influenza virus', in A. Billeau and N. Finter (eds) *Interferon 1: General and Applied Aspects*, Amsterdam: Elsevier, 1984, pp. 3–18. Interviews with Jean Lindenmann and David Tyrrell.

38 D. Tyrrell and I. Tamm, 'Prevention of virus interference by 2,5-dimethylbenzimidazole', *J. Immun.* 75 (1955), 43–9, p. 43; interview with Jean Lindenmann.
39 The plasma membrane of these cells had been ruptured by chemical treatment and, consequently, these red blood cells had lost their haemoglobin content (pigment).
40 H. Donald and A. Isaacs, 'Counts of influenza virus particles', *J. Gen. Microbiol.* 10 (1954), 457–64, p. 459.
41 Michael Lynch noted that the circumstantial character of the interpretative enterprise in laboratory science is *a fortiori* true for electron microscopy. See, Lynch, op. cit., pp. 10–12.
42 In the original experiment, as reproduced in Isaacs's notebook, an additional four groups were included each containing different concentrations of free inactivated virus or virus-coated ghosts.
43 In retrospect, Derek Burke, who was to join Isaacs's and Lindenmann's researches early in 1957, gave this impressionistic account of the performance of haemagglutination titrations or tests; see, D. Burke, 'Early days with interferon', *J. Interferon Res.* 7 (1987), 441–2, p. 442.
44 A. Isaacs, Laboratory Notebooks, 1946–65. National Library of Medicine (Bethesda), Film number: reel 91–28, Lab Note Book, S.O. Book 135, Code 28-72-0, 'Alick Isaacs Interference Expts. V', 25 September 1956.
45 Interview with Jean Lindenmann.
46 In principle it is impossible to recapture the irreproducable 'initial' observation which was constitutive in rendering the events anomalous.
47 Interview with Jean Lindenmann.
48 A. Isaacs, Laboratory Notebooks, 1946–65. National Library of Medicine (Bethesda), Film number: reel 91–28, Lab Note Book, S.O. Book 135, Code 28-72-0, 'Alick Isaacs Interference Expts. V', 6 November, 1956.
49 About half a year later Isaacs would write to Lindenmann that their filtration experiments had gone wrong. 'I have a shock for you. Ernst Fulton told me our filtration results were of no value since we had not used Broth. I had forgotten that Elford always insisted on Broth. We repeated filtrations in Broth and it goes through a 0.1μ membrane and an 0.048μ membrane!!! (previously both the virus and interferon activity had been held back by a 0.6μ membrane). Results are being confirmed at the moment but I am afraid we will have to publish a disclaimer soon. Of course this now fits with the centrifugation results but I feel rather foolish and very sorry to have misled you'; Alick Isaacs to Jean Lindenmann, letter dated 9 December 1957, Lindenmann correspondence, personal archives.
50 This list of trials and results is based on the following documents: A. Isaacs, Laboratory Notebooks, 1946–65. National Library of Medicine (Bethesda), Film number: reel 91–28, Lab Note Book, S.O. Book 321, Code 28–321, 'Alick Isaacs Interference Expts. VI', 10–12–1956/21–2–1957; and A. Isaacs, J. Lindenmann and R. Valentine, 'Virus interference. II. Some properties of interferon', *Proc. R. Soc.* 147 (1957), 268–73.
51 On the relative nature of expertise in science see also; T. Nickles, 'Justification and experiment', in D. Gooding, T. Pinch and S. Schaffer (eds) *The Uses of Experiment*, Cambridge University Press, 1989, pp. 300–1; and D. Gooding, *Experiment and the Making of Meaning*, Dordrecht: Kluwer Academic Publishers, 1990, pp. 29–69.
52 The use of the term tinkering was not to connote unskilled work but to emphasize the essential contingent nature of work at the frontier of science; see, K. D. Knorr-Cetina, *The Manufacture of Knowledge: An Essay on the Constructivist and Contextual Nature of Science*, Oxford: Pergamon Press, 1981, pp. 34; and F. L. Holmes, 'Manometers, tissue slices, and intermediary metabolism', in A. Clarke and J. Fujimura (eds) *The Right Tools for the Job*, Princeton: Princeton Univ. Press, 1992, pp. 169–70.

53 Since Alick Isaacs's personal laboratory archives were destroyed after his death in 1967, I was unable to recapture the process of drafting the early interferon papers and follow in detail the way both part of the continuous web of investigations was sealed off and ideas concerning this new interfering factor, 'interferon', further evolved.
54 Interview with Jean Lindenmann.
55 A. Isaacs and J. Lindenmann, 'Virus interference. I. The interferon', *Proc. R. Soc.* 147 (1957), 258–267; and A. Isaacs, J. Lindenmann and R. Valentine, 'Virus interference. II. Some properties of interferon', *Proc. R. Soc.* 147 (1957), 268–73.
56 A. Isaacs, J. Lindenmann and R. Valentine, 'Virus interference. II', op. cit. pp. 272–3.
57 Interviews with Derek Burke, Norman Finter and Joseph Sonnabend.
58 Ibid.
59 Ibid.
60 D. Burke, Laboratory Notebook, S.O. Book 321, Code 28-321, 'Interferon Book I', 4-3/26-9-1957, personal archives.
61 In order for the replication of the material realization of an experiment to be called successful it is required that 'the same actions are performed and the same experimental situations produced from the point of view of the daily language description of the material realization of the experiment'. See, H. Radder, 'Experimental reproducibility and the experimenter's regress', in D. Hull, M. Forbes and K. Okruklik (eds) *PSA, Volume I*, East Lansing, Michigan: Philosophy of Science Association, 1992, p. 65.
62 J. Lindenmann, D. Burke and A. Isaacs, 'Studies on the production, mode of action and properties of interferon', *Br. J. Exp. Path.* 38 (1957), 551–62, p. 559.
63 Interview with Derek Burke.
64 Interview with Jean Lindenmann.
65 J. Lindenmann, D. Burke and A. Isaacs, 'Studies on the production, mode of action and properties of interferon', op. cit., p. 561.
66 Interview with Jean Lindenmann.
67 R. Depoux and A. Isaacs, 'Interference between influenza and vaccinia viruses', *Br. J. Exp. Path.* 35 (1954), 415–18, p. 418.
68 J. Lindenmann, D. Burke and A. Isaacs, 'Studies on the production, mode of action and properties of interferon', op. cit., p. 558.
69 See the annual report from the Bacteriology and Virus Research Group of the NIMR, 10, 1956/'57, held at the NIMR Library Archives; and A. Isaacs, Laboratory Notebooks, 1946–65. National Library of Medicine (Bethesda), Film number: reel 91-28, Lab Note Book, S.O. Book 321, Code 28-321, 'Alick Isaacs Interference Expts. VI,' 21-3-/10-7-1957; and interview with Jean Lindenmann.
70 The 'real' *in vivo* situation is a chimera or something one cannot gain access to via the artificial and constructive nature of laboratory practice.
71 This observation is in line with Olga Amsterdamska's claim that 'the selection and formulation of problems in mission-oriented sciences must be considered within the wider contexts of their disciplinary goals, which are closely linked to the social contexts of practice'; O. Amsterdamska, 'Medical and Biological Constraints: Early Research on Variation in Bacteriology', *Soc. Stud. Sci.* 17 (1987), 657–87, p. 659.

2 Shaping a new field of research: investigating interferons(s)

1 J. Lindenmann und A. Isaacs, 'Versuche über Virus-Interferenz', *Schweiz. Z. Path. Bakt.* 20 (1957), 640–6, p. 646.

2 Jean Lindenmann to Alick Isaacs, letter dated 28 June 1957, Jean Lindenmann correspondence, personal archives.
3 F. M. Burnet, *Principles of Animal Virology*, New York: Academic Press, 1955.
4 Interview with John Beale; 'The Asiatic Flu', *Life*, 2 September 1957, 113–20.
5 Jean Lindenmann to Alick Isaacs, letter dated 28 June 1957, Jean Lindenmann correspondence, personal archives.
6 Jean Lindenmann to Alick Isaacs, letter dated 28 June 1957, Jean Lindenmann correspondence, personal archives.
7 Alick Isaacs to Jean Lindenmann, letter dated 1 July 1957, Jean Lindenmann correspondence, personal archives.
8 According to a 1941 article on pharmaceutical research there was a 'tendency amongst academic workers to view with apprehension whole-hearted alliance with the research of any one firm'. See, J. Liebenau, 'The British Success with Penicillin', *Soc. Stud. Sci.* 17 (1987), 69–86, p. 72.
9 Alick Isaacs to Jean Lindenman, letter dated 7 July 1957, Jean Lindenmann correspondence, personal archives.
10 See, for a detailed analysis of how the story of penicillin was intertwined with great issues in post-Second World War British culture: R. Bud, 'Penicillin and the new Elizabethans', *Br. J. Hist. Sci.* 31 (1998), 305–33.
11 D. Wilson, *Penicillin in Perspective*, London: Faber & Faber, 1976, p. 245; and L. G. Matthews, *History of Pharmacy in Britain*, London: E. & S. Livingstone, 1962, 330–3.
12 The publicly funded Medical Research Council played and still plays an important role in supporting and undertaking biomedical and clinical research in Britain. The NIMR was set up as the MRC's central institute for medical research in 1914. In 1960 up to 20 per cent of its research funds were allocated to finance the NIMR and the rest to finance research projects throughout the country. See, for the role of the MRC in shaping a national system of medical research in Britain: J. Austoker and L. Bryder (eds) *Historical Perspectives on the Role of the MRC*, Oxford: Oxford University Press, 1989.
13 Interview with John Porterfield.
14 Interview with Derek Burke.
15 J. Austoker and L. Bryder (eds) op. cit , p. 81.
16 J. Austoker and L. Bryder (eds) op. cit., pp. 234–9.
17 Alick Isaacs to Jean Lindenman, letter dated 26 June 1957, Jean Lindenmann correspondence, personal archives.
18 Jean Lindenmann to Alick Isaacs, letter dated 28 June 1957, Lindenmann correspondence, personal archives; and interview with Jean Lindenmann.
19 In 1964 Lindenmann would return to the University of Zürich as an associate professor at the Institute for Medical Microbiology. By then he had already left the field of interferon research. See, Jean Lindenmann to Alick Isaacs, letter dated 19 May 1964, Lindenmann Correspondence, Personal Archives; and interview with Jean Lindenmann.
20 Together with other WHO centres Isaacs's laboratory was tied up with characterizing an apparently new influenza virus strain, which was held responsible for a worldwide flu epidemic ('Asian flu') in 1957. See Alick Isaacs to Jean Lindenmann, letter dated 9 September 1957, Lindenmann Correspondence, Personal Archives; and Alfred Grafe, *A History of Experimental Virology*, Berlin: Springer-Verlag, 1991, p. 252.
21 Alick Isaacs to Jean Lindenmann, letters dated 17 October and 9 December 1957, Lindenmann correspondence, personal archives; J. Lindenmann, D. C. Burke and A. Isaacs, 'Studies on the production, mode of action and properties of interferon',

Br. J. Exp. Path. 38 (1957), 551–62; and, D. C. Burke and A. Isaacs, 'Further studies on interferon', Br. J. Exp. Path. 39 (1958), 78–84.
22 See the annual report from the Bacteriology and Virus Research Group of the NIMR, 10, 1956/57, held at the NIMR Library Archives.
23 See, Editorial, 'A lead towards virus chemotherapy?', Lancet i (1957), 931–2; and 'Interferon', Br. Med. J. i (1957), 1102.
24 Alick Isaacs to Jean Lindenmann, letter dated 23 October 1957, Lindenmann correspondence, personal archives.
25 Glaxo to MRC, letter dated 1 November, 1957, MRC Archives, File No. A814/17; and Benger Laboratories to MRC, letter dated 28 November 1957, MRC Archives, File No. A814/17.
26 MRC to Sir Charles Harrington, letters dated 6, 29 November and 5 December 1957, MRC Archives, File No. A814/17; and Sir Charles Harrington to MRC, letters dated 13 November, 7 and 10 December 1957, MRC Archives, File No. A814/17.
27 Statement of a senior MRC official during the discussion following the presentation of a preliminary version of this chapter to the Twentieth Medical History Group of the Wellcome Trust, held at the Royal College of Physicians, 11 February 1992.
28 NRDC to MRC, letter dated June 24 1958: '. . . applications covering the interferon invention have been filed in Germany under No. J14535 on March 11 1958, in Canada under No. 750330 on April 28 1958 and in the USA under No. 734106 on May 9 1958', MRC Archives, File No. A 813/104; Elkington and Fife, consulting chemists and chartered patent agents, to the MRC, letter dated 20 December 1957, MRC Archives, File No. A813/104.
29 See the invitation to the Annual Conversazione of the Royal Society on 15 May 1958, *The Annals of the Royal Society* I (1958), p. 10.
30 See newspaper article, 'Interferon May Aid Fight Against Flu', *The Daily Telegraph*, 16 May 1958.
31 Anonymous, 'Royal Society Conversazione', Br. Med. J. i (1958), 1229.
32 Alick Isaacs to Jean Lindenmann, letter dated 5 June 1958, Lindenman correspondence, personal archives.
33 D. Cantor, 'Cortisone and the Politics of Drama, 1949–55', in J. V. Pickstone (ed.) *Medical Innovations in Historical Perspective*, Hampshire: Macmillan Distribution, 1992, p. 173.
34 H. M. Marks, 'Cortisone, 1949: a year in the political life of a drug', Bull. Hist. Med. 66 (1992), 419–39.
35 The same 'penicillin syndrome' played a most vital part in the concurrent British efforts to develop cephalosporin C as a drug; see, Wilson, op. cit., pp. 252–3.
36 More rewarding and fashionable research topics for animal virologists in America in the late 1950s were the study of the biochemistry and genetics of animal viruses, quantitative tissue-culture studies and the characterization of new viruses, and the study of animal tumour viruses in conjunction with the possible virus eatiology of human cancer; A. Grafe, *A History of Experimental Virology*, Berlin: Springer-Verlag, 1991, pp. 164–97; M. Pollard (ed.) *Perspectives in Virology: A Symposium*, Texas: The University of Texas Medical Branch, 1958.
37 The researchers Temin and Rubin were closely associated with one of the protagonists of American animal virology, Renato Dulbecco. Grafe, op. cit., pp. 164–97; Pollard, op. cit. pp. 1–50; Interviews with Sam Baron, Robert Wagner, Norman Finter, David Tyrrell and Derek Burke.
38 I agree with Stephen Hall that it is impossible to reconstruct the story of interferon without taking into account Isaacs' manic-depressive disorder that manifested itself

for the first time in the fall of 1958; S. S. Hall, *A Commotion in the Blood; Life, Death and the Immune System*, New York: Henry Holt and Company, 1997.
39 Christopher Andrewes to Jean Lindenmann, letter dated 10 November 1958, Jean Lindenmann correspondence, personal archives; interviews with Susanna Isaacs-Elmhirst, Derick Burke and David Tyrrell.
40 See, for an in-depth analysis of the role of expectations and niches in technological developments, H. van Lente, *Promising Technology: The Dynamics of Expectations in Technological Developments*, Delft, PhD. diss., Eburon, 1993.
41 Interview with Derek Burke.
42 Henle *et al.* were already in the process of studying host cell–virus interactions in continuous lines of human cells, of malignant origin (e.g. human bone marrow cells derived from bone marrow of a leukaemic patient, so-called 'MCN cells'), which were chronically infected with virus (e.g. an avian virus that went by the name New Castle disease virus); G. Henle, F. Deinhardt, V. Bergs and W. Henle, 'Studies on persistent infections of tissue cultures: I. General aspects of the system', *J. Exptl. Med.* 108 (1958), 537–60.
43 F. Deinhardt, V. Berghs, G. Henle and W. Henle, 'Studies on persistent infections of tissue cultures: III. Some quantitative aspects of host cell–virus interactions', *J. Exptl. Med.* 108 (1958), 573–89; interviews with Monto Ho, Ion Gresser, Norman Finter and Kari Cantell.
44 M. Ho and J. Enders, 'An inhibitor of viral activity appearing in infected cell cultures' ,*Proc. N. A. S.* 45 (1959), 385–9.
45 Ibid.
46 Ibid., p. 389.
47 Interview with Monto Ho.
48 Further research on VIF was discussed at great length in a second article, which was accepted for publication in *Virology* about half a year later. As in the case of interferon, material conditions and the emerging scientific object, VIF, were strongly interwoven. Another striking feature is the use of the term 'VIF', which is far from consistent. Despite the statement of the authors that VIF stands for viral inhibitory fluid, more than once VIF seems to connote viral inhibitory factor (Ho's favourite translation); M. Ho and J. Enders, 'Further studies on an inhibitor of viral activity appearing in infected cell cultures and its role in chronic viral infection', *Virology* 9 (1959), 446–77.
49 Interview with Monto Ho.
50 Alick Isaacs to Jean Lindenmann, letters dated 17, 24 July and 1 September 1959, Jean Lindenmann, correspondence, personal archives; and interview with Derek Burke.
51 D. Burke and A. Isaacs, 'Some factors affecting the production of interferon', *Br. J. Exp. Path.* 39 (1958), 452–8, p. 456.
52 The Common Cold Unit was set up after the Second World War in the buildings of the former American Red Cross Hospital on the initiative of Andrewes and with support of the MRC and Ministry of Health as a centre for systematic studies on the aetiology and spread of common cold-inducing viruses such as the rhino and influenza viruses. See, D. Tyrrell, 'The Common Cold Unit 1946–1990: Farewell to a much-loved British institution', *PHLS Microbiology Digest* 6 (1991), 74–6.
53 Parainfluenza viruses were considered a subgroup of the Influenza viruses. These in turn were and are still counted as part of the Myxoviruses. Contrary to the influenza viruses the Parainfluenza subgroup did not appear to grow well in the chick embryo. The Sendai virus strain was held responsible for inducing pneumonia in infants; T. M. Bell, *An Introduction to General Virology*, London: William Heinemann Medical Books, 1965, 129–43.

54 D. Tyrrell, 'Interferon produced by cultures of calf kidney cell', *Nature* 184 (1959), 452–3, p. 453; and interviews with David Tyrrell, James Porterfield and Derek Burke.
55 Robert Wagner is definitely wrong when he retrospectively claims that the animal species specificity of interferon action was discovered in his laboratory. He confuses discovering with re-examining the hypothesis of species specificity as stated in his 1961 article in *Virology*; see, R. R. Wagner, 'Biological studies of interferon', *Virology*, 13 (1961), 323–37, p. 334; and, R. R. Wagner, 'Reminiscences of a virologist wandering in Serendi', *Arch. Virol.* 141 (1996), 787–97.
56 D. Tyrrell, 'Interferon produced by cultures of calf kidney cells', *Nature* 184 (1959), 452–3; interview David Tyrrell and John Porterfield.
57 Interview with David Tyrell.
58 Interview with Derek Burke.
59 A. Isaacs, 'Interferon: The prospects', *The Practitioner* 183 (1959), 601–5, p. 605.
60 Y. Nagano and Y. Kojima, 'Inhibition de l'infection vaccinale par un facteur liquide dans le tissu infecté par le virus homologue', *Compt. Rend. Soc. Biol. Filiales* 152 (1958), 1627–9.
61 Alick Isaacs to Jean Lindenmann, letter dated 10 September 1959, Jean Lindenmann correspondence, personal archives.
62 Minutes of the third meeting of the Scientific Committee on interferon, dated 26 October 1959, MRC Archives File No. 788/2/1.
63 The following section is a reconstruction largely based on: interviews with Robert Wagner, Derek Burke and Norman Finter; Minutes of the third meeting of the Scientific committee on interferon, dated 26 October 1959, MRC Archives File No. 788/2/1; and R. R. Wagner, 'Viral interference', *Bacteriol. Rev.* 24 (1960), 151–66, pp. 157–61.
64 Interview with Robert Wagner.
65 Interview with Robert Wagner
66 Interview with Robert Wagner.
67 R. R. Wagner and A. Levy, 'Interferon as a chemical intermediary in viral interference', *Ann. N.Y. Acad. Sci.* 88 (1960), 1308–18, p. 1310; and interviews with Robert Wagner, Sam Baron, Monto Ho and Hilton Levy.
68 Minutes of the fourth meeting of the Scientific Committee on interferon, dated 26 November 1959, MRC Archives File No. 788/2/1.
69 Alick Isaacs to Jean Lindenmann, letter dated 25 November 1959, Jean Lindenmann correspondence, personal archives.
70 Jean Lindenmann to Alick Isaacs, letter dated 2 December 1959, Jean Lindenmann correspondence, personal archives.
71 Alick Isaacs to Jean Lindenmann, letter dated 4 December 1959, Jean Lindenmann correspondence, personal archives.
72 Ibid.
73 Ibid.
74 Alick Isaacs, 'Nature and function of interferon', in Morris Pollard (ed.) *Perspectives in Virology*, Minneapolis: Burgess Publishing, 1960, p. 123.
75 A. Isaacs, 'Interferon', *Scientific American* 204 (1961), 51–7.
76 A. Isaacs, J. Lindenmann and R. Valentine, 'Virus interference. II. Some properties of interferon', *Proc. R. Soc.* 147 (1957), 268–73; A. Isaacs, 'Viral interference', *Symp. Soc. Gen. Microbiol.* No. 9 (1959), 102–21, p. 109; and R. R. Wagner and A. H. Levy, 'Interferon as a chemical intermediary in viral interference', *Ann. N.Y. Acad. Sci.* 88 (1960), 1308–18, p. 1308.
77 D. C. Burke, 'The Purification of Interferon', *Biochem. J.* 78 (1961), 556–64, p. 556; and M. R. Hilleman, 'Interferon in prospect and perspective', *J. Cell. Comp. Physiol.* 62 (1963), 337–53, p. 337.

206   *Interferon: The science and selling of a miracle drug*

78 In claiming that he and Emanuel Heller simultaneously discovered that interferon was a cellular product and not a product of the virus by showing in 1963 that the antibiotic actinomycin D (a known inhibitor of DNA-directed RNA synthesis of cellular proteins) inhibited the production of interferon, Robert Wagner not only disregards the work of his female fellow-scientist Joyce Taylor, but even more important the lengthy and complex transformation process through which interferon's identity as a cellular inhibitor of viral infection was shaped. Wagner's, Heller's and Taylor's experiments were neither more nor less than constitutive of interferon's genesis as a cellular protein; R. R. Wagner, 'Reminiscences of a virologist wandering in serendip', *Arch. Virol.* 141 (1996), 779–88, p. 783; E. Heller, 'Enhancement of Chikungunya virus replication and inhibition of interferon production by actinomycin D', *Virology* 21 (1963), 652–6; R. R. Wagner, 'Inhibition of interferon biosynthesis by actinomycin D', *Nature* 204 (1964), 49–51; and J. Taylor, 'Inhibition of interferon action by actinomycin', *Biochem. Biophys. Res. Comm.* 14 (1964), 447–51.
79 The point of modalities which are constantly added, modified and dropped in the transformation of a provisional claim into a scientific fact was first noticed by Fleck and developed in greater detail by Latour and Woolgar; L. Fleck, *Genesis and Development of a Scientific Fact*, Chicago: The University of Chicago Press, 1979, p. 118; Latour and Woolgar, *Laboratory Life: The Construction of Scientific Facts*, Princeton: Princeton University Press, 1986, 151–86.
80 E. De Maeyer, 'Interferon twenty years later', *Bulletin de L'Institut Pasteur* 76 (1978), 303–23, p. 303.
81 The table is based on author indexes and references from: M. Ho, 'Interferons', *N. Eng. J. Med.* 266 (1962), 1258–64; A. Isaacs, 'Interferon', *Adv. Virus Res.* 10 (1963), 1–39; N. B. Finter (ed.) *Interferons*, Amsterdam: North-Holland Publishing Company, 1966; and N. B. Finter (ed.) *Interferons and Interferon Inducers*, Amsterdam: North-Holland Publishing Company, 1973. See for the exponential growth in biomedical research in America in the 1960s, J. T. Patterson, *The Dread Disease*, Cambridge: Harvard University Press, 1987, p. 245.
82 C. Chany, 'Interferon-like inhibitor of viral multiplication from malignant cells', *Virology* 13 (1961), 485–92; E. De Maeyer and J. Enders, 'An interferon appearing in cell cultures infected with measles virus', *Proc. Soc. Exper. Biol. Med.* 107 (1961), 573–78; L. Glasgow and K. Habel, 'Role of interferon in vaccinia virus infection of mouse embryo tissue culture', *J. Exper. Med.* 115 (1962), 503–12; and M. Ho, 'Kinetic considerations of the inhibitory action of an interferon produced in chick cultures infected with sindbis virus', *Virology* 17 (1962), 262–75.
83 J. Lindenmann, 'L'Interferon', *Méd. Hyg.* 19 (1961), 945–6, p. 945; N. Gibbons (ed.) *Recent Progress in Microbiology*, Proceedings of the VIII International Congress for Microbiology, Montreal 1962, Toronto, University of Toronto Press, 1963, pp. 419–57; interviews with Jan Vilcek, Edward De Maeyer and Kari Cantell.
84 A. Isaacs, 'Interferon', *Adv. Virus. Res.* 10 (1963), 1–39, p. 2.
85 However, Isaacs's attempt to justify his and Lindenmann's priority claim regarding interferon could not prevent the American patent authorities from claiming in 1965 that the American researchers Lennette and Koprowski published a paper in the *Journal of Experimental Medicine* in 1946 which might have anticipated the work of Isaacs and Lindenmann. As such it threatened to undermine the possibility of a valid patent in the US, although eventually in March 1966 after lengthy negotiations between the NRDC and the American patent agency the basic US patent application on interferon was accepted. Meeting of the G.N.R.D. Patent Holdings Ltd, 18 February 1965, MRC Archives, File No. A 812/5; and, Minutes of the 33th meeting of the Scientific Committee on Interferon, dated 7 April 1966, MRC Archives File No. S788/2/4.

3 Interferon on trial

86 In cognitive sciences, e.g. psychology of science, the transformation process which interferon underwent would be argued to exemplify in a gradual fashion some variant of a classical gestalt switch (quite common are the rabbit-duck, pelican-antelope and Einstein-nude reversals). See, M. De Mey, *The Cognitive Paradigm*, Chicago: The University of Chicago Press, 1992, XVI–XVII, 89–93, 173–6.

1 A preliminary version of chapter 3 was published in *Medical History*; T. Pieters, 'Interferon and its first clinical trial': Looking behind the scenes', *Medical History* 37 (1993), 270–95.
2 Medical ethics were primarily defined by the hippocratic oath, which all newly qualified doctors were required to swear; J. S. Porterfield to T. Pieters, letter dated 9 November 1993.
3 In the 1950s toxicity testing required no more than an LD-50 test in a single species.
4 The description of the practice of drug testing is based on: Breckon, op. cit., pp. 121–41; A. Landsborough Thomson, *Half a Century of Medical Research Vol. II*, London: Her Majesty's Stationery Office, 1975, 226–43; W. Sneader, *Drug Development: From Laboratory to Clinic*, Chichester: John Wiley & Sons, 1986, pp. 75–86; D. Healy, *The Psychopharmacologists I*, London: Chapman and Hall, 1996, 113; and, information gathered during discussions following the presentation of a preliminary version of this chapter to the Twentieth Century Medical History Group of the Wellcome Trust, held at the Royal College of Physicians, 11 February 1992.
5 See, W. Breckon, *The Drugmakers*, Plymouth: The Bowering Press, 1972, pp. 121–41; and D. G. Grahame-Smith, 'Problems facing a regulatory authority', in J. F. Cavalla (ed.) *Risk–Benefit Analysis in Drug Research*, Lancaster: MTP Press Limited, 1981, 51–4; and E. M. Tansey, P. P. Catterall, D. A. Christie, S. V. Willhoft and L. A. Reynolds, *Wellcome Witnesses to Twentieth Century Medicine Vol I*, London: Wellcome Trust Occasional Publications, 1997.
6 The American pharmaceutical company Johnson & Johnson to the MRC, letter dated 25 June 1958, MRC Archives, File No. A 813/104.
7 Alick Isaacs to Johnson & Johnson, New Brunswick, New Jersey, US, letter dated 25 June 1958, MRC Archives, File No. A 813/104.
8 Alick Isaacs to Jean Lindenmann, letters dated 19 March, 13 May 1958, Jean Lindenmann correspondence, personal archives.
9 Internal Note MRC, dated 4 November 1958, MRC Archives, File No. A 814/17.
10 Liebenau pointed out how as a result of difficulties experienced when trying to coordinate things for insulin in the 1920s, the MRC decided to keep its distance in future dealings with drug companies as can be seen in the case of penicillin. Another factor that may have played a role in the MRC's reservations in the case of penicillin was the failed attempt in the 1920s to unite a number of prominent British firms – among others Boots, British Drug Houses and British Dyestuffs – in 'The Pharmaceutical Corporation': see, J. Liebenau, 'The MRC and the pharmaceutical industry: the model of insulin', in J. Austoker and L. Bryder (eds) *Historical Perspectives on the Role of the MRC*, Oxford: Oxford University Press, 1989, 179–80.
11 In 1955 the MRC and NRDC also initiated efforts to seek collaboration with British drug companies on the development of the possibly interesting antibiotic, cephalosporin. However, despite initial mutual enthusiasm, because of slow research progress most companies soon lost interest and it never came to a formal

Collaboration Agreement; see, D. Wilson *Penicillin in Perspective*, London: Faber & Faber, 1976, 251.
12 Interviews with Derek Burke, Norman Finter and David Tyrrell.
13 In the 1940s the image of international competition (with American pharmaceutical companies) also helped British academic workers to overcome part of their restraint towards British drug firms and start collaborating on the development of penicillin as a drug; see, J. Liebenau, 'The British success with penicillin', *Soc. Stud. Sci.* 17 (1987), 69–86, p. 75.
14 Internal Note MRC, dated 6 November 1958, MRC Archives, File No. A 814/17.
15 See, L. G. Matthews, *History of Pharmacy in Britain*, London: E. & S. Livingstone, 1962, 330–3; and, Wilson, op. cit., pp. 250–5.
16 Internal Note MRC, dated 4 November 1958, MRC Archives, File No. A 813/104.
17 L. E. Arnow, *Health in a Bottle; Searching for the Drugs that Help*, Philadelphia: J. B. Lippincott Company, 1970; and, W. Breckon, *The Drug Makers*, London: Eyre Methuen Ltd, 1972; and interviews with John Beale and Norman Finter.
18 For an in-depth historical analysis of the development of Ehrlich's chemotherapy theory which was largely based on the dictum 'If the law is true in chemistry that corpora non agunt nisi liquida, then for chemotherapy the principle is true that corpora non agunt nisi fixata'; see, J. Parascandola, 'The theoretical basis of Paul Ehrlich's chemotherapy', *J. Hist. Med.* 36 (1981), 19–43; and, P. Ehrlich, 'Chemotherapy', in F. Himmelweit (ed.) *The Collected Papers of Paul Ehrlich*, Vol. 3, London: 1914, 507.
19 D. Wilson, *Penicillin in Perspective*, London: Faber & Faber, 1976, 278–9.
20 Wilson, op. cit., pp. 278–82; and interviews with John Beale and Norman Finter.
21 Minutes of a meeting to discuss collaboration in a programme of work on interferon, held at the NIMR, 22 April 1959, MRC Archives, File No. S 788/1.
22 See, C. Andrewes, 'Alick Isaacs', *Biographical Memoirs of Fellows of the Royal Society*, 13 (1976), 205–21, p. 215; and interviews with Derek Burke and David Tyrrell.
23 Minutes of a meeting to discuss collaboration in a programme of work on interferon, held at the NIMR, 22 April 1959, MRC Archives, File No. S 788/1.
24 Elkington and Fife, consulting chemists and chartered patent agents, to Alick Isaacs, letters dated 9 March 1959, MRC Archives, File No. A 813/104.
25 The pharmaceutical companies Wellcome, Glaxo and ICI to the MRC and vice versa, letters dated May 1959, MRC Archives, File No. S 788/1.
26 Minutes of a meeting to discuss collaboration in a programme of work on interferon, held at the NIMR, 2 June 1959, MRC Archives, File No. S 788/1.
27 Liebenau, op. cit., p. 83.
28 Unfortunately due to the fact that the company archives have remained closed to outsiders up to this day the knowledge about the wheeling and dealing of the companies and their policies is limited; Liebenau, op. cit., p. 83.
29 Internal note MRC, dated 5 June 1959, MRC Archives, File No. S 788/1.
30 Minutes of a meeting to discuss collaboration in a programme of work on interferon, held at the NIMR, 2 June 1959, MRC Archives, File No. S 788/1.
31 Minutes of a meeting to discuss collaboration in a programme of work on interferon, held at the NIMR, 3 July 1959, MRC Archives, File No. A 812/5/1.
32 The executive body would be formalized in a GNRD Patent Holdings Ltd with a board made up of representatives of each of the collaborating firms and the NRDC, under an independent scientific chairman.
33 Proposed agreement with the NRDC on collaboration with some pharmaceutical firms, dated 10 July 1959, MRC Archives File No. A 812/5/1.

34 Minutes of the 8th meeting of the Scientific Committee on Interferon, dated 4 October 1960, MRC Archives, File No. S 788/2/1.
35 MRC to NRDC, letters, dated 10, 29 July 1959, MRC Archives, File No. A 812/5/1.
36 The formulation of a clause on the publication of research results indeed turned out to be a major obstacle in the negotiations. It would take until April 1961 to reach a formal Collaboration Agreement relating to interferon.
37 Minutes of the 1st and 2nd meeting of the Scientific Committee on interferon, 3, 9 September 1959, MRC Archives File No. S 788/2/1.
38 Arnow, *Health in a Bottle*; W. Breckon, *The Drug Makers*; and interview with Norman Finter.
39 Ibid.
40 Minutes of the 3rd and 4th meeting of the Scientific Committee on Interferon, dated 26 October, 26 November 1959, MRC Archives, File No. S 788/2/1.
41 Minutes of the 6th meeting of the Scientific Committee on Interferon, dated 6 April 1960, MRC Archives File No. 788/2/1.
42 Ibid.
43 The trachoma agent which is now considered to be part of the Chlamydiae family – a group of bacterial organisms – was still referred to as a virus in the early 1960s; interview with a senior MRC official who prefers not to be identified, following the presentation of a preliminary version of this chapter to the Twentieth Medical History Group of the Wellcome Trust, held at the Royal College of Physicians, 11 February 1992.
44 The collaboration agreement was signed on 24 October 1960. Memorandum and Articles of Association of GNRD Patent Holdings Ltd. MRC Archives File No. S 788/6.
45 Minutes of the 6th meeting of the Scientific Committee on Interferon, dated 6 April, 1960, MRC Archives File No. 788/2/1; A. Isaacs, 'Interferon', *Scientific American* 204 (1961), 51–7, p. 57; interviews with John Beale, Derek Burke, Norman Finter, and David Tyrrell.
46 Since its establishment in 1955 as a laboratory specially aimed at exercising laboratory control for the Ministry of Health over the safety of the American (Salk) vaccine against poliomyelitis, the NIMR's Immunological Products Control Laboratory had widened its activities. Other control work for the Ministry had been added to its functions and by the early 1960s it was involved in the safety testing of a large range of virus vaccines and other biologicals. Over the years a lot of knowledge and skill in testing the safety of biologicals had been accumulated. In asking the Hampstead people for advice the Scientific Committee on Interferon hoped to profit from this pooled experience; See, Landsborough Thomson, op. cit., pp. 250–2; and, J. Austoker and L. Bryder, 'The National Institute for Medical Research and related activities of the MRC', op. cit., pp. 56–7.
47 Brian Wynne was first in pointing out the important role of *ad hoc* rules and judgements in developmental practices such as clinical trials in: B. Wynne, 'Unruly technology: practical rules, impractical discourses and public understanding', *Social Studies of Science* 18 (1988), 147–67, p. 162.
48 Minutes of the 9th meeting of the Scientific Committee on Interferon, dated 5 December 1960, MRC Archives File, No. S 788/2/1.
49 Minutes of the 10th meeting of the Scientific Committee on Interferon, dated 16 February 1961, MRC Archives File No. S 788/2/1.
50 Minutes of the 11th meeting of the Scientific Committee on Interferon, dated 19 April 1961, MRC Archives File No. S 788/2/1.

210 *Interferon: The science and selling of a miracle drug*

51 Agreement for Collaboration on Interferon, 3 May 1961, MRC Archives, File No. A 812/5/1.
52 A. Isaacs, 'Interferon', *Scientific American* 204 (1961), 51–7.
53 Minutes of a meeting of Directors of GNRD Patent Holdings Limited, dated 19 June 1961, MRC Archives, File No. A 812/5/1.
54 Internal note MRC, dated 20 June 1961, MRC Archives, File No. A 812/5/1.
55 A. Isaacs, 'Interferon', *Scientific American* 204 (1961), 51–7.
56 Harington to NRDC, letter dated 21 June 1961, MRC Archives, File No. A 812/5/1.
57 Minutes of the 13th meeting of the Scientific Committee on interferon, dated 13 September 1961, MRC Archives, File No. S 788/2/1; Tyrrell, interview; and Finter, interview.
58 A. Isaacs, 'Interferon', *Scientific American* 204 (1961), 51–7; and, S. Baron and A. Isaacs, 'Interferon and natural recovery from virus diseases', *New Scientist* 243 (1961), 81–2; and, Internal note MRC, MRC Archives 9 March 1962, File No. A 742/14.
59 A. Isaacs, 'Interferon: A round unvarnish'd tale', *The Journal of Pharmacy and Pharmacology Supplement* 13 (1961), 57T–61T; and, Editorial, *The Chemist and Druggist*, 30 September 1961, 375.
60 Interferon report by The Wellcome Research Laboratories, December 1961, MRC Archives File No. S 788/2/1.
61 Officially self-testing was and still is frowned upon and condemned as 'sloppy' science: 'How can one possibly maintain objectivity in the reading of results if the experimenter and the human subject who will undergo the preliminary experiment are the same?'. However, in actual laboratory practice it was the normal thing for scientists to try out any new medicines or biomedical procedures on themselves first in order to get an indication as to whether a larger field experiment was warranted. Self-testing still takes place on a much larger scale in our laboratories than anyone dares to acknowledge. Minutes of meeting of the subcommittee on preliminary clinical trials in man at the MRC, Hampstead, 4 December 1961, MRC Archives File No. S 788/2/1.
62 See, for a personal account of the historical development of controlled trials in British medicine: R. Doll, 'Development of controlled trials in preventive and therapeutic medicine', *J. Biosoc. Sci.* 23 (1991), 365–78.
63 This description of the actual controls over new medicines in Britain around 1960 is based on: Breckon, op. cit., pp. 121–41; Landsborough Thomson, op. cit., 226–43; Sneader, op. cit. pp. 75–86; and information gathered during discussions following the presentation of a preliminary version of this chapter to the Twentieth Century Medical History Group of the Wellcome Trust, held at the Royal College of Physicians, 11 February 1992.
64 Minutes of the 15th meeting of the Scientific Committee on Interferon, dated 14 December 1961, MRC Archives File, No. S 788/2/1.
65 Minutes of the 15th meeting of the Scientific Committee on Interferon, dated 14 December 1961, MRC Archives File, No. S 788/2/1.
66 Minutes of the 16th meeting of the Scientific Committee on Interferon, dated 22 February 1962, MRC Archives File, No. S 788/2/2; and Tyrrell, interview.
67 Ibid.
68 Internal Note MRC, dated 4 April 1962, MRC Archives File No. A 813/127.
69 Alick Isaacs to Derek Burke, letter dated 5 April 1962, Derek Burke correspondence, Norwich, personal archives; and Internal Note MRC, dated 4 April 1962, MRC Archives File No. A 813/127.
70 Unfortunately the current editorial staff of the *Lancet* was unable to locate the 1962 *Lancet* editorial files dealing with the publication of the vaccinia trial

report from the Scientific Committee on Interferon; D. Sharp to T. Pieters, letter dated 25 March 1998, T. Pieters, personal archives.
71 Scientific Committee on Interferon. 'Effect of interferon on vaccination in volunteers', *Lancet* i (1962), 873–5, p. 875.
72 A. Isaacs, 'Interferon tried in man', *New Scientist* 285 (1962), 213–14.
73 Interviews with John Beale, Derek Burke, Norman Finter, David Tyrrell and Karl Fantes.
74 Interviews with John Beale, Derek Burke, Karl Fantes, Norman Finter and David Tyrrell.
75 Interview with Maurice Hilleman.
76 Minutes of the 17th meeting of the Scientific Committee on Interferon, dated 18 April 1962, MRC Archives File, No. S 788/2/2.
77 Interviews with Norman Finter and David Tyrrell.
78 Internal note MRC, MRC Achives File No. S 788/2; and MRC to Charles Harington, letter dated 29 May 1962, MRC Achives File No. S 788/2.
79 Minutes of the 16th meeting of the Scientific Committee on Interferon, dated 22 February 1962, MRC Archives File, No. S 788/2/2
80 D. Tyrrell, 'The Common Cold Unit 1946-1990: Farewell to a much-loved British institution', *PHLS Microbiology Digest* 6 (1991), 74–6.
81 In order to maintain a high public profile and attract enough volunteers for its regular common cold studies programme, regular press visits were organized to the Unit's laboratories and spacious centrally-heated volunteer huts; interview with David Tyrrell.
82 This impressionistic account of the 18th meeting of the Scientific Committee on Interferon is based on: Minutes of the 18th meeting of the Scientific Committee on Interferon, dated 2 July 1962, MRC Archives File, No. S 788/2/2; Internal note MRC, 11 July 1962, MRC Archives File, No. S 788/5; Internal note MRC, 16 August 1962, MRC Archives File, No. S788/2/2; I. Gresser, Production of interferon by suspensions of human leucocytes, *P.E.S.B.M.* 108 (1961) 799–803; Ion Gresser to John Enders, letter dated 26 June 1962, Ion Gresser personal archives. And interviews with John Beale, Derek Burke, Norman Finter and David Tyrrell.
83 Minutes of the 19th meeting of the Scientific Committee on Interferon, dated 25 September 1962, MRC Archives File, No. S 788/2/2; D. Tyrrell, Interim Report Clinical Trials with Interferon up to 24 September 1962, MRC Archives File, No. S 788/2/2; and, Internal Note MRC, 26 September 1962, MRC Archives File, No. S 788/2/2.
84 Minutes of the 19th meeting of the Scientific Committee on Interferon, dated 25 September 1962, MRC Archives File, No. S 788/2/2.
85 Peter Fairley, 'Whitehall Men Join Battle on Common Cold', *Evening Standard*, Britain, 5 October 1962.
86 Minutes of the 20th meeting of the Scientific Committee on Interferon, dated 4 December 1962, MRC Archives File, No. S 788/2/2.
87 Interviews with Norman Finter and John Beale.
88 Minutes of the 20th and 21th meeting of the Scientific Committee on Interferon, dated 4 December 1962 and 7 March 1963, MRC Archives File, No. S 788/2/2,3; Internal note MRC, 4 December 1962, MRC Archives File, No. S 788/1; D. Tyrrell, 'Experiments With Interferon', Third Draft of Interim Report to the Medical Research Council by the Scientific Committee on Interferon, MRC Archives File, No. S 788/2/2.
89 Interview with Hilleman.
90 See, for a detailed social-historical description and analysis of the development of biologicals at Merck, Sharp & Dohme between 1895 and 1995: L. Galambos and

J. E. Sewell, *Networks of Innovation*, Cambridge: University of Cambridge Press, 1995.
91 G. Lampson, A. Tytell, M. Nemes and M. Hilleman, 'Purification and characterization of chick embryo interferon', *P.S.E.B.M.* 112 (1963), 468–78.
92 Ibid.
93 Norman Finter to Alick Isaacs, letter dated 9 May 1963, MRC Archives File No. A 812/5/2.
94 Norman Finter to Alick Isaacs, letter dated 14 May 1963, MRC Archives File No. A 812/5/2.
95 Alick Isaacs to Norman Finter, letter dated 10 May 1963, MRC Archives File No. A 812/5/2.
96 Minutes of the 22th meeting of the Scientific Committee on Interferon, dated 21 May 1963, MRC Archives File No. S 788/2/3.
97 I agree with Stephen Hall that 'Isaacs's battle for intellectual normalcy, for a return to the wit and brilliance of his earlier career, is without doubt the saddest chapter of the interferon story'; see S. S. Hall, *A Commotion in the Blood: Life, death, and the immune system*, New York: Henry Holt and Company, 1997, p. 157.
98 When at the end of 1964 word came from the NIH that they were planning to repeat Tyrrell's trial against the common cold, the publication ban was lifted. Tyrrell was able to convince the other members of the Scientific Committee that because they had now further evidence that the British common cold trial had been unsuccessful due to the small dose and speed of removal of interferon inside the nose it would still be worthwhile to report the trial results to the scientific world; Minutes of the 28th meeting of the Scientific Committee on Interferon, dated 10 November 1964, MRC Archives File No. S 788/2/3; and, A Report to the Medical Research Council from the Scientific Committee on Interferon, 'Experiments with Interferon in Man', *Lancet* i (1965), 505–6.
99 Minutes of the 22th meeting of the Scientific Committee on Interferon, dated 21 May 1963, MRC Archives File No. S 788/2/3; and, Internal note MRC, 1 June 1963, MRC Archives File No. S 788/1.
100 Since Isaacs's correspondence got lost and access to Hilleman's correspondence was not granted it was not possible to address properly the interesting question of how both Isaacs and Hilleman developed similar ideas about the use of interferon inducers as an alternative means to exploit interferon's therapeutic potential and to what extent Hilleman's new line of research was related or influenced by Isaacs's work; M. Hilleman, 'Interferon in prospect and perspective', *J. Cell. Comp. Physiol.* 62 (1963), 337–53; Internal note MRC, MRC Archives, File, No. S 788/1; and interview with Maurice Hilleman.
101 Minutes of the 22th meeting of the Scientific Committee on Interferon, dated 21 May 1963, MRC Archives File No. S 788/2/3.
102 Interviews with Karl Fantes, Norman Finter and David Tyrrell.
103 Internal note MRC, 4 July 1963, MRC Archives File No. A 742/14.
104 In 1960 Medawar together with Burnet received the Nobel Prize 'for the discovery of acquired immunological tolerance'; A. M. Silverstein, *A History of Immunology*, San Diego, Academic Press, Inc, 1989, 344–5.
105 Internal note Himsworth, dated 11 July 1963, MRC Archives A 812/5/2.
106 J. Parascandola, 'The theoretical basis of Paul Ehrlich's chemotherapy', *J. Hist. Med.* 36 (1981), 19–43, p. 38; and interviews with John Beale and David Tyrrell.
107 Minute of a special meeting of the Board of G.N.R.D. Patent Holdings Ltd, dated 20 November 1963, MRC Archives File No. A 812/5/2.
108 Breckon, op. cit., p. 43.
109 Minute of a special meeting of the Board of G.N.R.D. Patent Holdings Ltd, dated 20 November 1963, MRC Archives File No. A 812/5/2.

110 Editorial, 'Interferon', *British Medical Journal*, 26 December 1964, 1612–13, p. 1613.
111 Interviews with John Beale, Derek Burke, Norman Finter and David Tyrrell.
112 Minutes of the 25th meeting of the Scientific Committee on Interferon, dated 8 January 1964, MRC Archives File No. S 788/2/3.
113 Helio Pereira to Jean Lindenmann, letter dated 8 January 1964, Jean Lindenmann personal archives; and, Internal note MRC, dated 7 January 1964, MRC Archives File No. A. 812/5/2.
114 Interviews with Robert Friedman, Robert Wagner and Joseph Sonnabend.
115 This very problem was discussed during an informal gathering of 31 interferon researchers at the 1964 meeting of the American Society for Microbiology in Washington, DC. According to Monto Ho they could not and would not produce an inclusive definition of what more precise conditions must be met for a substance to be called interferon; see M. Ho, 'Identification and "Induction" of Interferon', *Bacteriological Reviews* 28 (1964), 367–81, p. 367.
116 The ambiguous status of interferon is nicely illustrated by the following excerpt from Vilcek's 1969 monograph on interferon: 'The other day I had lunch with the Bacterial Geneticist working next door to my lab. While consuming his yoghurt (which he had for dessert) he asked what I was working on all the time. So I told him all about interferon synthesis requiring mRNA to be made thus being an induced protein which induces new mRNA and another protein which by binding to ribosomes inhibits translation of the viral mRNA but not the translation of other mRNA. 'That's very interesting', said the Bacterial Geneticist while finishing his yoghurt, 'but do you really think that interferon exists?'; J. Vilcek, *Interferon*, New York: Springer-Verlag, 1969, 111.
117 *Science* to Dr Jacqueline De Maeyer-Guignard, letter dated 21 January 1965, Edward de Maeyer personal archives; E. De Maeyer, 'Interferon twenty years later', *Bulletin De L'institut Pasteur* 76 (1978), 303–23, p. 304; Edward De Maeyer and Jacqueline De Maeyer-Guignard, interview.
118 Interviews with Sam Baron, Jean Lindenmann, Edward De Maeyer and Jan Vilcek.
119 We can see in Table 1, chapter 2, that the number of scientists working on interferon levelled off between 1964 and 1967.
120 Interview with Ion Gresser.

4 Managing differences

1 A preliminary version of chapter 5 was published in *Studies in History and Philosophy of Science*; T. Pieters, 'Managing differences in biomedical research: The case of standardizing interferons', *Stud. Hist. Phil. Sci.* 29 (1998), 31–79.
2 Quote is taken from statement by Pieter De Somer at the first session of the International Symposium on Standardization of Interferon and Interferon Inducers; see, F. T. Perkins and R. H. Regamey (eds) *Symposia Series in Immunobiological Standardization*, Vol. 14, Basel: S. Karger, 1970, 5.
3 Most studies in the history and sociology of science which deal with standardization as a key dimension of portability and transferability in the biomedical sciences are confined to discussing the role and effects of standardization; see, N. Oudshoorn, *The Making of the Hormonal Body*, Amsterdam: PhD. diss., Univ. Amsterdam: 1991, 72–80; J. H. Fujimura, 'Crafting science: standardized packages, boundary objects, and "translation"', in A. Pickering (ed.) *Science as Practice and Culture*, Chicago: The University of Chicago Press, 1992, 168–211; P. Gossel, 'A need for standard methods: The case of American bacteriology', in A. Clarke and J. Fujimura, *The Right Tools for the Job*, Princeton: Princeton University Press, 1992, 287–311; P. Faasse, *Experiments in Growth*, Amsterdam: PhD. diss., Univ.

Amsterdam: 1994, 92–111. With an exception to Theodore Porter's book *Trust in Numbers* the processes of realizing, maintaining and disseminating biological standards have received relatively little attention from science studies scholars to date; T. M. Porter, *Trust in Numbers*, Princeton: Princeton University Press, 1995, 29–32.
4. The notion of a 'gift culture' was borrowed from Theodore Porter; see, T. M. Porter, *Trust in Numbers*, Princeton: Princeton University Press, 1995, 226.
5. Cantell to Paucker, letter dated 24 September 1962, Cantell personal archives.
6. Interview with Kari Cantell.
7. Ibid.
8. Whenever possible, mostly on return from overseas meetings, biomedical researchers would carry the packages themselves or ask fellow scientists who happened to be in the area of a specific laboratory to pick up a package on their way home. For instance at the end of his stay in Philadelphia on his trip back to Helsinki Cantell took along with him a couple of bottles with virus strains he had worked with in Werner Henle's Department. It was regarded as a practical way to get around problems relating to logistics (customs, export licences, transport costs and damages). In particular during the summer months with a peak in congresses in both Europe and the USA there was a lively traffic of biological specimens between the continents. This observation is based on both correspondence and interviews of the various scientists under survey in this book.
9. Interviews with Kari Cantell and Ion Gresser.
10. Kari Cantell to Kurt Paucker, letter dated 23 October 1962, Cantell personal archives.
11. Interview with Kari Cantell.
12. Kurt Paucker to Kari Cantell, letter dated 3 November 1964, Cantell personal archives.
13. Interview with Kari Cantell.
14. Kari Cantell to Thomas Merigan, letter dated 30 October 1965, Cantell personal archives.
15. Kari Cantell to Thomas Merigan, letters dated 30 October 1965 and 11 July 1966, Cantell personal archives.
16. Thomas Merigan to Kari Cantell, letter dated 11 July 1966, Cantell personal archives.
17. Ibid.
18. Ibid.
19. Thomas Merigan to Kari Cantell, letter dated 17 November 1966, Cantell personal archives.
20. Interview with Thomas Merigan.
21. Ibid.
22. Jean Lindenmann to Alick Isaacs, letter dated 19 May 1964, Lindenmann personal archives.
23. Interview with Norman Finter.
24. Interviews with Sam Baron, Monto Ho, Thomas Merigan and George Galasso.
25. Ibid.
26. D. E. Green, 'An experiment in communication: The information exchange group', *Science* 143 (1964), 308–9; Anonymous, 'Information exchange group', *Nature* 204 (1964), 627; interviews with Sam Baron and George Galasso.
27. Interview with Sam Baron.
28. Minutes of the 28th meeting of the Scientific Committee on Interferon, dated 10 November 1964, MRC Archives File No. S788/2/3.
29. Internal note MRC, dated 10 November 1964, MRC Archives File No. S788/2/3.

30 Minutes GNRD Patent Holdings Ltd meeting, 10 December 1964, MRC Archives File No. A812/8/2; and Internal note MRC, dated 10 December 1964, MRC Archives File No. S788/9.
31 Sonnabend to Medawar, letter dated 9 November 1964, MRC Archives File No. S788/9; Medawar to NRDC, letter dated 9 November 1964, MRC Archives File No. S788/9; NRDC to Medawar, letter dated 10 November 1964, MRC Archives File No. S788/9; and Minutes GNRD Patent Holdings Ltd meeting, 10 December 1964, MRC Archives File No. A812/8/2.
32 IEG 6 membership list, dated 18 January 1965, Kari Cantell personal archives.
33 Sam Baron to members of IEG 6, letter dated 2 November 1966, Ion Gresser personal archives.
34 IEG 6, communications 1–220, Kari Cantell personal archives.
35 Interview with Sam Baron.
36 G. E. Wolstenholme and M. O'Connor (eds) *Interferon* (1967, *Ciba* Foundation Symposium), London: J & A Churchill, 1968, 260–1.
37 Minutes of the 26th, 27th, 28th and 29th meeting of the Scientific Committee on Interferon, dated 21 April, 7 July, 10 November 1964 and 25 January 1965, MRC Archives File No. S788/2/3.
38 Minutes of the 30th and 31th meeting of the Scientific Committee on Interferon, dated 4 May and 21 September 1965, MRC Archives File No. S788/2/4; and, Internal notes MRC, dated 5 May and 29 September 1965, MRC Archives File No. 788/2/4.
39 W. J. Kleinsmidt, J. C. Cline and E. B. Murphy, 'Interferon Production Induced by Statolon', *Proc. Natl. Acad. Sci. U.S.A.* 52 (1964), 741–4.
40 Professor Ernst Chain, who had received the Nobel Prize for the discovery of penicillin and its therapeutic effects in various infectious diseases, would not wait for Finter's assessments. On hearing from Isaacs that the interferon inducing substance Statolon was produced by a penicillium mould he immediately decided to start looking for other active interferon inducing agents in penicillium moulds, which were available in abundance in his laboratory at the Imperial College of Science and Technology in London. Both Chain and Isaacs hoped that the penicillium molds might lead the way into the field of chemotherapy against viruses just as they did into chemotherapy against bacteria. Minutes of the 30th and 31th meeting of the Scientific Committee on Interferon, dated 4 May and 21 September 1965, MRC Archives File No. S788/2/4; and, Internal notes MRC, dated 5 May and 29 September 1965, MRC Archives File No. 788/2/4; Minutes of the Scientific Steering Committee, dated 9 March, 1965, MRC Archives File No. S788/8/1; and interview with Norman Finter.
41 Unbeknownst to his colleagues, he attempted to take his own life on at least two occasions; interview with Susanna Isaacs-Elmhirst.
42 Internal note MRC, dated 14 October 1966, MRC Archives File No. S788/2/4; and N. B. Finter, *Interferons*, Amsterdam: North Holland Publishing Company, 1966, 264.
43 Minutes of the 34th meeting of the Scientific Committee on Interferon, dated 20 October 1966, MRC Archives File No. S788/2/4.
44 Minutes of the 34th meeting of the Scientific Committee on Interferon, dated 20 October 1966, MRC Archives File No. S788/2/4; and, Internal note MRC, dated 27 October, 27 October 1966, MRC Archives File No. S788/2/4; N. B. Finter, 'Interferon as an antiviral agent *in vivo*: Quantative and temporal aspects of the protection of mice against Semliki forest virus', *Brit. J. Exp. Pathol.* 47 (1966), 361–70.
45 Isaacs to Chany, draft letter, dated 27 October 1966, MRC Archives File No. S788/2/4.

46 Minutes of the 34th meeting of the Scientific Committee on Interferon, dated 20 October 1966, MRC Archives File No. S788/2/4; and, Internal note MRC, dated 27 October, 27 October 1966, MRC Archives File No. S788/2/4.
47 E. R. Falcoff, R. Falcoff, F. Fournier and C. Chany, 'Production en masse, purification partielle et caractérisation d'un interféron destiné a des essais therapeutiques humains', *Ann. Inst. Pasteur* 111 (1966), 562–84; A. K. Field, A. A. Tytell, G. P. Lampson, G. P. and M. Hilleman, 'Inducers of interferon and host resistance: II. Multistranded synthetic polynucleotide complexes', *Proc. Nat. Acad. Sci. U.S.A.* 58 (1967), 1004–10; W. Regelson, 'Prevention and treatment of Friend Leukemia Virus (FLV) infection by interferon-inducing synthetic polyanions', *Adv. Exp. Med. Biol.* 1 (1967), 316–32; and interviews with Kari Cantell, Maurice Hilleman and Thomas Merigan.
48 While it had been agreed in 1965 under pressure of the three pharmaceutical companies, who preferred to work independently in the area of interferon inducers, to leave this topic outside the Interferon collaboration, some MRC officers could be heard saying that it might be more advantageous to join forces on the study of inducers too; Internal note MRC, dated 11 September 1967, MRC Archives, File No. A812/17/1.
49 Tyrrell had been de facto chairman of the Scientific Committee on Interferon for the past few years and in practice Isaacs's scientific assistant, the virologist, Joseph Sonnabend, ran Isaacs's Laboratory for Research on Interferon at the NIMR. Medawar, the director of the NIMR, who had been thinking of disbanding the unproductive Laboratory for Research on Interferon for quite some time, seized the opportunity to reassimilate the Interferon Laboratory into what had been renamed the 'Division of Virology'. Medawar met hardly any opposition as he convinced worried MRC officers that this would not mean that all work at the NIMR on interferon would come to an end. On the contrary, he assured them that the NIMR's efforts in the field of interferon research would be continued 'at not less than its present intensity'. Medawar enforced his claim by pointing out that the most promising research work on interferon at the NIMR, studying how interferon acted at the molecular level, was currently performed by one of the staff members of the Biochemistry Division, the biochemist, Martin Kerr. Internal Note MRC dated 9 February 1967, MRC Archives, File No. A812/8/2; The MRC to Sir Peter Medawar, letter dated 14 February 1967, MRC Archives, File No. S788/13; and Sir Peter Medawar to the MRC, letter dated 17 February 1967, MRC Archives, File No. A812/8/2.
50 Interviews with Sam Baron, Kari Cantell, Norman Finter, Robert Friedman, Ion Gresser, Maurice Hilleman and Jan Vilcek.
51 The London meeting held on 19–21 April 1967, which was organized on behalf of the Ciba Foundation, was the first meeting after Smolinice completely devoted to the subject of interferon. With the idea of getting some new input from an expert outsider, Francis Crick was specially invited as a participant to the conference; see, G. E. Wolstenholme and M. O'Connor (eds, 1967, April, *London* Symposium) *Interferon*, London: J & A Churchill Ltd, 1968, 71–4.
52 Minutes of the 36th meeting of the Scientific Committee on Interferon, dated 28 June 1967, MRC Archives, No. S788/2/5; and interview with Norman Finter.
53 Gray Anderson to Kari Cantell, letter dated 19 September 1967, Cantell personal archives.
54 Minutes of the 36th meeting of the Scientific Committee on Interferon, dated 28 June 1967, MRC Archives, No. S788/2/5.
55 Minutes of the 36th meeting of the Scientific Committee on Interferon, dated 28 June 1967, MRC Archives, No. S788/2/5.

56 Minute of the 35th meeting of the Scientific Committee on Interferon, dated 3 March 1967, MRC Archives File No. S788/2/5.
57 Interim Report on Standards for Human, Mouse and Chick Embryo Interferons, Division of Biological Standards, dated September 1967, MRC Archives File No. S 788/2/5.
58 Minutes of the 36th meeting of the Scientific Committee on Interferon, dated 28 June 1967, MRC Archives File No. S788/2/5.
59 This impressionistic account of Finter's and Perkins' visit to Cantell is based on: N. Finter to J. Lindenmann, letters dated 7, 15, 28, 31 July and 11 August 1967, K. Cantell correspondence, personal archives; K. Cantell to N. Finter, letters dated 15 and 25 July and 10 August 1967, K. Cantell correspondence, personal archives; Minutes of the 36th meeting of the Scientific Committee on Interferon, dated 28 June 1967, MRC Archives File No. S788/2/5; Report of a visit made by Dr N. B. Finter and Dr F. T. Perkins to Helsinki to see Dr K. Cantell, Dr H. R. Nevanlinna and Dr H. Strander, MRC Archives File No. S788/2/5; H. Strander and K. Cantell, 'Production of interferon by human leukocytes in vitro', *Ann. Med. Exp. Fenn* 44 (1966), 265–73; K. Cantell, H. Strander, Gy. Hadhazy and H. R. Nevanlinna, 'How much interferon can be prepared in human leucocyte suspensions', in G. Rita (ed.) *The Interferons*, New York: Academic Press, 1968, 223–32; interviews with Kari Cantell, Norman Finter and Hans Strander.
60 In experimenting with a large number of different sorts of viruses from his laboratory stock Cantell figured out that a rare Finnish strain of Sendai virus in the back of his laboratory freezer was the best inducer of interferon.
61 If they spinned blood in the centrifuge, they were left with three layers. Formerly after removal of the top layer consisting of a thick, straw-coloured liquid known as plasma or 'serum' they got rid of the intermediate almost colourless layer of leucocytes, the so-called 'buffy coat', whereafter the bags with the remaining red cells were stored for further use. Instead of discarding all buffy coats they now routinely collected the leucocytes.
62 Interviews with John Beale, Karl Fantes and Norman Finter.
63 Internal notes MRC dated 9, 16 and 19 February 1968, MRC Archives File No A 812/5/3.
64 The minority of mostly European 'interferon guys' had a hard time defending themselves against the powerful American alliance of inducer supporters headed by the charismatic Maurice Hilleman. Ion Gresser to Sidney Farber, letter dated 17 January 1969, Gresser personal archives; and interviews with Kari Cantell, Ion Gresser, Maurice Hilleman, Monto Ho and Jan Vilcek.
65 Kari Cantell to Gray Anderson, letter dated 20 January 1969, Cantell personal archives; and Gray Anderson to Kari Cantell, letter dated 4 February 1969, Cantell personal archives.
66 Minutes of the 10th meeting of the Human Interferon Clinical Trials Working Party, dated 25 February 1969, MRC Archives File No. S788/15/2.
67 D. G. Evans, 'Conference Address', in F. T. Perkins and R. H. Regamey (eds) *Symposia Series in Immunobiological Standardization*, Basel: S. Karger, vol. 14, 1970, 2.
68 F. T. Perkins and R. H. Regamey (eds) *Symposia Series in Immunobiological Standardization*, Basel: S. Karger, vol. 14, 1970, 280–1
69 F. T. Perkins and R. H. Regamey (eds) *Symposia Series in Immunobiological Standardization*, Basel: S. Karger, vol. 14, 1970, 297.
70 F. T. Perkins and R. H. Regamey (eds) *Symposia Series in Immunobiological Standardization*, Basel: S. Karger, vol. 14, 1970, 291–5.
71 F. T. Perkins and R. H. Regamey (eds) *Symposia Series in Immunobiological Standardization*, Basel: S. Karger, vol. 14, 1970, 311.

218 *Interferon: The science and selling of a miracle drug*

72 Frank Perkins to Kari Cantell, letter dated 18 November 1969, Cantell personal archives.
73 Interview with Charles Chany.
74 First draft of the Minutes of the 14th meeting of the Human Interferon Working Party, dated 25 November 1969, MRC Archives File No. S788/15/2; Frank Perkins to Kari Cantell, letter dated 28 November 1969, Cantell personal archives; and Frank Perkins to Charles Chany, letter dated 28 November 1969, MRC Archives File No. S788/15 /2.
75 Charles Chany to Gray Anderson, letter dated 22 December 1970, Cantell personal archives.
76 Interview with Charles Chany.
77 Charles Chany to Gray Anderson, letters dated 22 December 1970 and 7 May 1971, Cantell personal archives.
78 George Galasso to David Tyrrell, letter dated 24 December 1970, MRC Archives File No. S788/15/2.
79 Gray Anderson to Charles Chany, letters dated 19 April and 12 May 1971, MRC Archives File No. S788/15/2; and interview with Charles Chany.
80 George Galasso to Gray Anderson, letter dated 11 February 1972, MRC Archives File No. S788/15/2.
81 George Galasso to Gray Anderson, letters dated 11 February and 12 April 1972, MRC Archives File No. S788/15/2; Gray Anderson to George Galasso, letter dated 15 March 1972, MRC Archives File No. S788/15/2.
82 With reference to this standard the poor results of the latest British common cold trial, which was performed in March 1970, were claimed to be due to the use of insufficient interferon, to the inadequate dosage and to an unexplained ten-fold decline in biological activity between the time of preparation and administration. This explanation could not prevent the respiratory trial from leaving its marks on the Interferon Collaboration per se. It played an instrumental role in Glaxo's and ICI's decision to abandon work on interferon, leaving only Burroughs Wellcome and the MRC in interferon research; Minute of the 17th meeting of the Human Interferon Clinical Trials Working Party, dated 14 April 1970, MRC Archives File No. S788/15/2.
83 Whereas the formal collaboration on interferon terminated on 2 May 1973, all parties agreed that the GNRD as the Patent Holdings Company should be continued in order to protect the property held – e.g. US interferon patent; Internal note MRC, dated 13 May 1971, MRC Archives File No. A 812/8 (ii); and, Minutes of the 46th meeting of the Scientific Committee on Interferon, dated 22 February 1973, MRC Archives File No. S788/2/6; Internal note MRC, 22 April 1973, MRC Archives File A812/2/1.
84 For instance, in Paris, upon receiving a sample of mouse interferon from Bill Stewart, who temporarily worked as a post-doctoral worker at the Rega Institute in Louvain, Belgium, Ion Gresser immediately wrote him a note to ask him whether the units indicated on the label were NIH units. 'The titer I gave you was in PDD-VSV units, which was about 1/5 of an NIH unit in Delaware but seems to be about the same (within 2-fold) as an NIH unit here in Leuven'; Ion Gresser to William E. Stewart II, letter dated 28 February 1972, Gresser personal archives; and William E. Stewart II to Ion Gresser, letter dated 2 March 1972, Gresser personal archives.
85 Interviews with Sam Baron, Norman Finter, Robert Friedman, Ion Gresser and Joseph Sonnabend. In 1965 Frederick Wheelock of the Western Reserve University in Ohio had published a highly controversial article in *Science* claiming that he had identified a new interferon-like inhibitor in cultures of human white blood cells, thus implicating that not only different species produced different interferons

but also that the same species produced different types of interferon. Without proper standards he had difficulty in substantiating this. Wheelock had to wait a couple of years before his study was taken seriously in the field of interferon research; E. F. Wheelock, 'Interferon-like virus-inhibitor induced in human leukocytes by phytohemagglutinin', *Science* 149 (1965), 319–21; and interview with Frederick Wheelock.

86 At the workshop all the recommendations of the 1969 Conference were reinforced by recommending the same set of research standards for acceptance as WHO International Reference Standards Internal memorandum on the occasion of an International Workshop on Interferon Standards held at Woodstock, Illinois, USA, dated September 1978, MRC Archives File No. S805/7.

5 About mice, malignancies and experimental therapies

1 A preliminary version of this chapter was published in: T. Pieters. 'History of the development of the interferons: from test-tube to patient', in R. Stuart-Harris and R. Penny (eds) *The Clinical Applications of Interferons*, London: Chapman & Hall, 1997.
2 R. A. Rettig, *Cancer Crusade*, Princeton: Princeton University Press, 1977, 18.
3 Rettig, op. cit., pp. 77–115, 281–315; and interview with Mathilde Krim.
4 Panem, op. cit., p. 16; and interview with Mathilde Krim.
5 S. Panem, *The Interferon Crusade*, Washington, DC: Brookings Institution, 1984, 16; and interview with Mathilde Krim.
6 P. Atanasiu and C. Chany, 'Action d'un interféron provenant de cellules malignes sur l'infection expérimental du hamster nouveau-né par le virus du polyoma', *Comptes Rendus Hebdomadaires Séances de L'Académie de Sciences* 251 (1960), 1687–9; and, C. Chany, 'An interferon-like inhibitor of viral multiplication from malignant cells (the viral autoinhibition phenomena)', *Virology* 13 (1961), 485–92.
7 J. Bader, 'Production of interferon by chick embryo cells exposed to Rous sarcoma virus', *Virology* 16 (1962), 436–43; R. Friedman, A. S. Rabson, W. Kirkham, 'Variation in interferon production by polyoma virus strains differing oncogenicity', *Proc. Soc. Exp. Biol. Med.* 112 (1963), 347–51; 'NCI scientists publish data on interferon as cancer-inhibitor', *The NIH Record* XVI, 17 (1964), 8. See, the session on viruses and interferon in *Viruses, Nucleic Acids and Cancer* [A Collection of Papers Presented at the Seventeenth Annual Symposium on Fundamental Cancer Research, 1963), Baltimore: Williams & Wilkins Co, 1963, 429–84.
8 J. D. Almeida, R. C. Hasselback and A. W. Ham, 'Virus-like particles in blood of two acute leukaemia patients', *Science* 142 (1963) 1487–9; and, E. J. Freireich and E. Frei, III, 'Recent advances in acute leukaemia', in L. M. Tocantins (ed.) *Progress in Hematology*, Vol. IV, New York: Grune and Stratton, 1964, 187–202.
9 Not until the late 1950s did research into the possible viral aetiology of cancer gain more respectability. For more than forty years the biomedical research community had persisted in the belief that mammalian cancer was not contagious, thereby relegating occasional claims about viruses that could cause tumours in mammals to the fringes of scientific research. The following developments played an important role in the change in attitude towards the idea of viruses causing cancer. First, after a virtual standstill in the reporting of viruses that were thought to cause cancer in mammals, a number of scientists from highly regarded research centers like the NIH came up with claims about the isolation of viruses that caused leukaemia and other cancers in laboratory animals like the rabbit, mice and rat. Second, virology as a whole had grown to maturity and was

increasingly recognized as a major scientific discipline. The isolation of the polio virus and the successful production of effective vaccines against the widely feared disease polio should also be mentioned. This medical breakthrough earned virologists a lot of credit both in and outside the scientific community. Third, it was noticed that prominent virus researchers like Jonas Salk, who played a major role in the development of a vaccine against polio virus, joined the club of supporters of the theory on the viral aetiology of cancer. Hence, by the early 1960s the idea that viruses could cause cancers at least in animals had gained a firm foothold; R.E. Shope, 'Evolutionary episodes in the concept of viral oncogenesis', *Perspectives on Biological Medicine* 9 (1966), 258–74; M. B. Shimkin, *Contrary to Nature*, Washington DC: US Department of Health, Education, and Welfare, National Institutes of Health, 1979, 213–23; K. Studer and D. Chubin, *The Cancer Mission: Social Contexts of Biomedical Research*, Beverly Hills: Sage Publications, 1980, 19–21; J. Patterson, *The Dread Disease*, Cambridge (MA): Harvard University Press, 1987, 59, 98, 186.

10 The reason to give up a potential bright research career at the world-famous Harvard Medical School for a relatively uncertain living in what was known as the French biomedical 'snake pit' was private in nature. During his one-year stay in 1961 as a post-doctoral research fellow at Charles Chany's virology laboratory at Hôpital St-Vincent-De-Paul in Paris he had fallen in love with both the city and a particular Parisian girl; interview with Ion Gresser.

11 I. Gresser to S. Farber, progress report period June–December 1965, dated 10 December 1965, I. Gresser personal archives; interview with Ion Gresser.

12 He only knew of one recent American study by the virologist Frederick Wheelock in the *New England Journal of Medicine* claiming that repeated administration of viruses to a patient with acute leukaemia had a temporary inhibitory effect on the disease process; F. E. Wheelock and J. H. Dingle, 'Observations on the repeated administration of viruses to a patient with acute leukaemia', *New. Eng. J. Med.* 271 (1964), 645–51.

13 R. R. Wagner, 'Interferon; A review and analysis of recent observations'. *American Journal of Medicine* 38 (1965), 726–37.

14 I. Gresser to S. Farber, letter dated 10 December 1965, Gresser correspondence, personal archives.

15 According to Gresser, for one or the other reason Farber treated him as his favourite adventurous son; Gresser, interview; and, Gresser to C. Chany, letter dated 19 March 1965, I. Gresser correspondence, personal archives.

16 N. Finter, 'A rich source of mouse interferon', *Nature* 204 (1964), 1114–15; N. Finter, 'Protection of mice by interferon against systemic virus infections', *Br. Med. J.* ii (1964), 981–5.

17 In 1963 in an desperate effort to save the life of his best friend's daughter, who suffered from acute leukaemia, Chany – as a strong believer in the virus aetiology of cancer – treated the girl with crude human interferon preparations. Since the patient with about 6 months life expectancy lived for another 1.5 years, he believed that the administration of interferon might have been beneficial; interviews with Charles Chany, Ernesto Falcoff and Ion Gresser.

18 E. Falcoff, R. Falcoff, F. Fournier and C. Chany, 'Production en masse, purification partielle et caractérisation d'un interferon destiné a des essais thérapeutiques humains', *Annales de l'Institut Pasteur* 5 (1966), 562–84; and interviews with Charles Chany, Ernesto Falcoff and Ion Gresser.

19 I. Gresser to N. Finter, letter dated 8 September 1965, Gresser correspondence, personal archives; N. Finter to I. Gresser, letter dated 14 September 1965, Gresser correspondence, personal archives; I. Gresser to S. Farber, progress report period

June-December 1965, dated 10 December 1965, I. Gresser personal archives; interviews with Norman Finter and Ion Gresser.
20 I. Gresser to S. Farber, progress report period June-December 1965, dated 10 December 1965, I. Gresser personal archives.
21 I. Gresser to S. Farber, progress report period December 1965 to June 1966, dated 2 September 1966, I. Gresser personal archives.
22 I. Gresser, J. Coppey, E. Falcoff and D. Fontaine, 'Action inhibitrice de l'interféron brut sur le développement de la leucémie de Friend chez la Souris', *C.R. Acad. Sci. Paris* 263 (1966), 586–8; I. Gresser, J. Coppey, E. Falcoff and D. Fontaine, 'Interferon and murine leukaemia. I. Inhibitory effect of interferon preparations on the development of friend leukaemia in mice', *P.S.E.B.M.* 124 (1966), 84–91; and, I. Gresser, D. Fontaine, J. Coppey, R. Falcoff and E. Falcoff, 'Interferon and murine leukaemia. II. Factors related to the inhibitory effect of interferon preparations on the development of Friend leukaemia in mice', *P.S.E.B.M.* 124 (1966), 91–4; and interview with Ion Gresser.
23 I. Gresser, J. Coppey, J. Falcoff and D. Fontaine, 'Action inhibitrice de l'interféron brut dur le développement de la leucémie de Friend Chez la Souris', *C.R. Acad. Sci. Paris* 263 (1966) 586–8; I. Gresser, J. Coppey, D. Fontaine-Brouty-Boyé, R. Falcoff, 'Interferon and murine leukaemia. III: Efficacy of interferon preparations administered after inoculation of friend virus', *Nature* 215 (1967), 174–5; I. Gresser, R. Falcoff, D. Fontaine-Brouty-Boyé, F. Zajdela, J. Coppey and E. Falcoff, 'Interferon and murine leukaemia. IV. Further studies on the efficacy of interferon preparations administered after inoculation of Friend virus', *P.S.E.B.M.* 126 (1967), 791–7; and, I. Gresser to S. Farber, letter dated 19 January 1967, I. Gresser correspondence, personal archives.
24 Interviews with Charles Chany, Ion Gresser and Ernesto Falcoff.
25 I. Gressser *et al.* (1967, April, Ciba Foundation Symposium) 'The effect of interferon preparations on Friend leukaemia in mice', in G. E. Wolstenholme and M. O'Connor (eds) *Interferon*, London: J & A Churchill Ltd, 1967, 240–8; Interviews with Sam Baron, Edward De Maeyer, Robert Friedman, Joseph Sonnabend and Robert Wagner.
26 M. VandePutte, J. Delafonteyne, J. Billeau and P de Somer, 'Influence and production of interferon in Rauscher virus infected mice', *Arch. Ges. Virusforsch.* 20 (1967), 235–45; and, E. F. Wheelock, 'Effect of Statolon on Friend virus leukaemia in mice', *P.S.E. B.M.* 124 (1967), 855–8; E. F. Wheelock and R. P. B. Larke, 'Efficacy of interferon in the treatment of mice with established Friend virus leukaemia', *P.S.E.B.M.* 127 (1968), 230–8. Interviews with Ion Gresser and Frederick Wheelock.
27 I. Gresser to J. Enders, letter dated 18 January 1968, I. Gresser correspondence, personal archives.
28 I. Gresser, C. Bourali, J. P. Lévy, D. Fontaine-Brouty-Boyé and M. Thomas, 'Cancérologie. Prolongation de la survie des souris inoculées avec des cellules tumourales et traitées avec des préparations d'interféron', *C.R. Acad. Sci.* 268 (1969), 994–7; I. Gresser, C. Bourali, J. P. Lévy, D. Fontaine-Brouty-Boyé and M. Thomas, 'Increased survival in mice inoculated with tumour cells and treated with interferon preparations', *Proc. Natl. Acad. Sci. USA* 63 (1969), 51–7; I. Gresser to F. Farber, progress report, period January-June 1968 (dated 1 July 1968), Gresser personal archives; and, interview with Ion Gresser.
29 Ibid.
30 Ibid.
31 I. Gresser, 'Prolongation de la survie des souris inoculées avec des cellules tumourales et traitées avec des préparation brutes d'interféron', *Symposium*

International sur L'interféron, Lyon, January 1969. Interviews with Ion Gresser, Kari Cantell, Charles Chany, Sam Baron and Robert Friedman.

32 I. Gresser, 'Prolongation de la survie des souris inoculées avec des cellules tumourales et traitées avec des préparation butes d'interferon', *Symposium International sur L'interféron*, Lyon, January 1969; and, H. Levy, L. Law and A. Rabson, 'Inhibition of tumour growth by polyinosinic-polycytidylic acid', *Proc. Nat. Acad. Sci. U.S.A.* 62 (1996), 357–63. Interviews with Hilton Levy, Ion Gresser and Kari Cantell.

33 'Anti-tumour action of chemical on mice suggests possible new cancer weapon', *The NIH Record* XXI, 2 (1969), 1, 7; and, I. Gresser to S. Farber, letter dated 17 January 1969, Ion Gresser correspondence, personal archives.

34 I. Gresser to S. Farber, letter dated 17 January 1969, Ion Gresser correspondence, personal archives; M. Hilleman, 'Double-stranded RNAs (PolyI:C) in the prevention of viral infections', *Arch. Intern. Med.* 126 (1970), 109–24. Interviews with Hilton Levy, Kari Cantell and Ion Gresser.

35 Interview with Kari Cantell.

36 'NIAID study seeks to explain role of interferon system in viral resistance', *The NIH Record* XX, 21 (1968), 1, 7.

37 K. Cantell to G. Galasso, dated 15 September 1969, Cantell personal archives; and, K. Cantell to K. Paucker, letter dated 12 November 1969, Cantell personal archives.

38 K. Paucker to K. Cantell, letter dated 12 August 1969, Cantell personal archives; and, K. Cantell to K. Paucker, letter dated 12 November 1969, Cantell personal archives; Cantell, interview; and Finter, interview. As soon as the British found out that the Americans were going to pay Cantell for his services they offered payment too: 'I am pleased to advise you that the Council has agreed to reimburse the institute the costs of the salaries, materials and equipment detailed in your letter (3390 pounds). For administrative reasons the scientific equipment, i.e. large capacity centrifuge will have to be regarded as property of the Council on loan for an unspecified period'; MRC to K. Cantell, letter dated 10 February 1970, MRC Archives File No. S788/15/2.

39 K. Cantell to K. Fantes, letter dated 12 February 1969, K. Cantell personal archives.

40 K. Cantell to K. Paucker, letter dated 12 November 1969, Cantell personal archives; K. Cantell to I. Gresser, letter dated 19 November 1969, Gresser personal archives; and K. Cantell, *The Story of Interferon: The Ups and Downs in the Life of a Scientist*, Singapore: World Scientific, 1998.

41 K. Cantell to K. Fantes, letter dated 27 February 1970, Cantell personal archives.

42 K. Cantell to K. Fantes, letter dated 25 May 1970, Cantell personal archives.

43 K. Cantell to N. Finter, letter dated 25 May 1970, Cantell personal archives; H. Strander to K. Cantell, letter dated 7 July 1970, Cantell personal archives.

44 Interviews with Kari Cantell and Hans Strander.

45 Interviews with Hans Strander and Ernest Borden.

46 K. Cantell to H. Strander, letter dated 12 September 1970, Cantell personal archives.

47 One of the researchers at the Tumor Institute claimed to have developed an enzyme test which could be used to identify virus-related cancers. Strander's selection of four cancers had all tested positive; H. Strander to K. Cantell, letter dated 2 December 1970, Cantell personal archives.

48 K. Cantell to I. Gresser, letter dated 19 February 1971, Cantell personal archives; interview with Strander.

49 H. Strander, K. Cantell, G. Carlström, and P. Jakobsson, 'Clinical and laboratory investigations on man: systemic administration of potent interferon in man', *J. Nat. Cancer. Inst.* 51 (1973), 733–42; and interview with K. Cantell.
50 Ibid.
51 K. Cantell to T. Merigan, letter dated 10 June 1971, Cantell personal archives.
52 H. Strander, K. Cantell, G. Carlström and P. Jakobsson, 'Clinical and laboratory investigations on man: Systemic administration of potent interferon in man', *J. Nat. Cancer. Inst.* 51 (1973), 733–42; and interview with K. Cantell.
53 K. Cantell to K. Kato, letter dated 6 September 1971, Cantell personal archives; H. Strander to K. Cantell, 2 September 1971, Cantell personal archives.
54 K. Cantell to N. Finter, letter dated 21 June 1971, Cantell personal archives.
55 K. Cantell to H. Strander, letter dated 29 december 1971, Cantell personal archives.
56 I. Gresser to H. Strander, letter dated 7 February 1972, I. Gresser and K. Cantell personal archives.
57 H. Strander to K. Cantell, letters dated 26 March and 4 July 1972, K. Cantell personal archives.
58 H. Strander, K. Cantell, G. Carlström and P. Jakobsson, 'Clinical and laboratory investigations on man: systemic administration of potent interferon in man', *J. Nat. Cancer. Inst.* 51 (1973), 733–42; I. Gresser to E. Frei III, letter dated 6 November 1972, I. Gresser personal archives; H. Strander to I. Gresser, letter dated 3 January 1973, I. Gresser personal archives.
59 H. Strander to I. Gresser, letter dated 15 November 1972, I. Gresser personal archives.
60 M. Krim to I. Gresser, letter dated 30 November 1971, I. Gresser personal archives.
61 In the fiscal year 1970, 22 contract were awarded, totalling 868.811 dollars. Evaluatory report of the Antiviral Substances Program, September 1976, NIAID Archives, DMID-ARB box #14; and, 'NIAID initiates antiviral drug program; awards contracts for interferon study', *The NIH Record* XXII, 18 (1970), 1, 4.
62 R. A. Rettig, *Cancer Crusade*, Princeton: Princeton University, 1977, 69.
63 S. E. Grossberg, 'The interferons and their inducers: Molecular and therapeutic considerations', *New. Engl. J. Med.* 287(2) (1972), 79–85.
64 R. W. Moss, *The Cancer Industry*, New York: Paragon House, 1989, 119–31. Interviews with Mathilde Krim, Joseph Sonnabend, Arthur Levine and Thomas Merigan.
65 During a visit to Russia in the summer of 1971 Galasso had made the surprising discovery that human leucocyte interferon was already available in Moscow area pharmacies for use as a nasal spray against influenza. This encouraged him in funding a number of clinical trials to test interferon as a potential antiviral drug. For instance, the NIAID financially supported Tom Merigan's and David Tyrrell's collaborative effort to test interferon on respiratory viruses in volunteers at the Common Cold Unit in Salisbury (UK). In 1973 the latter attempt would result in claims in both the *New York Times* and *Lancet* that interferon was effective against the common cold; 'Three US scientists visit Soviet Union; find antiviral research programs differ', *The NIH Record* XXIII, 18 (1971), 1, 4; Minutes Antiviral Substances Workshop, NIH, dated 1 December 1971, NIAID archives; 'Substance in body said to block colds', *New York Times*, 24 March 1973; 'Scientists report interferon – Naturally occurring antiviral protein – Effective against common cold', The NIH record, 24 April 1973, p. 8; T. Merigan, S. Reed, T. Hall and D. Tyrrell, 'Inhibition of respiratory virus infection by locally applied interferon', *Lancet* i (1973), 563–7; T. Merigan to J. Gold

(SKF), letter dated 4 September 1972, Tyrrell personal archives; and interview with George Galasso.
66 S. Baron, 'The defensive and biological roles of the interferon system', in N. Finter (ed.) *Interferon and Interferon Inducers*, Amsterdam: North-Holland Publishing Company, 1973, 267–92; E. Richards, *Vitamin C and Cancer: Medicine or Politics?*, New York: St Martin's Press, 1991, 21–2; and interviews with Sam Baron, George Galasso and Joseph Sonnabend.
67 M. Krim to I. Gresser, letter dated 20 March 1974, Gresser personal archives; and interviews with Kari Cantell and Mathilde Krim.
68 Interview with Mathilde Krim; I. Gresser to S. Farber, letter dated 7 November 1969, Gresser personal archives.
69 M. Krim to I. Gresser, letter dated 20 March 1974, Gresser personal archives; Krim, interview; and Gresser, interview. Krim and Gresser met in the fall of 1971, when Krim was in the process of interesting the NIH in supporting work on the production of interferon for clinical trials in cancer patients; M. Krim to I. Gresser, letter dated 30 November 1971, Gresser personal archives.
70 I. Gresser (unpublished manuscript, 1973), 'Cancer and immunology or the resurrection of Sir Colenso Ridgeon'; Gresser loved to quote his assistant Pernilla Lindahl who defined interferon as a substance 'that does what you want it to do'; I. Gresser, 'Interferon therapy: obvious and not so obvious applications', *Acta Medica Scandinavica* 197 (1975), 49–53, 52: and interviews with Ion Gresser and Mathilde Krim.
71 From the 1920s up to the 1960s immunologists would describe their specialty in terms of the study of specific defence mechanisms. The concept of antibody-related specificity as Stephen Hall aptly expressed it 'remained magnetic north to all immunologists, the orientation against which everyone set their research compasses'. Those who pioneered non-specific mechanisms were hardly taken serious by mainstream immunologists; Stephen Hall, op. cit., p. 170. The growing emphasis on local defensive processes and substances was not unique to immunotherapy but was also visible in other areas of pharmacotherapy. For instance in the early 1970s pharmacologists began to talk about aspirin as an opponent of defensive local hormone(s), so-called prostaglandines; See, H. O. J. Collier, 'The story of Aspirin', in M. J. Parnham and J. Bruinvels (eds) *Selections from Discoveries in Pharmacology*, Amsterdam: Elsevier Science Publishers, 1987.
72 Robert Teitelman notes that with Robert Good's arrival in New York: he and immunotherapy became famous overnight. *Time* put him on the cover, and *The New York Times Magazine* published a long admiring profile. 'Today almost every puzzling disease in the medical handbook has become the target of the new immunological weapons', said the magazine; R. Teitelman, *Gene Dreams: Wall Street, Academia and the Rise of Biotechnology*, New York: Basic Books, 1989, p. 31.
73 I. Löwy, *Between Bench and Bedside*, Cambridge, Mass.: Harvard University Press, 1996, 110.
74 Interviews with Ion Gresser and Mathilde Krim.
75 H. Strander to K. Cantell, letters dated 8 and 13 March, 12 April and 22 September 1973, Cantell personal archives; H. Strander to I. Gresser, letters dated 8 August, 22 September, 8 October, 29 December 1973, Gresser personal archives; M. Krim, 'Interferon as an antiviral and anticancer agent', *Clinical Bulletin* 5 (1975), 34–6; H. Strander, P. Jakobsson, G. Carlström and K. Cantell, 'Administration of potent interferon to patients with malignant diseases, *Cancer Cytology* 13 (1974), pp. 18–9; and interviews with Kari Cantell, Mathilde Krim and Hans Strander.

76 Interviews with Hans Strander and Kari Cantell.
77 I. Löwy, *Between Bench and Bedside*, Cambridge, Mass.: Harvard University Press, 1996, 59.
78 H. Strander to I. Gresser, letter dated 29 December 1973, Strander personal archives; and interviews with Mathilde Krim and Hans Strander.
79 Interviews with Hans Strander and Mathilde Krim.
80 M. Krim to D. Habif, letter dated 19 July 1974, Cantell personal archives; and interview with Mathilde Krim.
81 Merigan, an infectious disease specialist, developed his plans for trials with interferon in cancer patients suffering from disseminated virus infection (e.g. shingles) simultaneously with Cantell in 1971. Only after Merigan had obtained a high visibility with his promising common cold trial in Salisbury (UK) did he receive sufficient financial support from the NIAID for realizing his trial plans at Stanford (CA). In the course of 1975 he would ask permission for a trial involving human hepatitis infection, which was very common among the large gay community in the San Francisco area. After initial problems about the costs of the interferon required – the costs for the interferon were estimated at a little less than 100,000 dollars for eight patients – the green light was given in June 1975; T. Merigan to G. Galasso, letters dated 16 April, 10 May 1974 and 21 March 1975, NIAID Archives, DMID-ARB box #14; G. Galasso to T. Merigan, letters dated 22 April and 4 June 1975, NIAID Archives, DMID-ARB box #14.
82 Interviews with Kari Cantell, George Galasso, Mathilde Krim, Arthur Levine, Thomas Merigan and Joseph Sonnabend.
83 As director of the NCI Rauscher was part of what was dubbed the 'college of cardinals' which used to meet twice a month – consisting of the directors of the various institutes of the NIH. They held in their hands space, positions and budget. Individually and collectively they were a very influential group; interview with Frank Rauscher.
84 See the report of the International Workshop on Interferon in the Treatment of Cancer, held in New York: March 31, April 1 and 2, 1975, Sonnabend personal archives; H. Strander and K. Cantell to Timothy O'Connor, telegram dated 13 January 1975, Cantell personal archives; M. Krim to J. Sonnabend, letter dated 15 January 1975, Sonnabend personal archives; Panem, op. cit., p. 17; and interviews with Mathilde Krim and Hans Strander.
85 M. Krim to J. Sonnabend, letter dated 15 January 1975, J. Sonnabend personal archives.
86 Interview with Jan Vilcek.
87 M. Krim to K. Cantell, letter dated 13 March 1975, K. Cantell personal archives.
88 I. Gresser to M. Krim, letter dated 16 April 1975, Ion Gresser personal archives; Krim, interview; Sonnabend, interview; Levine, interview; Cantell, interview; Gresser, interview; and, G. Rosen *et al.*, 'High dose methotrexate with citrovorum factor and rescue and adriamycin in childhood osteogenic sarcoma', *Cancer* 33 (1974), 1151–63.
89 Kari, Cantell, *Interferonin Tarina. Tiedemiehen Elämää- Iloja Ja Suruja. (The Interferon Story: The Ups and Downs of the Life of a Scientist)*, Helsinki: Finland, WSOY, 1993, 131.
90 Cantell, *Interferonin Tarina*, op. cit., p. 133.
91 Interviews with Kari Cantell and Hans Strander.
92 See report of the International Workshop on Interferon in the Treatment of Cancer, held in New York: March 31, April 1 and 2, 1975, Sonnabend personal archives; and interviews with Kari Cantell, Mathilde Krim, Thomas Merigan and Joseph Sonnabend.

93 L. Glasgow, 'Interrelationships of interferon and immunity during viral infections', *J. Gen. Phys.* 56 (1970), 212–27; and, E. de Maeyer, 'Interferon twenty years later', *Bulletin de L'Institut Pasteur* 76 (1978), 303–23.
94 See the report of the International Workshop on Interferon in the Treatment of Cancer, held in New York: March 31, April 1 and 2, 1975, Sonnabend personal archives; E. De Maeyer and J. De Maeyer-Guignard, interview; and Gresser, interview.
95 Few American pharmaceutical companies expressed interest in working on interferon, for the greater part because of profitability assessments. Profitability of a drug like interferon was often examined from four standpoints: the patentability of the compound (the British had obtained most of the interferon patents), the developmental costs (producing small quantities of semi-purified interferon was already beset with difficulties and immensely costly), the size of the market and the likely prize of the product (treating one patient for a couple of months already seemed to cost thousands of dollars) and the difficulty and expense of acquiring FDA approval (going by past experience with human cell-produced products, a lot of work was expected); M. Hilleman, interview.
96 M. Krim to I. Gresser, letter dated 20 May 1975, Ion Gresser personal archives.
97 M. Krim, 'Interfering with cancer', first draft for *Nature*, News and Views (April, 1975), J. Sonnabend, personal archives.
98 M. Krim to I. Gresser, letter dated 20 May 1975, Gresser personal archives.
99 M. Krim, A. S. Levine, T. Merigan and J. Vilcek, 'Interfering with Cancer', *Nature* 255 (1975), 372–4.
100 M. Krim to I. Gresser, letter dated 23 April 1975, Gresser personal archives.
101 M. Krim to K. Cantell, letter dated 11 April 1975, Cantell personal archives.
102 Minutes of the meeting of the National Cancer Institute's National Cancer Advisory Board, 17–18 November 1975, NCAB/NCI Archives.
103 Interview with Arthur Levine.
104 Minutes of the meeting of the National Cancer Institute's Division of Cancer Treatment Board of Scientific Counselors, 10, 11 November 1975, Board of Scientific Counselors/NCI Archives.
105 G. Kolata, 'Dilemma in cancer treatment', *Science* 209 (1980), 792–4.
106 T. C. Chalmers, 'The clinical trial', *Milbank Memorial Fund Quarterly Health and Society* 41 (1972), 753–8; E. Gehan and E. Freireich, 'Non-randomized controls in cancer clinical trials', *New England Journal of Medicine* 290 (1974), 198–203; and, Löwy, op. cit., p. 59.
107 Minutes of the meeting of the National Cancer Institute's Division of Cancer Treatment Board of Scientific Counselors, 10, 11 November, 1975, Board of Scientific Counselors/NCI Archives.
108 Ibid.
109 Rettig, op. cit., p. 297.
110 Minutes of the meeting of the National Cancer Institute's National Cancer Advisory Board, 17–18 November 1975, NCAB/NCI Archives.
111 A committee consisting of both NCI and NIAID officials, under the chairmanship of Alan Rabson, was installed to regulate the distribution of the $1 million worth of interferon for use in clinical studies. Among others Tom Merigan was to receive a portion for a prospective randomized trial in patients with lymphomas suffering from shingles; Minutes of the March 8–9 and October 25–26 meetings of the Board of Scientific Counselors of the Division of Cancer Treatment of NCI, NCI Archives; 'Interferon Working Group', *The Cancer Letter* 2(49) (1976), 1; and interview with Thomas Merigan.
112 Minutes of the meeting of the National Cancer Institute's National Cancer Advisory Board, 17–18 November 1975, NCAB/NCI Archives.

113 M. Krim to K. Cantell, letter dated 2 March 1976, Cantell personal archives.
114 H. Strander to A. Levine, letter dated 21 April 1976, Strander personal archives
115 V. DeVita to H. Strander, letter dated 15 April 1976, Strander personal archives.
116 H. Strander to N. Finter, letter dated 9 August 1976, Strander personal archives.
117 A. Levine to H. Strander, letter dated 14 July 1976, Strander personal archives; Report on Site Visit to Karolinska Hospital Stockholm, dated 14 July 1976, Strander personal archives.
118 S. Panem and J. Vilcek, 'Will interferon ever cure cancer?', *The Atlantic*, December 1982.
119 Minutes of the meeting of the National Cancer Institute's Division of Cancer Treatment Board of Scientific Counselors, 25–26 October 1976, Board of Scientific Counselors/NCI Archives. At least the site visit report might have played a role in Strander never being able to get his osteosarcoma data published in an international refereed medical journal.
120 Panem, op. cit., p. 21; and interview with Frank Rauscher.
121 Ibid.

6 Interferon, audiences and cancer

1 Preliminary versions of chapter 6 were published in: T. Pieters. 'About media, audiences and marketing medicines: The interferons', in M. Gijswijt-Hofstra, G. M. van Heteren and E. M. Tansey (eds) *Biographies of Remedies: Drugs, Medicines and Contraceptives in Dutch and Anglo-American Healing Cultures*, Amsterdam: Rodopi, 2002, 229–43; T. Pieters, 'Hailing a miracle drug: the interferon', in C. Osborne and W. de Blecourt (ed.) *Cultural Approaches to the History of Medicine*, Londen: Palgrave, 2003, 212–32.
2 The actors to be analysed are not only those individually and collectively 'present', articulate and committed to action but also those implicated by the actions of others. As we will see the cancer patients as potential users of interferon are the implicated actors examined in this chapter.
3 Panem, *The Interferon Crusade*, Washington, DC: The Brookings Institution, 1984.
4 E. De Maeyer, 'Interferon twenty years later, *Bull. Inst. Pasteur* 76 (1978), 303–23, p. 316.; R. Friedman, 'Interferon research in the Red Queen's Kingdom', *Arch. Pathol.* 98 (1974), 73–6; R. Friedman, 'Guest editorial: Interferon and cancer', *J. Natl. Cancer Inst.* 60 (1978), 1191–4; and interviews with Derek Burke, Joseph Sonnabend, Kari Cantell and George Galasso.
5 Panem, *The Interferon Crusade*.
6 A. Elzinga and A. Jamison, 'Changing policy agendas in science and technology', in S. Jasanoff *et al.* (eds) *Handbook of Science and Technology Studies*, London: Sage Publications, 1995, 588–9; J. Patterson, *The Dread Disease*, Cambridge (MS): Harvard University Press, 1987, 256–7.
7 I. Löwy, op. cit., p. 112.
8 R. Teitelman, *Gene Dreams: Wallstreet, Academia and the Rise of Biotechnology*, New York: Basic Books, 1989, 29; and, E. Richards, *Vitamin C and Cancer: Medicine or Politics?*, New York: St. Martin's Press, 1991, 207.
9 Interview with Jordan Gutterman.
10 J. Gutterman, 'Immunotherapy for recurrent malignant melanoma: The efficacy of BCG in prolonging the postoperative disease-free interval and survival', in M. Sela (ed.) *The Role of Non-Specific Immunity in the Prevention and Treatment of Cancer*, Rome: Pontificia Academia Scientiarum, 1977, 57–65.
11 Report of the International Workshop on Interferon in the Treatment of Cancer, March 31, April 1 and 2 1975 New York: Sonnabend personal archives;

J. U. Gutterman et al., 'Chemoimmunotherapy of advanced breast cancer: Prolongation of remission and survival with BCG?', *Br. Med. J.* ii (1976), 774–7.
12 Panem, op. cit., p. 21; and interviews with Mathilde Krim and Jordan Gutterman.
13 With the American defeat in South-East Asia, the metaphors of war began to turn against the American Cancer Establishment. 'By comparison with the fight against polio', claimed a former FDA commisioner in 1978, 'the war on cancer is a medical Vietnam'; Donald Kennedy quoted in J. C. Petersen and G. E. Markle Petersen, 'Expansion of conflict in cancer controversies', in L. Kriesberg (ed.) *Research in Social Movements, Conflicts and Change*, 4, 1981, 151–69, p. 152.
14 Richards, op. cit., pp. 46–47; R. Moss, *The Cancer Industry*, New York: Paragon House, 1989; and, R. Proctor, *Cancer Wars*, New York: Basic Books, 1995, and Patterson, pp. 256–68.
15 Interview with Mathilde Krim.
16 Panem, op. cit., pp. 21–2.
17 J. Gutterman to K. Cantell, letter dated 8 June 1976, Cantell personal archives; K. Cantell to J. Gutterman, letter dated 26 August 1976, Cantell personal archives.
18 Interview with Kari Cantell.
19 In 1976 the annual Finnish production was about 100 billion units of crude interferon (c-IF) or 50 billion units (with 50 per cent recovery) of partially purified preparations (p-IF); Report on the production of human leucocyte interferon, dated 21 September 1977, Cantell personal archives; T. Merigan to K. Cantell, letter dated 8 September 1979, Cantell personal archives; and interviews with Kari Cantell and Jordan Gutterman.
20 In the 1950s the NCI developed guidelines for the screening of chemical agents of potential value for chemotherapy of cancer. Among other things compounds were considered active in animals when minimum standards of activity were met at the maximum-tolerated dose. According to Löwy this approach was later transferred to the clinics; Löwy, op. cit., p. 55.
21 Edelhart, *Interferon: The New Hope for Cancer*, op. cit., p. 36.
22 Report of the NIAID/NCI workshop on Clinical Trials with Interferon, dated March 21, 22 and 23, 1978 at the NIH, DMID-ARB box #14.
23 In terms of formal clinical trial methodology in oncology these exploratory studies were considered 'Phase I' trials. Basically these were toxicity screening studies to determine the maximum tolerated dose and a first step in the established testing trajectory for testing potential new cancer drugs (followed by the larger, longer and stricter 'Phase II' trials screening for clinical activity – and 'Phase III' trials – determination of relative efficacy). Phase I trials also served the purpose of getting a first indication of whether or not a new compound was of any clinical value in cancer therapy; Report of the NIAID/NCI workshop on Clinical Trials with Interferon, dated March 21, 22 and 23, 1978 at the NIH, DMID-ARB box #14; interview with Arthur Levine; M. Buyse, M. Staquet and R. Sylvester (eds) *Cancer Clinical Trials*, Oxford: Oxford University Press, 1984.
24 H. Greenberg, M. Richard, R. Pollard, L. Lutwick, P. Gregory, W. Robinson and T. Merigan, 'Effect of human leucocyte interferon on Hepatitis B virus infection in patients with chronic active hepatitis', *New England Journal of Medicine* 295 (1976), 517–22.
25 K. Fantes, 'Purification and physico-chemical properties of interferons', in N. Finter (ed.) *Interferons and Interferon Inducers*, Amsterdam: North Holland Publishing Company, 1973, 171–99; and, J. O'Malley and W. A. Carter, 'Human interferons: characterization of the major molecular components', *Journal of the ReticuloEndothelial Society* 23 (1978) 299–305.

26 US federal researchers estimated that there were approximately 200,000 hepatitis carriers in the United States and worldwide some 100 million people. With no cure available it was considered an important research area by NIAID officials, although the escalating hepatitis epidemic among gay men was still non-existent as far as the NIAID was concerned; L. Garrett, *The Coming Plague*, New York: Farrar, Straus and Giroux, 1994, 273; and, D. Burke, 'The Status of Interferon', *Scientific American* 236 (1977), 42–50.
27 Report of the NIAID/NCI workshop on Clinical Trials with Interferon, dated March 21, 22 and 23, 1978 at the NIH, DMID-ARB box #14.
28 Panem, op. cit., p. 30; interview with Alfons Billiau. In 1976 Gresser had already submitted an article to *Science* concerning the question of kidney, liver and spleen toxicity and disease causation by interferon in mice. The paper was rejected because of insufficient evidence that the toxic effects were related to interferon and not to impurities in the preparations. '. . . without data to associate a specific material in the preparation with the progressive glomerulonephritis (kidney toxicity), the findings presented are really only a starting point to isolate the cause . . .'; I. Gresser to K. Cantell, letter with enclosed *Science* referee comment dated 22 September 1976, K. Cantell personal archives.
29 Panem, op. cit., p. 30.
30 Report of the NIAID/NCI workshop on Clinical Trials with Interferon, dated March 21, 22 and 23, 1978 at the NIH, DMID-ARB box #14.
31 The ACS is one of the biggest fundraising organizations in the US, which by relying on 2.5 million volunteers raised $85 million in 1979, plus an additional $35 million from legacies; J. Patterson, *The Dread Disease; Cancer and Modern Culture*, Cambridge (MA): Harvard University Press, 1987, 268–9; and interview with Frank Rauscher.
32 Interviews with Mathilde Krim, Jordan Gutterman and Frank Rauscher.
33 K. Krim to I. Gresser, letter dated 31 August 1978, I. Gresser personal archives; Edelhart, op. cit., 38–9.
34 'Test planned on substance used in cancer treatment', *The Washington Post*, 30 August 1978; 'Natural body substance: $2 million test on cancer retardant', *The Chicago Tribune*, 30 August, 1978; J. Hixson, 'Interferon: The cancer drug we have ignored', *The New Yorker*, 4 September 1978, 59–64; 'New cancer weapon?', *Newsweek*, 18 September 1978, 90–1.
35 Krim, interview; Interview with Frank Rauscher, 16 October 1992, Stamford CT.
36 'New Cancer Weapon?', *Newsweek*, 18 September 1978, 90–1.
37 Discussing the interleukine-2 hype in the mid-1980s Ilana Löwy noted similar connections: Löwy, op. cit., p. 158.
38 See, for extensive analyses of the rise of the information metaphor in molecular biology, immunology and medicine: E. Fox-Keller, *Refiguring Life*, New York: Columbia University Press, 1995; and, E. Martin, *Flexible Bodies*, Boston: Beacon Press, 1994.
39 W. E. Stewart II, *The Interferon System*, Vienna: Springer-Verlag, 1979, 322.
40 J. Hixson, 'Interferon: The cancer drug we have ignored', *New York Magazine*, 4 September 1978, 59.
41 K. White, 'Interferon's future: For common colds, transplants?', *Medical Tribune* (US), 10 November 1978.
42 See, for detailed social histories of the recombinant DNA debate, the development of international regulatory policy for genetic engineering and the subsequent rise of industrial genetics: S. Krimsky, *Genetic Alchemy*, Cambridge (MA): The MIT Press, 1982; S. Wright, *Molecular Politics*, Chicago: The University of

Chicago Press, 1994; S. Krimsky, Biotechnics and Society, New York: Praeger, 1991; R. Teitelman, *Gene Dreams*, New York: BasicBooks, 1989.

43 According to Panem only a few other proteins, like insulin, growth hormone and several proteins identified for use as human vaccines, qualified for a demonstration project; Panem, op. cit., p 26.

44 Hall, op. cit., p. 184.

45 Weissman, interview. See, for a detailed personal account of the cloning of the gene for leucocyte interferon: C. Weissman, 'The cloning of interferon and other mistakes', in I. Gresser (ed.) *Interferon 3*, London: Academic Press, 1981, 101–34.

46 The public announcement, early in September 1978, of the cloning of the first medically significant human gene to make human insulin hit the headlines and was said to open up a new and most exciting era in biology, only added to the public interest in interferon: Victor Cohn, 'Scientists in California create gene to make human insulin', *The Washington Post*, 7 September 1978.

47 Interviews with Mathilde Krim, Joseph Sonnabend and Thomas Merigan.

48 The large drug companies had begun investing in genetic technology with the objective of protecting products and markets from being undermined by the new technology and out of economical need for new growth and above-average returns on investment. See, for a detailed account of dynamics of the industrial interest in interferon: Panem, op. cit., pp. 25–8, 58–74.

49 See, for a thorough analysis of the changing nature of the flow of research information and materials in interferon research owing to the growing interaction between academic and industrial research activities: Panem, op. cit., pp. 75–81.

50 The following impressionistic account of the Second International Workshop on Interferons, April 22–24, 1979 at the Sloan Kettering Memorial Cancer Center in New York is based on: Anonymous, 'Can interferons cure cancer?' *Lancet* i (1979), 1171–2; T. Merigan, 'Human interferon as a therapeutic agent', *New. Engl. J. Med.* 300 (1979), 42–3; K. Cantell, 'Why is interferon not in clinical use today?', in I. Gresser (ed.) *Interferon*, London: Academic Press, 1979, 2–28; J. L. Marx, 'Interferon (I): On the threshold of clinical application', *Science* 204 (1979), 1183–6; J. L. Marx, 'Interferon (II): Learning about how it works', *Science* 204 (1979), 1293–5; A. Rosenfeld, 'If IF works it could . . .', July 1979, *Life*, 55–62; J. Gutterman *et al.*, 'Leucocyte-interferon-induced regression in human metastatic breast cancer, multiple myeloma and malignant lymphoma', *Ann. Int. Med.* 93 (1980), 399–406; F. Balkwill, 'What future for the interferons?', *New Scientist* (1980), 230–2; and interviews with Sam Baron, Kari Cantell, Norman Finter, Robert Friedman, Mathilde Krim and Frank Rauscher.

51 The fact that interferon's chemical formula and structure had remained a mystery for more than 20 years is not exceptional in the history of biochemistry. We only have to compare it with the more than 30 years it cost to produce insulin's or penicillin's chemical structure; M. Bliss, *The Discovery of Insulin*, London: Faber and Faber, 1982; G. L. Hobby, *Penicillin: Meeting the Challenge*, New Haven: Yale University Press, 1985.

52 Interviews with Charles Chany and Kathy Zoon.

53 Stewart *et al.*, 'Interferon nomenclature: Report from the Committee on Interferon Nomenclature', in Gresser (ed.) *Interferon 2*, London: Academic Press, 1980, 97–9.

54 In 1979 virtually all the human interferon available worldwide was still supplied by the Finnish Red Cross. Their production facility had an annual production output of 2.5×10^{11} standard interferon units (roughly 1 g of pure human leucocyte interferon), using leucocytes from 90,000 Finnish blood donors. The production costs amounted to about 5 to 10 million dollars per gram. The

Finnish supply of partially purified interferon preparations ('p-IF', containing 10–20 million units per millilitre) only sufficed for annually enlisting 200 patients with neoplastic or other chronic diseases (employing a daily dose of 3×10^6 standard interferon units) in clinical trials; Cantell, interview.
55 *E. coli* or 'the common lab's workhorse' was and still is widely used in laboratory experiments for bacteriological and genetic studies. The bacteria originate from the intestines of humans and many animals; interview with Charles Weissman.
56 This particular human lymphoblastoid cell line originated from a young Ugandan girl named Namalwa who suffered from a fatal kind of African lymphoma (a Burkitt's tumor). A piece of her tumor taken for diagnostic purposes in the early 1970s was sent to and stored in George Klein's laboratory for tumor biology at the Karolinska Institute. Strander and Cantell had figured out that somehow this particular cell line not only grew readily under laboratory conditions but also yielded high amounts of human interferon when induced with Sendai virus. With their help the Wellcome researcher Norman Finter had been able to obtain a sample of the Namalwa cell line. Despite the knowledge that in the early 1960s the use of malignant cell lines for the production of vaccines had been forbidden for safety reasons (the risk that one of the residual biological agents in the preparation might induce cancer was considered too high), the Wellcome team had given it a try, but only after overcoming fierce opposition from management circles. Finter was able to convince them that by employing state-of-the-art purification and detection techniques – to get rid of all suspicious biological material and in particular genetic material – and performing extensive animal safety testing, he would be able to persuade the regulatory authorities into accepting the final interferon product. The doubts about the high investments associated with the safety aspects were taken away by pointing out the fact that the lymphoblastoid cells had the great practical and commercial advantage over leucocyte cells that they could be grown in continuous cultures on a very large scale using conventional and relatively cheap fermentor technology. It turned out to be a lucky venture; interviews with Kari Cantell, Norman Finter and John Beale.
57 Important in the change in attitude toward the use of transformed (tumorous) cells in the production of a biological like interferon was the consensus achieved at the 1978 meeting at Lake Placid sponsored by the United States Bureau of Biologics of the FDA, that if such materials could be made nucleic-acid free, cautious initiation of a limited trial in humans was warranted. Wellcome indeed claimed their interferon material to be nucleic-acid free; T. Merigan, 'Human interferon as a therapeutic agent', *New England Journal of Medicine* 300 (1979), 42–3.
58 A. Rosenfeld, 'If IF works it could . . .', *Life Magazine*, July 1979, 55; and F. Hauptfuhrer, 'Will interferon kill cancer? Finnish Dr Kari Cantell is helping the world find out', *People* (US weekly), 2 July 1979.
59 K. Krim to I. Gresser, letter dated 5 October 1978, Ion Gresser personal archive.
60 Minutes of the October 16–17, 1978 Meeting of the NCI's Division of Cancer Treatment Board of Scientific Counselors, NCI Archives Volume 1.
61 Pepper Committee Hearings, *The Cancer Letter*, dated 29 June 1979.
62 Pepper Committee Hearings, *The Cancer Letter*, dated 29 June 1979. Pepper was a firm believer in interferon's potential as an anti-cancer agent as is illustrated by his introduction of Strander: 'Doctor, we are very much honoured to have you here as a recognized pioneer as I understand it, of the development of interferon. As I indicated this morning, I wish to the Lord that I had known about interferon before my wife came to the end to see whether or not it would have this miraculous effect upon her illness. It seems to me one of the most promising substances there is to offer hope and aid to cancer victims . . .'. Pepper several times

indicated that his political support of NCI's BRM programme was dependent on the extent to which this programme would focus attention to interferon; Frontiers in Cancer Research for the Elderly, Comm. Pub. No.96–188, Strander personal archives.
63 Pepper Committee Hearings, *The Cancer Letter*, dated 29 June 1979.
64 Pepper Committee Hearings, *The Cancer Letter*, dated 29 June 1979; and, NCI Memorandum, dated 21 June 1979, NCI Archives D-7906–3418.
65 Gutterman should have known better considering the paragraph on toxicity in his September 1980 article dealing with interferon therapy. His non-toxicity claim is all the more remarkable when taking into account Constance McAdam's article (she was at the time associate director of nursing at the M.D. Anderson Hospital and Tumor Institute) in the *American Journal of Nursing*. She indicated that because of it's side-effects patients receiving interferon required serious nursing assessment and close monitoring. Krim's silence is also surprising when taking into account the following remark by Krim in a letter to Cantell. 'Dr Oettgen has started treating some patients at our hospital. . . . Although he started at doses 10 times less than Levy and Arthur Levine (remember?) say can be tolerated, he already found severe toxicity, particularly immediate very high fever. He had to discontinue one patient'; K. Krim to K. Cantell, letter dated 21 May 1979, Cantell personal archives; C. McAdams, Interferon; The Penicillin of the Future?, *American Journal of Nursing*, April 1980, 714–7; J. Gutterman *et al.*, 'Leukocyte interferon-induced tumor regression in human metastatic breast cancer, multiple myeloma, and malignant lymphoma', *Annals of Internal Medicine* 93 (1980), 399–406, p. 402.
66 Congressional inquiry received and answered by phone relating to interferon, NCI Archives No. 00–78032816. In 1980 and 1981, 80 per cent of the BRMP funds would be allocated for studies of interferon; in Background on Interferon, ACS News Service, NCI Archives No. 0000–001171; Panem, op. cit., p. 52.
67 'Interferon: The Body's Own Wonder Drug', *Saturday Review*, 13 October 1979; 'A wonder drug in the making', *Business Week*, 19 November 1979; Hal Lancaster, 'Potent protein; Medical researchers say the drug interferon holds great promise', *The Wallstreet Journal*, 6 December 1979; 'The Race is on for miracle drug', *The Observer*, 30 December 1979; 'Interferon trial: Early results promising', *Medical World News*, 10 December 1979, 27.
68 Richards, op. cit., pp. 206–15.
69 Edelhart, op. cit., p. 146.
70 Interferon: The body's own wonder drug', *Saturday Review*, 13 October, 1979; and 'Interfering with Cancer', *Scientific American*, April 1979.
71 'Race is on for miracle drug', *The Observer*, 30 December 1979.
72 Hal Lancaster, 'Potent protein; Medical researchers say the drug interferon holds great promise', *The Wallstreet Journal*, 6 December 1979.
73 This is nicely illustrated by a cover story of *Newsweek* magazine entitled 'DNA's new miracles'; 'By turning bacteria into living factories scientists can cure disease and create new forms of life'; *Newsweek*, 17 March 1980.
74 N. Wade, 'Cloning gold rush turns basic biology into big business', *Science* 208 (1980), 688–92; S. Andreopoulos, 'Sounding board; and, gene cloning by press conference', *New England Journal of Medicine* 302 (1980), 743–6.
75 T. Taniguichi, M. Sakai, Y. Fujii-Kuriyama, M. Muramatsu, S. Kobayashi and T. Sudo, 'Construction and identification of a bacterial plasmid containing the human fibroblast interferon gene sequence', *Proc. Jpn. Acad.* 55B (1979), 461–9.
76 N. Wade, op. cit., pp. 688–92; Andreopoulos, op. cit., 743–6.

77 'Medical breakthrough reported', *Los Angeles Time*, 21 January 1980; 'L'interferon: Enjeu d'une competition mondiale scientifique et industrielle', *Le Monde*, 6 February 1980; 'Cancer treatment available soon', *The Guardian*, 20 March 1980; 'The big if – interferon', *The Listener*, 29 February 1980; 'Interferon: The IF drug for cancer', *Time*, 31 March 1980; 'Interferon duurste stof ter wereld', *Telegraaf*, 29 May 1980; 'The making of a miracle drug', *Newsweek*, 28 January 1980; *The Reader's Digest* monthly magazine with a worldwide circulation of about 50 million copies in 15 different languages published the cover story 'Interferon, the new miracle drug?' in January 1980; K. Cantell, personal correspondence 1979–1981; Interview with K. Cantell, 4 May 1992, Helsinki; Interview with J. Sonnabend, 13 November 1993, New York.

78 'Medical breakthrough reported', *Los Angeles Time*, 21 January 1980; 'Cancer treatment available soon', *The Guardian*, 20 March 1980; 'Interferon: The IF drug for cancer, *Time*, 31 March 1980; 'At only $100 million a gram, this "miracle" has a future', *Science Digest*, April 1980.

79 Wade, op. cit., pp. 688–92; 'At only $100 million a gram, this "miracle" has a future', *Science Digest*, April 1980.

80 Panem, op. cit., p. 99.

81 For privacy reasons I can only refer to the letter files I have studied in Cantell's personal archives.

82 See, for a detailed account of the cortisone case: D. Cantor, 'Cortisone and the politics of drama, 1949–55, in J. V. Pickstone (ed.) *Medical Innovations in Historical Perspective*, Houndmills: MacMillan, 1992, 165–99; and interview with Norman Finter.

83 'Wonder drug hope for Miss Anelli', *Sunday Mirror*, 1 June 1980; 'Dad's wonder drug plea', *Sunday Mirror*, 8 June 1980; and 'Drug brings hope for tumor boy Daniel', *The Daily Telegraph*, 12 April 1980.

84 Thanks to Susanna Isaacs Elmhirst I was able to get hold of a copy of both the script and the videotape of the *TV Eye* documentary 'Cancer – the new weapon'.

85 The use of the penicillin analogy was reiterative in the sense that journalists as well as interferon researchers made frequent use of it in their performances; K. Cantell, 'Why is interferon not in clinical use today?', in I. Gresser (ed.) *Interferon*, London: Academic Press, 1979, 2–28, p. 3.

86 'Publicity on interferon has caused great distress, specialists say', *The Times*, 10 June 1980; J. Gordon McVie, 'Medicine and media', *BMJ*, 28 June 1980, 161; and, R. Ridgway, 'Interferon: The hopes and the reality', *BMA News Review*, September 1980, 18–22.

87 Department of Education and Science to Sir John Eden (MP/House of Commons), letter dated 12 June 1980, MRC Archives File No. S806/5.

88 A flourishing black market rapidly developed in often dubious interferon samples, fuelled by those rich, famous and desperate enough to try anything like the dying exiled Shah of Iran.

89 MRC to ICRF, letter dated 23 June 1980, MRC Archives File No. D1009/40.

90 Internal note MRC, dated 23 June 1980, MRC Archives File No. D1009/40.

91 Interview with J. Petricciani, 6 November 1993, Cambridge, MA; and, interview with D. Tyrrell, 21 May 1990, Salisbury.

92 Internal note MRC, dated 27 June 1980, MRC Archives File No. D1009/40; Press notice, dated 13 May 1980, MRC Archives File No. D1009/40.

93 Editorial, 'What not to say about interferon', *Nature* 285 (1980), 603–4.

94 Edelhart, op. cit., pp. 1–9; and, 'Interferon: The hopes and the reality', *BMA News Review*, September 1980, 18–22.

95 'Interferon results promising, ACS will Commit Additional $3.4 Million', *The Cancer Letter*, 22 February 1980.

234 *Interferon: The science and selling of a miracle drug*

96 Ibid.
97 'Interferon results cool expectations', *The Cancer Letter*, 6 June 1980.
98 'Interferon: 'Hysteria out of hand', Asco warns against expecting benefit', *The Cancer Letter*, 20 June 1980.
99 V. Cohn, 'Leading U.S. cancer doctors agree to issue warnings on interferon', *The Washington Post*, 15 June 1980.
100 Edelhart, op. cit., p. 6.
101 H. M. Schmeck, 'Interferon: Studies put cancer use in doubt', *New York Times*, 27 May 1980; 'Is it a wonder cure for cancer or the most expensive flop in history?', *The Daily Star*, 19 June 1980.

7 Yet another twist: marketing interferon as a helpful neighbour

1 Preliminary versions of chapter 7 were published in the *British Medical Journal*: T. Pieters, 'Marketing medicines through randomised controlled trials: The case of interferon, *BMJ* 317 (1998), 1231–3; and, T. Pieters, 'What constitutes therapeutic success?: The interferons', in J. Lindenmann and W. D. Schleuning, *Interferon: The Dawn of Recombinant Protein Drugs*, Berlin: Springer, 1999.
2 See, G. Taubes, 'Use of placebo controls in clinical trials is disputed', *Science* 267 (1995), 25.
3 A number of case studies dealing with post-war therapeutic research and development in medicine, which were published in the 1990s, support the view that therapeutic evaluation is an inherently social and cultural process: see, E. Richards, *Vitamin C and Cancer*; A. Clarke and T. Montini, 'The many faces of RU486: Tales of situated knowledges and technological contestations', *Science, Technology and Human Values* 18 (1993), 42–78; N. Oudshoorn, *Beyond the Natural Body*; I. Löwy, *Between Bench and Bedside*; and, H. M. Marks, *The Progress of Experiment; Science and Therapeutic Reform in the United States, 1900–1990*, Cambridge: Cambridge University Press, 1997.
4 Krim and her scientific staff's letter in the *New York Times* was a direct reply to Harold Schmeck's article of 24 August in the very same newspaper: see, M. Krim, W. E. Stewart II, F. Sanders and L. Lin, 'Interferon therapy', *New York Times*, 17 June 1980, p. C5.
5 According to a *JAMA* reporter the conference had an air of 'upstage scientific excitement and backstage financial intrigue'; R. Johnson, 'Interferon: cloudy but intriguing future', *JAMA* 245 (1981), 109–16, p. 109.
6 Johnson, op. cit., p. 115.
7 An indirect research strategy to determine the structure of natural proteins commonly employed by biochemists was to try synthesizing the entity from pure amino acids on the basis of available structural data. When a substance was synthesized that matched the biological activity of the purified natural entity the structure of the natural protein might then be matched. However, interferon appeared too complex a molecule for this approach to be successful. E. Knight, 'Purification and characterization of interferons', in I. Gresser (ed.) *Interferon 2*, London: Academic Press, 1980, 1–12, p. 2.
8 See for the historical reconstruction of the cloning of interferon; C. Weissmann, 'The cloning of interferon and other mistakes', in I. Gresser (ed.) *Interferon 3*, London: Academic Press, 1981, 101–34; and Hall, op. cit., pp. 178–208.
9 Interview with Sidney Pestka.
10 S. Krown, 'Prospects for the treatment of cancer with interferon', in J. Burchenal and H. Oettgen (eds) *Cancer; Achievements, Challenges, and Prospects for the 1980s*, New York: Grune & Stratton, 1981, 367–79; Johnson, 'Interferon: cloudy but intriguing future', *JAMA* 245 (1981), 109–16; P. Newmark, 'Interferon: decline

and stall', *Nature* 291 (1981), 105–6; M. Sun, 'Interferon: no magic bullet against cancer', *Science* 212 (1981), 141–2.
11 Biological Response Modifiers Program Review during the February 1981 meeting of NCI's Board of Scientific Counselors, minutes of the Board, dated 12–13 February, NCI Archives; Johnson, op. cit. pp. 109–16; Newmark, op. cit., pp. 105–6; Sun, op. cit. pp. 141–2. Due to the high degree of species specificity of interferon, the drug industry depended on monkeys for preclinical studies of the biosynthetic human interferons (basically aimed at proving that a preparation was active and non-toxic *in vivo*). As head of the Dutch Centre for Primate Research, Huub Schellekens played a central role in the preclinical testing of the first -DNA-produced human interferon preparations; H. Schellekens *et al.*, 'Comparative antiviral efficiency of leukocyte and bacterially produced human alpha interferon in rhesus monkeys', *Nature* 292 (1981), 775–6.
12 F. Balkwill, 'Interferon: a progress report', *New Scientist*, 25 March 1982, 783–5, p. 784.
13 National Cancer Institute Program on Interferon and Other Biological Response Modifiers, dated October 1980, NCI Archives File No. AR-8000-007320; *The Cancer Newsletter*, 24 April 1981, p. 1.
14 Edelhart, op. cit., p. 49; L. Epstein, 'Interferon as model lymphokine, *Fed. Proc.* 40 (1981), 56–61; B. Feder, 'Technology; complexities cloud interferon', *New York Times*, 23 April 1981; Epstein, interview.
15 C. Fenyvesi, 'Beyond interferon', *The Washington Post*, 14 June 1981, p. 28.
16 Interview with Ernest Borden.
17 See, for a detailed account of the development of what Ilana Löwy dubbed the 'trialist ethos' in clinical oncology: Löwy, op. cit., pp. 36–83.
18 Fenyvesi, op. cit., p. 28.
19 Minutes of American Society of Clinical Oncology, 18th Annual Meeting, St. Louis, Missouri, 25–27 April 1982, NCI Archives No. 8204-001971; K. Sikora and H. Smedley, 'Interferon and cancer', *British Medical Journal* 286 (1983), 739–40.
20 R. Walgate, 'Side effect scare hits French trials', *Nature* 300 (1982), 97–8. According to a February 1983 ACS press release interferon was one of the most studied natural substances in medicine, 'as researchers explore its benefits, potentials and limitations'; Interferon Update ACS, dated 11 February 1983, NCI Archives DC-8301-006691. See, on overzealous treatment practices; 'Cancer a progress report', *Newsweek*, 2 November 1981.
21 'France halts interferon study', *New York Times*, 4 November 1982.
22 T. Gup and J. Neuman, 'Experimental drugs cause pain, deaths', *The Boston Sunday Globe*, 18 October 1981.
23 See, S. Rosenberg, *The Transformed Cell*, London: Phoenix, 1992, 236–7.
24 See, Moss, op. cit., p. 22.
25 Ibid.
26 Minutes of NCI's Board of Scientific Counselors meeting, dated 27–28 January 1983, NCI Archives.
27 T. Powledge, 'Interferon on trial', *Bio/Technology* 2 (1984), 214–28.
28 'Million dollar cold cure', *The Sunday Times*, 3 October 1982, pp. 22–30; and, G. Scott *et al.*, 'Prevention of rhinovirus colds by human interferon alpha-2 from Escherichia coli', *Lancet* ii (1982), 186–7.
29 J. Alper, 'First there was interferon', *New York Times*, 18 November 1984, F. 13.
30 In 1983 Hoffmann-La Roche and Schering-Plough reportedly allocated 15 per cent of their research budgets – more than 40 million dollars each – to interferon; Powledge, op. cit., pp. 214–28; and interviews with Norman Finter and Leon Gauci.

31 Powledge, op. cit., pp. 214–28; A. Wycke, 'Molecules and markets', *The Economist*, 7 February 1987; and interviews with Leon Gauci and J. P. Warbeer.
32 Interviews with John Petricciani and Leon Gauci.
33 Interviews with John Petricciani and Leon Gauci. See, for a similar cases of drugs looking for diseases: R. Vos, *Drugs Looking for Diseases*; N. Oudshoorn, *Beyond the Natural Body*.
34 'Interferon may help AIDS victims', *New Scientist*, 3 November 1983; 'Interferon tested on sclerosis', *New York Times*, 21 November 1981; and, N. Finter and R. K. Oldham (eds) *Interferon (vol. 4); In Vivo and Clinical Studies*, Amsterdam: Elsevier, 1985; 'Update on interferon', News Service American Cancer Society, dated 11 February 1983, NCI Archives File No. DC8301-006691; and, 'Intron A', press information videotape produced by Schering-Plough in 1986 on the occasion of the market introduction of their interferon product.
35 B. G. Leventhal, 'Treatment of virus-associated tumours and papillomas with interferons', in N. Finter and R. Oldham (eds) *Interferon (vol. 4); In Vivo and Clinical Studies*, Amsterdam: Elsevier, 1985, 326–35; and interviews with Kari Cantell and Hans Strander.
36 Hall, op. cit., pp. 203–5.
37 J. Quesada, E. Hersh and J. Gutterman, 'Hairy cell leukaemia: Induction of remission with alpha interferon', *Blood* 62 (1983), 207a.
38 J. Quesada, J. Reuben, J. Manning, E. Hersh and J. Gutterman, 'Alpha interferon for induction of remission in Hairy Cell leukaemia', *N. Engl. J. Med.* 310 (1984), 15–18; and interviews with Leon Gauci and John Petricciani.
39 Powledge, op. cit., p. 227.
40 'Anticancer interferon available soon', *Hospital Doctor*, 22 September 1983. In contrast to interferon products which were still in the investigational stage and limited in their clinical use (Phase I, II and III testing), licensed products which would be available for general commercial distribution and use by the medical profession must meet a variety of extra regulatory requirements, at least in the United States, as prescribed by the Good Manufacturing Practice (GMP) provisions and the General Provisions for Licensed Biologicals (GPLB) Code of Federal Regulations; J. Petricciani, E. Esber, H. Hopps and A. Attallah, 'Manufacture and safety of interferons in clinical research', in P. Came and W. Carter (eds) *Interferons and their Applications*, Berlin: Springer-Verlag, 1984, 357–70, p. 359.
41 A. Hecht, 'Interferon: Trying to live up to its Press', *Hospital Doctor*, 22 September 1983; H. Hopps, K. Zoon, J. Djeu and J. Petricciani, 'Interferons for clinical use: purity, potency and safety', in N. Finter and R. Oldham (eds) *Interferon (vol. 4); In Vivo and Clinical Studies*, Amsterdam: Elsevier, 1985, 121–33.
42 Interviews with John Petricciani and Kathy Zoon; and, Hopps *et al.*, op. cit., pp. 121–33.
43 US Department of Health and Human Services, news announcement, dated 4 June 1986, NCI Archive No. 8606–001949; Editorial, 'Interferon drugs join cancer fight', *Clinical Pharmacy*, 15 March 1986.
44 See, P. Temin, *Taking Your Medicine*, Cambridge (MS): Harvard University Press, 1980; Richards, op. cit., pp. 58–9 and 224–5.
45 See, Temin, op. cit., p. 143.
46 'Interferon licensed to treat hairy cell leukaemia', *New Scientist*, 12 June 1986.
47 F. Young, 'The reality behind the headlines', in FDA Anonymous, *New Drug Development in the United States*, Rockville (MD), 1990.
48 Interview with John Petricciani.
49 Interview with a Dutch Glaxo-Wellcome worker who prefers not to be identified. Kari Cantell's Finnish leucocyte interferon 'cocktail' met with the same difficulties from the drug regularity authorities: H. Hage, 'Controlerende instanties moeten

soepeler worden met interferon', *Toegepaste Wetenschap TNO*, October 1986. Finally on 21 November 1997, the FDA authorised Wellcome to manufacture, ship and sell its product Interferon alpha-n1, lymphoblastoid on the American market.
50 See, for the annual number and titles of interferon-related publications the Dindi data base, Cologne.
51 S. Baron, F. Dianzani, G. Stanton and W. Fleischmann (eds) *The Interferon System: A Current Review to 1987*, Austin, TX: The University of Texas Press, 1987; S. Baron *et al.*, *Interferon: Principles and Medical Applications*, Austin, TX: The University of Texas Press, 1992; C. Wallis, 'What's become of interferon', *Time*, 1 June 1985; F. Balkwill, 'Interferons: From common colds to cancer, *New Scientist*, 14 March 1985, pp. 26–8; F. Balkwill, *Cytokines in Cancer Therapy*, Oxford: Oxford University Press, 1989.
52 See, for a focus on what constitutes today's state-of-the-art molecular medicine the website http://www.isinet.com/hot/ (last visited 10 March 2004).
53 Interview with Leon Gauci; and, L. Gauci, 'Interferon drug development: A history truely consistent with the discovery process', paper presentation at the International Meeting From Clone to Clinic, March 1990, Amsterdam; R. K. Oldham, 'Interferon: A model', in I. Gresser (ed.) *Interferon 6*, London: Academic Press, 1985, pp. 127–41; D. Barnes, 'Biologics gain influence in expanding NCI programme', *Science* 237 (1987), 848–50.
54 A. Foerstner, 'How we make our own wonder drugs', *Chicago Tribune*, 20 January 1985.
55 H. Johnson *et al.*, 'How interferons fight disease', *Scientific American*, May 1994, 40–7.
56 'Interferon in Prospect', A three-part film series from Schering Corporation, USA, that was made available in 1983 to clinical investigators all over the world.
57 Interview with Leon Gauci.
58 H.C. Thomas, F. Cavalli, and M. Talpaz (eds) 'Thirty years of interferon', *Interferons Today and Tomorrow* 5 (1987) (this journal is published by Mediscript, London); H. Kirchner (ed.) 'Update on interferons', *Progress in Oncology* 2 (1986) (Abstracting journal published by Mediscript, London); H. Kirchner (ed.) 'Update on Interferons', *Progress in Virology* 1 (1987) (Abstracting journal published by Mediscript, London); H. K. Silver (ed.) *Interferons in Cancer Treatment*, Mississauga: Medical Education Services (Canada) INC, 1986. ICON or Interferon Communication Network was established in 1980 by Schering-Plough. In 1988 ICON contained about 15,000 items related to interferon which had been selected by a group of five physicians payed for by Schering. The Schering-Plough team produced their own abstracts of the papers and on special request customers could obtain the original publications. Interview with C. v. Helsen.
59 P. E. Came and W. A. Carter (ed.) *Interferons and their Applications*, Berlin: Springer-Verlag, 1984; C. Pinsky (ed.) 'Biological response modifiers', *Seminars in Oncology* 13 (1986), 131–227; D. Parkinson (ed.) 'The expanding role of interferon-alfa in the treatment of cancer', *Seminars in Oncology* 21 (1994), 1–37.
60 My distinction between effectiveness and efficacy is derived from clinical epidemiology; see, S. Pocock, *Clinical Trials*, Chichester: John Wiley & Sons, 1983.
61 See, for a similar kind of phenomenon in the treatment of reproductive disorders in women: J. Van Dyck, *Manufacturing Babies and Public Consent*, New York: New York University Press, 1995, 124.
62 See, for interview with Schellekens; 'Intron A', press information videotape produced by Schering-Plough in 1986 on the occasion of the market introduction of their interferon product.
63 According to an editorial by Ronald Penny the new journal title reflected the real breadth of focus in this ever-expanding field. In his view the number of factors

238 *Interferon: The science and selling of a miracle drug*

involved grew 'almost at the rate of action of the cytokines themselves': R. Penny, *Interferons and Cytokines* 15 (1990), 3.
64 Pinsky, op. cit., pp. 131–227; D. Parkinson (ed.) 'The expanding role of interferon-alfa in the treatment of cancer', *Seminars in Oncology* 21 (1994), 1–37; R. Stuart Harris and R. Penny, *Clinical Applications of the Interferons*, London: Chapman & Hall Medical, 1997.
65 D. Fischer et al., *The Cancer Chemotherapy Handbook* (5th edn), St. Louis: Mosby, 1997.
66 A. J. Elsworth et al., *Mosby's 2004–2005 Medical Drug Reference*, St. Louis: Mosby, 2004.
67 See, for FDA product approval information on biologics the FDA web site: www.FDA.gov/cber/efoi/approve.htm and www.FDA.gov/cber/establish.htm.
68 According to December 1997 *Pharma Business Signals* (Competitive Intelligence and Strategic Issues/Trends) and 2004 BioPortfolio Limited AS Insights.
69 'MS drug could fight Sars', BBC News, 24 July 2003; 'Interferon promising drug for Sars', CBC News World Edition, 24 December 2003; 'Interferon protects against SARS virus, in monkeys.' Reuters Health, 23 February 2004.
70 A. Shiell and G. Salkeld, 'The economic aspects of interferon', in R. Stuart-Harris and R. Penny (eds) *The Clinical Applications of the Interferons*, London: Chapman & Hall, 1997, pp. 376–90; R. B. Forbes, A. L. Norman Waugh and R. J. Swingler, 'Population based cost utility study of interferon beta-1b in secondary progressive multiple sclerosis', *BMJ* 319 (1999), 1529–33.
71 See, *Guidance on beta interferon and glatiramer acetate for the treatment of multiple sclerosis*, Technology appraisal guidance No. 32, January 2002 of the National Institute for Clinical Excellence in Britain; and the 2000 Dutch government report on the use of interferon beta in multiple sclerosis.
72 See, for a detailed account of the increasingly important social and cognitive role of clinical trials in twentieth-century scientific medicine: H. M. Marks, *The Progress of Experiment*.

8 Interferons in retrospect and prospect

1 This impressionistic account of a meeting to inform multiple sclerosis patients of the newly registered therapeutic drug, interferon beta, was based on a series of meetings held between May and December 1997 as part of a national government-sponsored information campaign in the Netherlands.
2 See, for a historical analysis of development and change in the relations between 'science' and 'the public': S. Shapin, 'Science and the public', in R. C. Olby, G. N. Cantor, J. R. Christie and M. J. Hodge (eds) *Companion to the History of Modern Science*, London: Routledge, 1990, 990–1007.
3 See, for an exemplary philosophical introduction to intervention and representation in natural science: I. Hacking, *Representing and Intervening*, Cambridge: Cambridge University Press, 1983.
4 See, R. M. Burian, 'How the choice of expcrimental organism matters: Epistemologiocal reflections on an aspect of biological practice', *Journal of the History of Biology* 26 (1993), 351–67, p. 365.
5 For instance, in 1997 the executive director for cancer research at Merck Research Laboratories, Alan Oliff, admittted that 'the fundamental problem in drug discovery for cancer is that the model systems are not predictive at all': see, Trisha Gura, 'Systems for identifying new drugs are often faulty', *Science* 278 (1997), 1041–2.
6 See, for a similar kind of argument: Löwy, op. cit., p. 114.
7 At least until the 1970s the gift culture remained largely intact. But in the early 1980s the emerging profit potential in molecular biology, began to strain and rupture

the informal traditions of scientific exchange. In the 1990s the once relatively open exchange of biological materials has become more and more formalized and selective. Increasingly research materials and techniques are regarded as commodities and therefore no longer freely exchanged as part of a gift culture but rather formally acquired for money.
8 The American researchers Jonas Salk and Maurice Hilleman serve as most notable examples of this new generation of scientific entrepreneurs in biomedicine: see, L. Galambos and J. E. Sewell, *Networks of Innovation: Vaccine Development At Merck, Sharp & Dohme, and Mulford 1895–1995*, Cambridge: Cambridge University Press, 1995; and, J. S. Smith, *Patenting the Sun; Polio and the Salk Vaccine*, New York: William Morrow and Company, 1990.
9 D. Cantor, 'Cortisone and the Politics of Drama, 1949–1955', in J. Pickstone (ed.) *Medical Innovations in Historical Perpsective*, London: MacMillan, 1992, 165–84, p. 173; H. M. Marks, 'Cortisone, 1949: A year in the political life of a drug', *Bulletin of the History of Medicine* 66 (1992), 419–39; A. Yoshioka, 'Streptomycine in postwar-Britain: A cultural history of a miracle drug', in M. Gijswijt-hofstra, G. M. Van Heteren and E. M. Tansey, *Biographies of Remedies*, Amsterdam: Editions Rodopi, 2002, 203–28.
10 R. Bud, 'Penicillin and the new Elizabethans', *British Journal of the History of Science* 31 (1998), 305–33, p. 305.
11 See, for a comprehensive account of medicine and the counter-culture in Britain and America in the second half of the twentieth century: M. Saks, 'Medicine and the counter culture', in R. Cooter and J. Pickstone (eds) *Medicine in the 20th Century*, Amsterdam: Harwood Publishers, 2000, 113–23.
12 See, Richards, op. cit., pp. 45–49.
13 Bert Hansen convincingly argues that the high media visibility of therapeutic discoveries in America originates from as early as the 1880s: B. Hansen, 'New images of a new medicine: Visual evidence for the widespread popularity of therapeutic discoveries in America after 1885', *Bulletin of the History of Medicine* 73 (1999), 629–78.
14 Karpf's pioneering study examines the dynamics of mediating health and medicine by the mass media from the 1930s up to the 1980s: A. Karpf, *Doctoring the Media*, London: Routledge, 1988, pp. 9–31.
15 See, R. Porter, *The Greatest Benefit to Mankind; A Medical History of Humanity from Antiquity to the Present*, London: Harper Collins Publishers, 1997, 718.
16 See, Cantor, op. cit., pp. 165–84; Yoshioka, op. cit., pp. 203–28.
17 Despite the many parallels between the research and development trajectories of the interferons and another closely related family of recombinant protein drugs, the interleukins, in the 1980s and 1990s, I would argue that it was precisely this magic public imprint that helped to transform the former into blockbuster compounds and relegated the latter to the status of orphan drugs or worse to the laboratory shelf; I. Löwy, *Between Bench and Bedside*.
18 See, for a similar shift in the way immunity was dealt with in American culture: E. Martin, *Flexible Bodies*, Boston: Beacon Press, 1994, p. 61. With his paper on interferon biosemiotics the interferon researcher Yoshioka Kawade represents a most radical view on biological complexity and the role of the cytokine regulatory-communication network: Y .Kawade, 'A biosemiotic view of interferon: Toward a biology of really living organisms', in J. Lindenmann and W. D. Schleuning (eds) *Interferon: The Dawn of Recombinant Protein Drugs*, Berlin: Springer, 1999.
19 See, for an insightful analysis of the innovation crisis in the international pharmaceutical industry at the end of twentieth century: J. Drews, *In Quest of Tomorrow's Medicines*, New York: Springer, 1999.

20 See, for an exemplary study of the use of methaphors in biology: E. F. Keller, *Refiguring Life; Metaphors of Twentieth-Century Biology*, New York: Columbia University Press, 1995, 94–114.
21 L. Lasagna, 'The Nature of Evidence', in J. D. Cooper (ed.) *The Philosophy of Evidence: Vol 3 Philosophy and Technology of Drug Assessment*, Washington, DC: Interdisciplinary Communications Program, 1972, 24.
22 During the 1998 celebration of the 50th anniversary of the publication in the *British Medical Journal* of the world's 'first' randomized trial, the 1948 streptomycin trial, tackling this methodological and assessment problem was regarded as the challenge to medical science in the twenty-first century: 'The randomised controlled trial at 50 issue', *British Medical Journal* 317 (1998), 1167–262.
23 See, for an exemplary case-study; J. Van Kammen, 'Who represents the users? Critical encounters between women's heath advocates and scientists in contraceptive R&D', in N. Oudhoorn and T. Pinch (eds) *How Users Matter; The Co-construction of Users and Technology*, Cambridge (MA.): The MIT Press, 2003, 151–72. The new way of how consumers may matter to drug discovery and development is most vividly publicly expounded in the 1992 Hollywood film 'Lorenzo's oil' directed by George Miller.
24 See, for MS-related sites on the Internet: www.nmss.org; www.mssociety.org.uk; www.mssociety.ca; for links to other international MS societies http://www.mswebpals.org/links.htm; and for newsgroups http://www.news2mail.com/alt/support/mult-sclerosis.html (accessed 20 February 2004).
25 See, S. Epstein, 'Activism, drug regulation, and the politics of therapeutic evaluation in the AIDS Era: A Case Study of ddc and the "surrogate markers" debate', *Social Studies of Science* 27 (1997), 691–726. A more radical and disturbing view on the dependencies between science, medicine and industry at the beginning of the twenty-first century can be found in: C. Medawar and A. Hardon, *Medicines Out of Control? Antidepressants and the Conspiracy of Goodwill*, Amsterdam: Aksant, 2004.
26 See for information on the 2nd Cytokines and Beyond conference that took place at the end of April 2004; http://www.gtcbio.com/confpage.asp?cid=1 (accessed 1 April 2004).
27 See, Thomas Nickles, 'Philosophy of science and history of science', *Osiris* 10 (1995), 139–63, p. 154.

Bibliography

Abraham J. (1995) *Science, Politics and the Pharmaceutical Industry*, London: UCL Press.
Almeida J. D., Hasselback R. C. and Ham A. W. (1963) 'Virus-like particles in blood of two acute leukemia patients', *Science*, 142: 1487–9.
Alper J. (1984) 'First there was Interferon', *The New York Times*, 18 November, F. 13.
Amsterdamska O. (1987) 'Medical and biological constraints: early research on variation in bacteriology', *Social Studies of Science*, 17: 657–87.
Andreopoulos S. (1980) 'Sounding board; and, gene cloning by press conference', *New England Journal of Medicine*, 302: 743–6.
Andrewes C. H. (1942) 'Interference by one virus with the growth of another in tissue-culture', *British Journal of Experimental Pathology*, 23: 214–20.
Andrewes C. H. (1944) 'Virus diseases of man: A review of recent progress', *British Medical Bulletin*, 2: 265–9.
Andrewes C. H. (1967) 'Alick Isaacs', *Biographical Memoirs of Fellows of the Royal Society*, 13: 205–21.
Arnow L. E. (1970) *Health in a Bottle; Searching for the Drugs that Help*, Philadelphia: J.B. Lippincott Company.
Atanasiu P. and Chany C. (1960) 'Action d'un interféron provenant de cellules malignes sur l'infection expérimental du hamster nouveau-né par le virus du polyoma', *C.R.H.S.A.S.*, 251: 1687–9.
Austoker J. and Bryder L. (eds) (1989) *Historical Perspectives on the Role of the MRC*, Oxford: Oxford University Press.
Bader J. (1962) 'Production of interferon by chick embryo cells exposed to Rous sarcoma virus', *Virology*, 16: 436–43.
Balducci D. and Penso G. (1963) *Tissue Cultures in Biological Research*, Amsterdam: Elsevier Publishing Company.
Balkwill F. (1980) 'What future for the interferons?', *New Scientist*, 85: 230–2.
Balkwill F. (1982) 'Interferon: a progress report', *New Scientist*, 93: 783–5.
Balkwill F. (1985) 'Interferons: from common colds to cancer', *New Scientist*, 105: 425–31.
Balkwill F. (1989) *Cytokines in Cancer Therapy*, Oxford: Oxford University Press.
Baron S. and Isaacs A. (1961) 'Interferon and natural recovery from viral diseases', *New Scientist*, 243: 81–2.
Baron S., Dianzani F., Stanton G. and Fleischmann W. (eds) (1987) *The Interferon System: A Current Review to 1987*, Austin, TX: The University of Texas Press.
Bell T. M. (1965) *An Introduction to General Virology*, London: William Heinemann Medical Books.

Berg M. and Mol A. (eds) (1998) *Differences in Medicine*, Durham: Duke University Press.
Beveridge W. and Burnet F. (1946) *The Cultivation of Viruses and Rickettsiae in the Chick Embryo*, Special Report Series, No. 256 of the Medical Research Council, London: His Majesty's Stationery Office.
Billiau A. and Finter N. (eds) (1984) *Interferon I; General and Applied Aspects*, Amsterdam: Elsevier, 1984.
Bliss M. (1982) *The Discovery of Insulin*, London: Faber and Faber.
Bijker W., Hughes T. P. and Pinch T. J. (eds) (1987) *The Social Construction of Technological Systems*, Cambridge (MA): The MIT Press.
Bijker W. (1995) *Of Bicycles, Bakelites and Bulbs: Toward a Theory of Sociotechnical Change*, Cambridge (MA): MIT Press.
Booth C. C. (1989) 'Clinical research', in J. Austoker and L. Bryder (eds) *Historical Perspectives on the Role of the MRC*, Oxford: Oxford University Press, pp. 234–9.
Breckon W. (1972) *The Drug Makers*, London: Eyre Methuen Ltd.
Bud R. (1998) 'Penicillin and the new Elizabethans', *British Journal of the History of Science*, 31: 305–33.
Buddingh G. (1952) 'Chick-embryo technics', in T. Rivers (ed.) *Viral and Rickettsial Infections of Man*, Philadelphia: J.B. Lippincott Company.
Burian R. M. (1993) 'How the choice of experimental organism matters: Epistemologioal reflections on an aspect of biological practice', *Journal of the History of Biology*, 26: 351–67.
Burke D. C. and Isaacs A. (1957) 'Studies on the production, mode of action and properties of interferon', *British Journal of Experimental Pathology*, 38: 551–62.
Burke D. C. and Isaacs A. (1958) 'Further studies on interferon', *British Journal of Experimental Pathology*, 39: 78–84.
Burke D. C. and Isaacs A. (1958) 'Some factors affecting the production of interferon', *British Journal of Experimental Pathology*, 39: 452–8.
Burke D. C. (1961) 'The purification of interferon', *Journal of Biochemistry*, 78: 556–64.
Burke D. C. and Buchan A. (1965) 'Interferon production in chick embryo cells; I. Production by ultraviolet-inactivated virus', *Virology*, 26: 28–35.
Burke D. C. (1977) 'The status of interferon', *Scientific American*, 236: 42–50.
Burke D. C. (1987) 'Early days with interferon', *Journal of Interferon Research*, 7: 441–3.
Burnet F. M. (1936) *The Use of the Developing Egg in Virus Research*, Special Report Series, No. 220 of the Medical Research Council, London: His Majesty's Stationery Office.
Burnet F. M. (1955) *Principles of Animal Virology*, New York, Academic Press.
Buyse M., Staquet M. and Sylvester R. (eds) (1984) *Cancer Clinical Trials*, Oxford: Oxford University Press.
Came P. and Carter W. (eds) (1984) *Interferons and their Applications*, Berlin: Springer-Verlag.
Canetti E. (1994) *Nachträge aus Hampstead: Aus den Aufzeichnungen 1954–1971*, München: Carl Hanser Verlag.
Cantell K. (1979) 'Why is interferon not in clinical use today?', in I. Gresser (ed.) *Interferon*, London: Academic Press, pp. 2–28.
Cantell K. (1998) *The Story of Interferon: The Ups and Downs in the Life of a Scientist*, Singapore: World Scientific.
Cantor D. (1992) 'Cortisone and the politics of drama', 1949–55, in J. V. Pickstone (ed.) *Medical Innovations in Historical Perspective*, Hampshire: Macmillan, pp. 165–84.

Carr G. (1998) 'The pharmaceutical industry', *The Economist*, 21 February.
Casper M. and Berg M. (1995) 'Constructivist perspectives on medical work', *Science Technology & Human Values*, 20: 395–407.
Chalmers T. C. (1972) 'The clinical trial', *Milbank Memorial Fund Quarterly/ Health and Society*, 41: 753–8.
Chany C. (1961) 'An Interferon-like inhibitor of viral multiplication from malignant cells (the viral autoinhibition phenomena)', *Virology*, 13: 485–92.
Clarke A. and Fujimura J. (eds) (1992) *The Right Tools for the Job*, Princeton: Princeton University Press.
Clarke A. and Montini T. (1993) 'The many faces of RU486: Tales of situated knowledges and technological contestations', *Science, Technology and Human Values*, 18: 42–78.
Cockburn W. C. (1991) 'The international contribution to the standardization of biological substances. I. Biological standards and the League of Nations 1921–1946', *Biologicals* 19: 161–9.
Cohn V. (1980) 'Leading U.S. cancer doctors agree to issue warnings on interferon', *The Washington Post*, 15 June.
Collins H. M. (1985) *Changing Order*, London: Sage Publications.
Collins H. M. and Pinch T. (1993) *The Golem: What Everyone should know about Science*, Cambridge: Cambridge University Press.
DavenPort Hines R. P. T. and Slinn J. (1992) *Glaxo: A History to 1962*, Cambridge: Cambridge University Press.
Davis P. (1997) *Managing Medicines*, Buckingham: Open University Press.
De Chadarevian S. and Kamminga H. (eds) (1998) *Molecularizing Biology and Medicine*, Amsterdam: Harwood.
Delbrück M. and Luria S. (1943) 'Interference between bacterial viruses', *Archives of Biochemistry*, 1: 111–41.
Depoux R. and Isaacs A. (1954) 'Interference between influenza and vaccinia viruses', *British Journal of Experimental Pathology*, 35: 415–18.
Doll R. (1991) 'Development of controlled trials in preventive and therapeutic medicine', *Journal of Biosocial Science*, 23: 365–78.
Donald H. and Isaacs A. (1954) 'Counts of influenza virus particles', *Journal of General Microbiology*, 10: 457–64.
Drews J. (1999) *In Quest of Tomorrow's Medicines*, Berlin: Springer.
Dulbecco R. (1952) 'Production of plaques in monolayer tissue cultures by single particles of an animal virus', *Proceedings of the National Academy of Sciences U.S.A.*, 38: 74752.
Van Dyck J. (1995) *Manufacturing Babies and Public Consent*, New York: New York University Press.
Edelhart M. (1981) *Interferon: The New Hope for Cancer*, Reading: AddisonWesley Publishing Company.
Elsworth A. J. *et al.* (2004) Mosby's 2004–2005 Medical Drug Reference, St. Louis: Mosby.
Elzen B. (1988) *Scientists and Rotors; The Development of Biochemical Ultracentifuges*, Enschede: PhD thesis.
Enders J. (1960) 'A consideration of the mechanisms of resistance to viral infection based on recent studies of the agents of measles and poliomyelitis', *Transactions and Studies of the College of Physicians of Philadelphia*, 28: 6879.
Editorial (1957) 'A lead towards virus chemotherapy?', *Lancet*, i: 931–2.

Epstein L. B. (1981) Interferon as model lymphokine, *Federal Proceedings*, 40: 5661–72.

Epstein S. (1997) 'Activism drug regulation, and the politics of therapeutic evaluation in the AIDS era: A case study of ddc and the "surrogate markers" debate', *Social Studies of Science*, 27: 691–726.

Faasse P. (1994) *Experiments in Growth*, Amsterdam: PhD thesis.

Fairley P. (1962) 'Whitehall men join battle on common cold', *Evening Standard* (Britain) 5 October.

Falcoff E., Falcoff R., Fournier F. and Chany C. (1966) 'Production en masse, purification partielle et charactérisation d'un interferon destiné a des essais thérapeutiques humains', *Annales de l'Institut Pasteur*, 5: 256–84.

Fazekas de St. Groth S., Isaacs A. and Edney M. (1952) 'Multiplication of influenza virus under conditions of interference', *Nature*, 170: 573–4.

Fenyvesi C. (1981) 'Beyond interferon', *The Washington Post*, 14 June.

Findlay G. M. and MacCallum F. O. (1937) 'An interference phenomenon in relation to yellow fever and other viruses', *Journal of Pathology and Bacteriology*, 44: 405–24.

Finter N. (1964) 'A rich source of mouse interferon', *Nature*, 204: 1114–15.

Finter N. (1964) 'Protection of mice by interferon against systemic virus infections', *British Medical Journal*, ii: 981–5.

Finter N. (ed) (1966) *Interferons*, Amsterdam: North-Holland Publishing Company.

Finter N. (ed) (1973) *Interferons and Interferon Inducers*, Amsterdam: North Holland Publishing Company.

Finter N. and Oldham R. (eds) (1985) *Interferon (vol 4); In Vivo and Clinical Studies*, Amsterdam: Elsevier.

Fleck L. (1979) *Genesis and Development of a Scientific Fact*, Chicago: The University of Chicago Press.

Fleissner E. (1986) 'Salvador Luria', in L. Levidow (ed.) *Science as Politics*, Radical Science Series, no. 20, London: Free Association Books.

Freeman R. (2000) *The Politics of Health in Europe*, Manchester: Manchester University Press.

Freireich E. J. and Frei E. (1964) 'Recent advances in acute leukemia', in L. M. Tocantins (ed.) *Progress in Hematology*, Vol IV, New York: Grune and Stratton, pp. 187–202.

Fried S. (1998) *Bitter Pills*, New York: Bantam Books.

Friedman R., Rabson A. S. and Kirkham W. (1963) 'Variation in interferon production by polyoma virus strains differing oncogenicity', *Proceedings of the Society of Experimental Biology and Medicine*, 112: 347–51.

Friedman R. (1974) 'Interferon research in the Red Queen's Kingdom', *Archives of Pathology*, 98: 73–6.

Friedman R. (1978) 'Guest editorial: Interferon and cancer', *Journal of the National Institute of Cancer*, 60: 1191–4.

Fujimura J. H. (1996) *Crafting Science: A Sociohistory of the Quest for the Genetics of Cancer*, Cambridge (MA): Harvard University Press.

Fulton F. and Armitrage P. (1951) 'Surviving tissue suspensions for influenza virus titration', *Journal of Hygiene*, 49: 247–63.

Fulton F. and Isaacs A. (1953) 'Influenza virus multiplication in the chick chorioallantoic membrane', *Journal of General Microbiology*, 9: 119–31.

Galambos L. and Sewell J. E. (1995) *Networks of Innovation: Vaccine Development At Merck, Sharp & Dohme, and Mulford 1895–1995*, Cambridge: Cambridge University Press.

Galambos L. and Sturchi J. L. (1997) 'The transformation of the pharmaceutical industry in the twentieth century', in J. Krige and D. Pestre (eds) *Science in the Twentieth Century*, Amsterdam: Harwood Academic Publishers.

Gard S. (1944) 'Tissue immunity in mouse poliomyelitis', *Acta Medica Scandinavica*, 119: 27–46.

Garrett L. (1994) *The Coming Plague*, New York: Farrar, Straus and Giroux.

Gehan E. and Freireich E. (1974) 'Non-randomized controls in cancer clinical trials', *New England Journal of Medicine*, 290: 198–203.

Geison G. L. (1995) *The Private Science of Louis Pasteur*, Princeton: Princeton University Press.

Gibbons N. (ed.) (1963) *Recent Progress in Microbiology*, Proceedings of the VIII International Congress for Microbiology, Montreal 1962, Toronto, University of Toronto Press, pp. 419–57.

Glasgow L. and Habel K. (1962) 'Role of interferon in vaccinia virus infection of mouse embryo tissue culture', *Journal of Experimental Medicine*, 115: 503–12.

Glasgow L. (1970) 'Interrelationships of interferon and immunity during viral infections', *Journal of General Physiology*, 56: 212–27.

Golinski J. (1998) *Making Natural Knowledge: Constructivism and the History of Science*, Cambridge: Cambridge University Press.

Gooding D., Pinch T. and Schaffer S. (eds) (1989) *The Uses of Experiment*, Cambridge: Cambridge University Press.

Gooding D. (1990) *Experiment and the Making of Meaning*, Dordrecht: Kluwer Academic Publishers.

Goodman J. and Walsh V. (2001) *The Story of Taxol, Nature and Politics in the Pursuit of an Anti-cancer Drug*, Cambridge: Cambridge University Press.

Grafe A. (1991) *A History of Experimental Virology*, Berlin: Springer-Verlag.

Grahame-Smith D. G. (1981) 'Problems facing a regulatory authority', in J. F. Cavalla (ed.) *Risk-Benefit Analysis in Drug Research*, Lancaster, MTP Press Limited.

Green D. E. (1964) 'An experiment in communication: the information exchange group', *Science*, 143: 308–9.

Greenberg H., Richard M., Pollard R., Lutwick L., Gregory P., Robinson W. and Merigan T. (1976) 'Effect of human leucocyte interferon on hepatitis B virus infection in patients with chronic active hepatitis', *New England Journal of Medicine*, 295: 517–22.

Gresser I., Coppey J., Falcoff E. and Fontaine D. (1966) 'Action inhibitrice de l'interféron brut sur le développement de la leucémie de Friend chez la Souris', *Critical Reviews of the Academy of Sciences (Paris)*, 263: 586–8.

Gresser I., Coppey, J., Falcoff E. and Fontaine D. (1966) 'Interferon and murine leukemia. I. inhibitory effect of interferon preparations on the development of Friend leukemia in mice', *Proceedings of the Society for Experimental Biology and Medicine*, 124: 84–91.

Gresser I., Fontaine D., Coppey J., Falcoff R. and Falcoff E. (1966) 'Interferon and murine leukemia. II. Factors related to the inhibitory effect of interferon preparations on the development of Friend leukemia in mice', *Proceedings of the Society for Experimental Biology and Medicine*, 124: 91–4.

Gresser I., Coppey J., Fontaine-Brouty-Boyé D. and Falcoff R. (1967) 'Interferon and murine leukemia. III: Efficacy of interferon preparations administered after inoculation of Friend virus', *Nature*, 215: 174–5.

Gresser I., Falcoff R., Fontaine-Brouty-Boyé D., Zajdela F., Coppey J. and Falcoff E. (1967) 'Interferon and murine leukemia. IV. Further studies on the efficacy of interferon preparations administered after inoculation of Friend virus', *Proceedings of the Society for Experimental Biology and Medicine*, 126: 791–7.

Gresser I., Bourali C., Lévy J. P., Fontaine-Brouty-Boyé D. and Thomas M. (1969) 'Cancérologie. prolongation de la survie des souris inoculées avec des cellules tumorales et traitées avec des préparations d'interferon', *C.R. Acad. Sci.*, 268: 994–7.

Gresser I., Bourali C., Lévy J. P., Fontaine-Brouty-Boyé D. and Thomas M. (1969) 'Increased survival in mice inoculated with tumor cells and treated with interferon preparartions', *Proceedings of the National Academy of Sciences USA*, 63: 51–7.

Gresser I., Bourali C., Lévy J. P., Fontaine-Brouty-Boyé D. and Thomas M. (1969) 'Cancérologie. Prolongation de la survie des souris inoculées avec des cellules tumorales et traitées avec des préparations d'interferon', *C.R. Acad. Sci.*, 268: 994–7.

Gresser I. (1975) 'Interferon therapy: Obvious and not so obvious applications', *Acta Medica Scandinavica*, 197: 49–53.

Gresser I. (ed) (1980) *Interferon 2*, London: Academic Press.

Griesemer J. and Gerson E. M. (1993) 'Colloboration in the museum of vertebrate zoology', *Journal of the History of Biology*, 26: 185–203.

Grossberg S. E. (1972) 'The interferons and their inducers: molecular and therapeutic considerations', *New England Journal of Medicine*, 287(ii): 79–85.

Gup T. and Neumann J. (1981) 'Experimental drugs cause pain and death', *The Boston Sunday Globe*, 18 October.

Gura T. (1997) 'Systems for identifying new drugs are often faulty', *Science*, 278: 1041–2.

Gutterman J. et al. (1976) 'Chemoimmunotherapy of advanced breast cancer: Prolongation of remission and survival with BCG', *British Medical Journal*, ii: 774–7.

Gutterman J. et al. (1980) 'Leucocyte-interferon-induced regression in human metastatic breast cancer, multiple myeloma and malignant lymphoma.' *Annals of Internal Medicine*, 93: 399–406.

Hacking I. (1983) *Representing and Intervening*, Cambridge: Cambridge University Press.

Hale K. L. (1994) Interferon: the evolution of a biological therapy, *M. D. Anderson Oncolog*, 39: 4–25.

Hall S. S. (1997) *A Commotion in the Blood; Life, Death and the Immune System*, New York: Henry Holt and Company.

Hansen B. (1999) 'New images of a new medicine: Visual evidence for the widespread popularity of therapeutic discoveries in America after 1885', *Bulletin of the History of Medicine*, 73: 629–78.

Hauptfuhrer F. (1979) 'Will interferon kill cancer? Finnish Dr Kari Cantell is helping the world find out', *People* (US weekly) 2 July.

Hecht A. (1981) 'Interferon: Trying to live up to its press', *FDA Consumer*, June.

Heller E. (1963) 'Enhancement of chikungunya virus replication and inhibition of interferon production by actinomycin D', *Virology*, 21: 652–56.

Van Helvoort T. (1993) *Research Styles in Virus Studies in the Twentieth Century: Controversies and the Formation of Consensus*, Maastricht University: PhD. thesis.

Van Helvoort T. (1994) 'History of virus research in the twentieth century: The problem of conceptual continuity', *History of Science*, xxxii: 185–235.

Henle W. (1950) 'Interference phenomena between animal viruses: A review', *Journal of Immunology* 64: 203–35.

Henle G., Deinhardt F., Bergs V. and Henle W. (1958) 'Studies on persistent infections of tissue cultures: I. General aspects of the system', *Journal of Experimental Medicine*, 108: 537–60.

Henle W. and Henle G. (1984) 'The road to interferon: Interference by inactivated influenza virus', in A. Billeau and N. Finter (eds) *Interferon 1: General and Applied Aspects*, Amsterdam: Elsevier, pp. 3–18.

Higby G. J. and Stroud E. C. (eds) (1997) *Inside Story of Medicines: A symposium*, Madison (WI): The American Institute of the History of Pharmacy.

Hilleman M. (1963) 'Interferon in Prospect and Perspective', *Journal of Cellular and Comparative Physiology*, 62: 337–53.

Hilleman M. (1970) 'Double-stranded RNAs (PolyI: C) in the prevention of viral infections', *Archives of International Medicine*, 126: 109–24.

Hixson J. (1978) 'Interferon: The cancer drug we have ignored', *New York Magazine*, 4 September, pp. 59–64.

Ho M. and Enders J. (1959) 'An inhibitor of viral activity appearing in infected cell cultures', *Proceedings of the National Academy of Sciences U.S.A.*, 45: 385–9.

Ho M. and Enders J. (1959) 'Further studies on an inhibitor of viral activity appearing in infected cell cultures and its role in chronic viral infections', *Virology*, 9: 446–77.

Ho M. (1962) 'Interferon', *New England Journal of Medicine*, 266: 258–64.

Ho M. (1962) 'Kinetic considerations of the inhibitory action of an interferon produced in chick cultures infected with Sindbis virus', *Virology*, 17: 262–75.

Ho M. (1964) 'Identification and "induction" of interferon', *Bacteriology Review* 28: 367–81.

Ho M. (1987) 'An early interferon: "Viral inhibitory factor"', *Journal of Interferon Research*, 7: 455–7.

Hobby G. L. (1985) *Penicillin: Meeting the Challenge*, New Haven: Yale University Press.

Holmes F. (1985) *Lavoisier and the Chemistry of Life*, Madison: Wisconsin Press.

Holmes F. (1987) 'Scientific writing and scientific discovery', *Isis,* 78: 220–35.

Holmes F. L. (1992) 'Manometers, tissue slices, and intermediary metabolism', in A. Clarke and J. Fujimura (eds) *The Right Tools for the Job*, Princeton: Princeton University Press, pp. 169–70.

Holub M. (1982) *Interferon, or on Theater*, Ohio: The Field Translation Series 7, Oberlin College.

Hoskins J. M. (1967) *Virological Procedures*, London: Butterworths.

Hughes S. (1977) *The Virus: A History of the Concept*, New York: Science History Publications.

Isaacs A. (1948) 'Laboratory methods used in investigating influenza', *Glasgow Medical Journal*, 29: 357–61.

Isaacs A. and Edney M. (1950) 'I. Quantitative aspects of interference', *Australian Journal of Experimental Biology*, 28: 219–30.

Isaacs A. and Lindenmann J. (1957) 'Virus interference. I. The Interferon', *Proceedings of the Royal Society*, 147: 258–67.

Isaacs A. and Valentine R. (1957) 'The structure of influenza virus filaments and spheres', *Journal of General Microbiology*, 16: 195–204.

Isaacs A., Lindenmann J. and Valentine R. (1957) 'Virus interference. II. Some properties of interferon', *Proceedings of the Royal Society*, 147: 268–73.

Isaacs A. (1959) 'Viral interference', *Symposium of the Society for General Microbiology*, 9: 102–21.

Isaacs A. (1959) 'Interferon: The prospects', *The Practitioner*, 183, 601–5.

Isaacs A. (1960) 'Metabolic effects of interferon on chick fibroblasts', *Virology*, 10: 144–5.
Isaacs A. (1961) 'Interferon', *Scientific American*, 204: 51–7.
Isaacs A. (1962) 'Interferon tried in man', *New Scientist*, 285: 213–14.
Isaacs A. (1963) 'Interferon', *Advances in Virus Research*, 10: 1–39.
Jasanoff S., Markle G. E., Petersen J. C. and Pinch T. (eds) (1995) *Handbook of Science and Technology Studies*, Thousand Oaks and London: Sage.
Johnson R. (1981) 'Interferon: Cloudy but intriguing future', *Journal of the American Medical Association*, 245: 109–16.
Kahn P. L., Yang E. J., Egan J. W., Higinbotham H. N. and Weston J. F. (1982) *Economics of the Pharmaceutical Industry*, New York, Praeger Publishers.
Van Kammen J. (2003) 'Who represents the users? Critical encounters between women's heath advocates and scientists in contraceptive R&D', in N. Oudhoorn and T. Pinch (eds) *How Users Matter; The Co-construction of Users and Technology*, Cambridge (MA.): The MIT Press, pp. 151–172.
Karpf A. (1988) *Doctoring the Media: The Reporting of Health and Medicine*, London: Routledge.
Kawade Y. (1999) 'A biosemiotic view of interferon: Toward a biology of really living organisms', in J. Lindenmann and W. D. Schleuning (eds) *Interferon: The Dawn of Recombinant Protein Drugs*, Berlin: Springer.
Kay L. (1993) *The Molecular Vision of Life*, Oxford: Oxford University Press.
Keller E. F. (1995) *Refiguring Life; Metaphors of Twentieth-Century Biology*, New York: Columbia University Press.
Kevles D. J. (1993) 'Renato Dulbecco and the new animal virology: Medicine, methods, and molecules', *Journal of the History of Biology*, 26: 409–42.
Kleinsmidt, J. C. Cline W. J. and Murphy E.B. (1964) 'Interferon production induced by Statolon', *Proceedings of the National Academy of Sciences U.S.A.* 52: 741–4.
Knorr-Cetina K. (1981) *The Manufacture of Knowledge*, Oxford: Pergamon Press.
Kohler R. (1994) *Lords of the Fly: Drosophila Genetics and the Experimental Life*, Chicago: Chicago University Press.
Krim M., Stewart II W. E., Sanders F. and Lin L. (1980) 'Interferon therapy', *New York Times*, 17 June, p. C5.
Krimsky S. (1982) *Genetic Alchemy*, Cambridge (MA): The MIT Press.
Krimsky S. (1991) *Biotechnics and Society*, New York: Praeger.
Krown S. (1981) 'Prospects for the treatment of cancer with interferon', in J. Burchenal and H. Oettgen (eds) *Cancer; Achievements, Challenges, and Prospects for the 1980's*, New York: Grune & Stratton, pp. 367–79.
Lampson G., Tytell A., Nemes M. and Hilleman M. (1963) 'Purification and characterization of chick embryo interferon', *Proceedings of the Society for Experimental Biology and Medicine*, 112: 468–78.
Lancaster H. (1979) 'Potent protein; medical researchers say the drug interferon holds great promise', *The Wallstreet Journal*, 6 December.
Landsborough Thomson A. (1973) *Half a Century of Medical Research Vol I*, London: Her Majesty's Stationery Office.
Landsborough Thomson A. (1975) *Half a Century of Medical Research Vol. II*, London: Her Majesty's Stationery Office.
Lasagna L. (1972) 'The nature of evidence', in J. D. Cooper (ed.) *The Philosophy of Evidence: Vol 3 Philosophy and Technology of Drug Assessment*, Washington, DC: Interdisciplinary Communications Program.

Latour B. and Woolgar S. (1986) *Laboratory Life: The Construction of Scientific Facts*, Princeton: Princeton University Press.
Latour B. (1987) *Science in Action*, Milton Keynes: Open University Press.
Latour B. (1988) *The Pasteurization of France*, Cambridge (MA): Harvard University Press.
Latour B. (1999) *Pandora's Hope; Essays on the Reality of Science Studies*, Cambridge (MA): Harvard University Press.
Law J. (1987) 'On the social explanation of technical change: The case of the Portugese maritime expansion', *Technology and Culture*, 28: 227–52.
Law J. (ed.) (1991) *A Sociology of Monsters: Essays on Power, Technology and Domination*, London: Routledge.
Le Grand H. E. (ed.) (1990) *Experimental Enquiries: Historical, Philosophical and Social Studies of Experimentation in Science*, Dordrecht: Kluwer Academic Publishers.
Lennette E. and Koprowski H. (1945) 'Interference between viruses in tissue culture', *Journal of Experimental Medicine*, 83: 195–219.
Lennette E. (1951) 'Interference between animal viruses', *Annual Review of Microbiology*, 5: 277–94.
Van Lente H. (1993) *Promising Technology: The Dynamics of Expectations in Technological Developments*, Delft: PhD thesis.
Leuret F. and Bon H. (1957) *Modern Miraculous Cures*, London: Peter Davies.
Levine A. J. (1992) *Viruses*, New York: W.H. Freeman and Company.
Levy H., Law L. and Rabson A. (1969) 'Inhibition of tumor growth by polyinosinic-polycytidylic acid', *Proceedings of the National Academy of Sciences USA*, 62: 357–63.
Liebenau J. (1987) *Medical Science and Medical Industry*, Baltimore: The Johns Hopkins University Press.
Liebenau J. (1987) 'The British success with penicillin', *Social Studies of Science*, 17: 69–86.
Lindenmann J., Burke D. and Isaacs A. (1957) 'Studies on the production, mode of action and properties of interferon', *British Journal of Experimental Pathology*, 38: 551–62.
Lindenmann J. and Isaacs A. (1957) 'Versuche über virus-interferenz', *Schweiz. Z. Path. Bakt.*, 20: 640–6.
Lindenmann J. (1960) 'Neuere aspekte der virus-interferenz', *Zeitschrift für Hygiene und Infektionskrankheiten*, 146: 369–97.
Lindenmann J. (1961) 'L'Interferon', *Médicine et Hygiène*, 19: 945–6.
Lindenmann J. (1981) 'Induction of chick interferon: Procedures of the original experiments', *Methods in Enzymology*, 78: 181–8.
Lindenmann J. (1982) 'From interference to interferon: a brief historical introduction', *Philosophical Transactions of the Royal Society of London*, series B, 299: 3–5.
Lindenmann J. (1995) 'The National Institute for Medical Research, Mill Hill: Personal recollections from 1956/57', *Archives of Virology* 140: 1687–91.
Lindenmann J. (1998) 'On Toine Pieters'; 'Shaping a New Biological Factor', *Studies in History and Philosophy of Science*, 29.
Lindenmann J. (1998) 'How to avoid making a fortune in medicine', *Nature*, 394: 844–5.
Loutfy, M. R., Blatt, L. M. *et al.* (2003) 'Interferon Alfacon-1 plus cortocosteroids in severe acute respiratory syndrome', *Journal of the American Medical Association* 290: 3222–8.
Löwy I. (1996) *Between Bench and Bedside*, Cambridge (Mass): Harvard University Press.

Lynch M. (1985) *Art and Artifact in Laboratory Science: A Study of Shop Work and Talk in a Research Laboratory*, London: Routledge and Kegan Paul.
McAdams C. (1980) Interferon; The penicillin of the future?, *American Journal of Nursing*, April, pp. 714–17.
Mackenzie D. A. (1981) *Statistics in Britain: 1865–1930*, Edinburgh: Edinburgh University Press.
Mackenzie D. (1990) *Inventing Accuracy: A Historical Sociology of Nuclear Missile Guidance*, Cambridge: The MIT Press.
De Maeyer E. and Enders J. (1961) 'An interferon appearing in cell cultures infected with measles virus', *Proc. Soc. Exper. Biol. & Med.*, 107: 573–78.
De Maeyer E. (1978) 'Interferon twenty years later', *Bulletin de L'Institut Pasteur*, 76: 303–23.
Mahoney J. T. (1959) *The Merchants of Life. An Account of the American Pharmaceutical Industry*, New York: Harper.
Mann C. C. and Plummer M. L. (1991) *The Aspirin Wars*, New York: Knopf.
Marks H. M. (1992) 'Cortisone, 1949: A year in the political life of a drug', *Bulletin of the History of Medicine*, 66: 419–39.
Marks H. M. (1997) *The Progress of Experiment; Science and Therapeutic Reform in the United States, 1900–1990*, Cambridge: Cambridge University Press.
Martin E. (1994) *Flexible Bodies*, Boston: Beacon Press.
Marx J. L. (1979) 'Interferon (I): On the threshold of clinical application', *Science*, 204: 1183–6.
Marx J. L. (1979) 'Interferon (II): Learning About How it Works', *Science*, 204: 1293–5.
Matthews L. G. (1962) *History of Pharmacy in Britain*, London: E.& S. Livingstone.
Medawar C. and Hardon A. (2004) *Medicines Out of Control? Antidepressants and the Conspiracy of Goodwill*, Amsterdam: Aksant.
Merigan T., Reed S., Hall T. and Tyrrell D. (1973) 'Inhibition of respiratory virus infection by locally applied interferon, *Lancet*, i: 563–7.
Merigan T. (1979) Human interferon as a therapeutic agent, *New England Journal of Medicine*, 300: 42–3.
De Mey M. (1992) *The Cognitive Paradigm*, Chicago: The University of Chicago Press.
Moore T. J. (1999) *Prescription for Disaster*, New York: Dell.
Mooser H. and Lindenmann J. (1957) 'Homologe interferenz durch hitzeinaktiviertes, an erythrozyten absorbiertes influenza-B-virus', *Experientia*, XIII: 147–8.
Moss R. W. (1989) *The Cancer Industry*, New York: Paragon House.
Nagano Y. and Kojima Y. (1958) 'Inhibition de l'infection vaccinale par un facteur liquide dans le tissu infecté par le virus homologue', *Compt. Rend. Soc. Biol. Filiales*, 152: 1627–9.
Newmark P. (1981) 'Interferon: decline and stall', *Nature*, 291: 105–6.
Nickles T. (1989) 'Justification and experiment', in D. Gooding, T. Pinch and S. Schaffer (eds) *The Uses of Experiment*, Cambridge University Press, pp. 299–334.
Nickles T. (1992) 'Good science as bad history: From order of knowing to order of being', in E. McMullin (ed.) *The Social Dimensions of Science*, Indiana: The University of Notre Dame Press, pp. 85–129.
Nickles T. (1995) 'Philosophy of science and history of science', *Osiris*, 10: 139–63.
Olby R. C., Cantor G. N., Christie J. R. and Hodge M. J. (eds) (1990) *Companion to the History of Modern Science*, London: Routledge.
O'Malley J. and Carter W. A. (1978) 'Human interferons: Characterization of the major molecular components', *Journal of the ReticuloEndothelial Society*, 23: 299–305.

Oudshoorn N. (1994) *Beyond the Natural Body: An Archaeology of Sex Hormones*, London: Routledge.

Oudshoorn N. (2003) *The Male Pill*, Durham: Duke University Press.

Oudshoorn N. and Pinch T. (eds) (2003) *How Users Matter; The Co-construction of Users and Technology*, Cambridge (MA.): The MIT Press.

Panem S. and Vilcek J. (1982) 'Will interferon ever cure cancer?', *The Atlantic*, December.

Panem S. (1984) *The Interferon Crusade*, Washington, DC: The Brookings Institution.

Parascandola J. (1981) 'The theoretical basis of Paul Ehrlich's chemotherapy', *Journal of the History of Medicine*, 36: 19–43.

Parker R. (1950) *Methods of Tissue Culture*, New York: Paul B. Hoeber, Inc.

Parker R. (1962) *Methods of Tissue Culture*, New York: Paul B. Hoeber, Inc.

Parkinson D. (ed.) (1994) 'The expanding role of interferon-Alfa in the treatment of cancer', *Seminars in Oncology*, 21: 1–37.

Parnham M. J. and Bruinvels J. (eds) (1987) *Selections from Discoveries in Pharmacology*, Amsterdam: Elsevier Science Publishers.

Patterson J. T. (1987) *The Dread Disease*, Cambridge: Harvard University Press.

Payer L. (1992) *Disease Mongers*, New York: John Wiley.

Perkins F. T. and Regamey R. H. (eds) (1970) *Symposia Series in Immunobiological Standardization vol. 14*, Basel: S. Karger.

Petersen J. C. and Markle G. E. (1981) 'Expansion of conflict in cancer controversies', in L. Kriesberg (ed.) *Research in Social Movements, Conflicts and Change*, 4.

Peyer H. C. (1996) *Roche: A Company History*, Basel: Editiones Roche.

Pickering A. (1984) *Constructing Quarcks: A sociological History of Particle Physics*, Chicago: Chicago University Press.

Pickering A. (ed) (1992) *Science as Practice and Culture*, Chicago: The University of Chicago Press.

Pickering A. (1995) *The Mangle of Practice*, Chicago: The University of Chicago Press.

Pieters T. (1993) 'Interferon and its first clinical trial': Looking behind the scenes', *Medical History*, 37: 270–95.

Pieters T. (1997) 'Shaping a new biological factor, 'The Interferon', in room 215 of the National Institute for Medical Research, 1956/57', *Studies in History and Philosophy of Science*, 28: 27–73.

Pieters T. (1997) 'History of the development of the interferons: from test-tube to patient', in R. Stuart-Harris and R. Penny (eds) *The Clinical Applications of Interferons*, London: Chapman & Hall.

Pieters T. (1998) 'Managing differences in biomedical research: The case of standardizing interferons', *Studies in History and Philosophy of Biological & Biomedical Sciences*, 29: 31–79.

Pieters T. (1998) 'Marketing medicines through randomised controlled trials: The case of interferon', *British Medical Journal*, 317: 1231–3.

Pieters T. (1999) 'What constitutes therapeutic success?: The Interferons', in J. Lindenmann and W. D. Schleuning (eds) *Interferon: The Dawn of Recombinant Protein Drugs*, Berlin: Springer.

Pieters T. (2002) 'About media, audiences and marketing medicines: The interferons', in M. Gijswijt-Hofstra, G. M. van Heteren and E. M. Tansey (eds) *Biographies of remedies: Drugs, Medicines and Contraceptives in Dutch and Anglo-American Healing Cultures*. Amsterdam: Rodopi, pp. 229–43.

Pieters T. (2003) 'Hailing a miracle drug: the interferon', in C. Osborne and W. de Blecourt (eds) *Cultural Approaches to the History of Medicine*, Houndmills: Palgrave, pp. 212–32.

Pieters T. (2003) 'New molecules, markets and changing drug regulatory practices at the end of the twentieth century', in J. Abrahams (ed.), *Regulation of the Pharmaceutical Industry*, Houndmills (UK): Palgrave MacMillan, pp. 146–160.

Pieters T. and Snelders S. (in preparation) Seige's cycle: The careers of psychotropic drugs in mental health care since 1869 – A programmatic essay.

Pinch T. (1986) *Confronting Nature: the Sociology of Solar-Neutrino Detection*, Dordrecht: D. Reidel Publishing Company.

Pocock S. (1983) *Clinical Trials*, Chichester: John Wiley & Sons.

Pollard M. (ed.) (1958) *Perspectives in Virology: A Symposium*, Texas: The University of Texas Medical Branch.

Pollard M. (ed.) (1960 *Perspectives in Virology*, Minneapolis: Burgess Publishing.

Porter R. (1997) *The Greatest Benefit to Mankind; A Medical History of Humanity from Antiquity to the Present*, London: Harper Collins Publishers,.

Porter T. M. (1995) *Trust in Numbers: The Pursuit of Objectivity in Science and Public Life*, Princeton: Princeton University Press.

Powledge T. (1984) 'Interferon on trial', *Bio/technology*, 2: 214–28.

Proctor R. (1995) *Cancer Wars: How Politics Shapes What We Know & Don't Know About Cancer*, New York: Basic Books.

Van de Putte M., Delafonteyne J., Billeau J. and de Somer P. (1967) 'Influence and production of interferon in Rauscher virus infected mice', *Arch. Ges. Virusforsch.*, 20: 235–45.

Quesada J., Hersh E. and Gutterman J. (1983) 'Hairy cell leukemia: induction of remission with alpha interferon', *Blood*, 62: 207a.

Quesada J., Reuben J., Manning J., Hersh E. and Gutterman J. (1984) 'Alpha interferon for induction of remission in hairy cell leukemia', *New England Journal of Medicine*, 310: 15–18.

Radder H. (1992) 'Experimental reproducibility and the experimenter's regress', in D. Hull, M. Forbes and K. Okruklik (eds) *PSA, Volume I*, East Lansing, Michigan: Philosophy of Science Association.

Rettig R. A. (1977) *Cancer Crusade*, Princeton: Princeton University Press.

Rheinberger H. J. (1992) 'Experiment, difference, and writing: I. Tracing protein synthesis', *Studies in History and Philosophy of Science*, 23: 305–31.

Rheinberger H. J. (1992) 'Experiment, difference, and writing: II. The laboratory production of transfer RNA', *Studies in History and Philosophy of Science*, 23: 389–422.

Richards E. (1991) *Vitamin C and Cancer: Medicine or Politics*, New York: St. Martin's Press.

Rita G. (ed.) (1968) *The Interferons*, New York: Academic Press.

Robbins F. and Enders J. (1950) 'Tissue culture techniques in the study of animal viruses', *American Journal of the Medical Sciences*, 220: 316–38.

Rosen G. et all (1974) 'High dose methotrexate with citrovorum factor and rescue and adriamycin in childhood osteogenic sarcoma', *Cancer*, 33: 1151–63.

Rosenberg S. (1992) *The Transformed Cell*, London: Phoenix.

Rosenfeld A. (1979) 'If IF works it could . . .', *Life*, July, pp. 55–62.

Rudwick M. J. S. (1985) *The Great Devonian Controversy*, Chicago: The University of Chicago Press.

Saks M. (2000) 'Medicine and the counter culture', in R. Cooter and J. Pickstone (eds) *Medicine in the 20th Century*, Amsterdam: Harwood Publishers, pp. 113–23.

Sanders M. and Lennette E. H. (1968) *Medical and Applied Virology, Proc. Soc. Int. Symp.*, St. Louis: Warren H. Green Inc.

Sapp J. (1990) *Where the Truth Lies: Franz Moewus and the Origins of Molecular Biology*, Cambridge: Cambridge University Press.

Schellekens H. *et al.* (1981) 'Comparative antiviral efficiency of leukocyte and bacterially produced human alpha interferon in rhesus monkeys', *Nature*, 292: 775–6.

Schmeck H. M. (1980) 'Interferon: studies put cancer use in doubt', *New York Times*, 27 May.

Schwartzman D. (1976) *Innovation in the Pharmaceutical Industry*, Baltimore: Johns Hopkins University Press.

Scott G. *et al.* (1982) 'Prevention of rhinovirus colds by human interferon alpha-2 from escherichia coli', *Lancet*, ii: 186–7.

Sela M. (ed.) (1977) *The Role of Non-Specific Immunity in the Prevention and Treatment of Cancer*, Rome: Pontificia Academia Scientiarum.

Shapin S. and Schaffer S. (1985) *Leviathan and the Airpump; Hobbes, Boyle and the Experimental Life*, Princeton: Princeton University Press.

Shapin S. (1990) 'Science and the Public', in R. C. Olby, G. N. Cantor, J. R. Christie and M. J. Hodge (eds) *Companion to the History of Modern Science*, London: Routledge, pp. 990–1007.

Shope R. E. (1966) 'Evolutionary episodes in the concept of viral oncogenesis', *Perspectives in Biology and Medicine* 9: 258–74.

Sikora K. and Smedley H. (1983) 'Interferon and Cancer', *British Medical Journal*, 286: 739–40.

Silvermann M. and Lee P. R. (1974) *Pills Profits & Politics*, Berkeley: California Press.

Silverstein A. M. (1989) *A History of Immunology*, San Diego: Academic Press, Inc.

Smith J. S. (1990) *Patenting the Sun; Polio and the Salk Vaccine*, New York: William Morrow and Company.

Sneader W. (1985) *Drug Discovery: the Evolution of Modern Medicines*, Chichester: John Wiley & Sons.

Sneader W. (1986) *Drug Development: From Laboratory to Clinic,* Chichester: John Wiley & Sons.

Spilker B. (1989) *Multinational Drug Companies: Issues in Drug Discovery and Development*, New York: Raven Press.

Star S. L. (1989) *Regions of the Mind: Brain Research and the Quest for Scientific Certainty*, Stanford (CA): Stanford University Press.

Stephens T. and Brynner R. (2001) *Dark Remedy*, Cambridge (MA): Perseus Publishing.

Stewart II W. E. (1979) *The Interferon System*, Vienna: Springer-Verlag.

Strander H. and Cantell K. (1966) 'Production of interferon by human leukocytes in vitro', *Ann. Med. exp. Fenn*, 44: 265–73.

Strander H., Cantell K., Carlström G. and Jakobsson P. (1973) 'Clinical and laboratory investigations on man: systemic administration of potent interferon in man', *Journal of the National Cancer Institute*, 51: 733–42.

Stuart Harris R. and Penny R. (1997) *Clinical Applications of the Interferons*, London: Chapman & Hall Medical.

Studer K. and Chubin D. (1980) *The Cancer Mission: Social Contexts of Biomedical Research*, Beverly Hills: Sage publications.

Sun M. (1981) 'Interferon: no magic bullet against cancer', *Science*, 212: 141–2.

Swann J. P. (1988) *Academic Scientists and the Pharmaceutical Industry*, Baltimore: The Johns Hopkins University Press.
Taniguichi T., Sakai M., Fujii-Kuriyama Y., Muramatsu M., Kobayashi S. and Sudo T. (1979) 'Construction and identification of a bacterial plasmid containing the human fibroblast interferon gene sequence', *Proc. Jpn. Acad.* 55B: 461–9.
Tansey E. M. (1997) *Catterall P.P., Christie, D.A., Willhoft S.V. and Reynolds, L.A., Wellcome Witnesses to Twentieth Century Medicine Vol I*, London: Wellcome Trust Occasional Publications.
Taubes G. (1995) Use of placebo controls in clinical trials is disputed, *Science*, 267: 24–7.
Taylor J. (1964) 'Inhibition of interferon action by actinomycin', *Biochemical and Biophysical Research Communication*, 14: 447–51.
Teitelman R. (1989) *Gene Dreams: Wall Street, Academia and the Rise of Biotechnology*, New York: Basic Books.
Temin P. (1980) *Taking Your Medicine: Drug Regulation in the United States*, Cambridge: Harvard University Press.
Thompson P. (1988) *The Voice of the Past*, Oxford: Oxford University Press.
Tyrrell D. and Tamm I. (1955) 'Prevention of virus interference by 2,5-dimethylbenzimidazole', *Journal of Immunology*, 75: 43–9.
Tyrrell D. (1959) 'Interferon produced by cultures of calf kidney cells', *Nature*, 184: 452–3.
Tyrrell D. (1987) 'Personal memories of the early days', *Journal of Interferon Research*, 7: 443–4.
Tyrrell D. (1991) 'The Common cold unit 1946–1990: Farewell to a much-loved British institution', *PHLS Microbiology Digest*, 6: 74–6.
Vilcek J. (1969) *Interferon*, New York: Springer-Verlag.
Vos R. (1991) *Drugs Looking for Diseases. Innovative Drug Research and the Development of the Beta Blockers and the Calcium Antagonists*, Amsterdam: Kluwer Academic Publishers.
Wade N. (1980) 'Cloning gold rush turns basic biology into big business', *Science*, 208: 688–92.
Wagner R. R. (1960) 'Viral interference', *Bacteriology Review*, 24: 151–66.
Wagner R. R. (1961) 'Biological studies of interferon', *Virology*, 13: 323–37.
Wagner R. R. (1963) 'The interferons: Cellular inhibitors of viral infection', *Annual Review of Microbiology*, 17: 285–94.
Wagner R. R. (1964) 'Inhibition of interferon biosynthesis by actinomycin D', *Nature*, 204: 49–51.
Wagner R. R. (1965) 'Interferon; A review and analysis of recent observations'. *American Journal of Medicine*, 38: 726–37.
Wagner R. R. (1996) 'Reminiscences of a virologist wandering in Serendip', *Archives of Virology*, 141: 787–97.
Walgate R. (1982) 'Side effect scare hits French trials, *Nature*, 300: 97–8.
Walker H. D. (1971) *Market Power and Price Levels in the Ethical Drug Industry*, Bloomington: Indiana University Press.
Weatherall M. (1990) *In Search of a Cure*, Oxford: Oxford University Press.
Weber M. M. (1999) *Die Entwickelung der psychopharmakologie im Zeitalter der Naturwissenschaftlichen Medizin*, München:Urban & Vogel Munich.
Weintraub M. and Northington F. K. (1986) 'Drugs that wouldn't die'. *Journal of the American Medical Association*, 255: 2327–8.
Weissman C. (1981) 'The cloning of interferon and other mistakes', in I. Gresser (ed.) *Interferon 3*, London: Academic Press.

Wheelock. E. F and Dingle J. H. (1964) 'Observations on the repeated administration of viruses to a patient with acute leukemia', *New England Journal of Medicine*, 271: 645–51.
Wheelock E. F. (1967) 'Effect of statolon on friend virus leukemia in mice', *Proceedings of the Society of Experimental Biology and Medicine*, 124: 855–8.
Wheelock E. F. and Larke R. P. B. (1968) Efficacy of interferon in the treatment of mice with established friend virus leukemia, *Proceedings of the Society for Experimental Biology and Medicine*, 127: 230–8.
Wilson D. (1976) *Penicillin in Perspective*, London: Faber & Faber.
Wolstenholme G. E. and O'Connor M. (eds) (1968) *Interferon*, London: J & A Churchill.
Wright S. (1994) *Molecular Politics: Developing American and British Regulatory Policy for Genetic Engineering 1972–1982*, Chicago: Chicago University Press.
Wycke A. (1987) 'Molecules and Markets', *The Economist*, 7 February.
Wycke A. (1997) *21st-Century Miracle Medicine*, New York: Plenum Trade.
Wynne B. (1988) 'Unruly technology: Practical rules, impractical discourses and public understanding', *Social Studies of Science*, 18: 147–67.
Yoshioka A. (2002) 'Streptomycine in postwar-Britain: A cultural history of a miracle drug', in M. Gijswijt-hofstra, G. M. Van Heteren and E. M. Tansey (eds) *Biographies of Remedies*, Amsterdam: Editions Rodopi, pp. 203–28.

Index

ABC (American television network) 140
academic setting vs. commercial setting 145
Achievement 1961 (BBC) 70
ACS *see* American Cancer Society
Adamson R. (NCI) 130
Advances in Virus Research in 1963 (Isaacs) 55
AIDS 165
AIDS epidemic: USA 167
AIDS-crisis 189
All Things Considered (American radio programme) 140
Alper, J. 163
alpha: beta and gamma terminology 145
alpha interferon 159, 162, 164–65
American Cancer Society (ACS) 51, 140, 144, 146, 148, 156, 161–2
  grant for clinical testing 140
American penicillin 'syndrome' *see* penicillin syndrome
American Society for Microbiology 95
American Society of Clinical Oncology (ASCO) 156
Andrewes, C.H. ('Cha') 9–12, 15, 18, 28, 36, 51
animals
  experimental work 10, 31–2, 66, 84, 112–6
  safety tests 71
  virology 11, 32
Annual Conversazione of the Royal Society 37
anti-cancer therapies: American politics 134–6
anti-tumour effects of interferon 111–6, 119–120, 124–5, 138, 165–6, 172
anti-tumour agent: USA 7, 120–6
antitumour claims 121, 143, 148

anti-viral drugs: search 34, 121
anti-viral effects of interferon 2, 32, 35, 37–8, 68, 111, 120, 138, 172, 173
'Antiviral penicillin' 1, 36–9, 61, 154, 173, 182
Asiatic flu epidemic (1957) 36
Aspirin 4
Atanasiu, P. 111

bacteriology 9–11
Balkwill, F. 160–1
Baron, S. 95–7
BBC (British television network) 38, 70
BCG (Bacillus Calmette-Guérin) 123, 135–6
Benger Laboratories (British drug company) 37
Behringwerke Company (German drug company) 122
Berlex Laboratories 171
Beta (fibroblast) interferon 145, 171, 172, 176–7
big science 145, 180
Biogen 144, 150, 152, 159
biography 4
biological response modifiers (BRM) 147, 149, 161, 169, 172, 187
  naturalizing interferons as 168
biological standards 91–2, 100–1, 105–6, 108
biologicals 4, 59
  problematic developmental track record 60–1
  biosynthetic 163–4, 168, 187
biomedical model of disease and treatment 121
biomedicine: patient-activist interventions 189

biotechnology companies: genetic boutiques, 143–4, 150–1, 159–160, 163–4
billion dollar molecules 2, 173
black market: coping with imminent 154–6
blockbuster drugs 164
Board of the Patent Holdings Company 69
bone cancer: osteosarcoma 118–20, 125–6, 132–3
Borali, C. 114
breast cancer 138, 165
Bristol-Myers Company (American drug company) 144
Britain: 1960's drug testing culture 59
British Interferon Collaboration 59–64, 81, 89, 91
British Medical Journal (BMJ) 37, 38, 76, 89
British Pharmaceutical Conference 71–72
British Scientific Committee on Interferon Research 67–8, 101
British/Swiss collaboration 12–18
BRM. See biological response modifiers
Bud, R. 182
Burian, R. 180
Burke, D. 29–30, 36, 37, 41–2, 45, 48, 49, 55, 86
Burnet, F.M. 17
Burroughs Wellcome Company *see* Wellcome

Cahagan, Douglas H. (American actor) 147
cancer(s)
 as a multi-factorial molecular disease 188
 breast 138, 165
 bone (Osteosarcoma) 118–20, 125–6, 132–3
 immunotherapy 2, 122–3, 135–6, 159–61, 168–72
 interferon as cure for 1, 141, 147–9, 151–4
 leukaemia 116
 lymphoma (non-Hodgkin's) 117, 118, 120, 138, 172
 kidney cancer 172
 malignant melanoma 117, 118, 172
 multiple myeloma 117, 172
 prostate 165
 viruses as cause of 111
Cancer Chemotherapy Handbook 172
cancer treatment
 see multi modality approach
 NCI Division of Cancer Treatment 130–1
Cantell, K. 3, 126, 152
 Finnish interferon 101, 103, 104, 137
 human cancer trials with Strander 118–20
 British laboratory inspection 102
 leukaemia trials in children 116–17
 plans for clinical trials 101
 relationship with Paucker 92–3
cardiac toxicity 162
CBS (American television network) 140
cellular immunity 48–9, 56, 126–7, 142–3, 146, 161, 168–9, 187
cephalosporin C 60
Cetus 144
Chany, C. 98–101, 106–7, 111
Chaproniere, D. 12
Chemical Industries-Pharmaceuticals *see* ICI
chemotherapeutic approach 61, 99, 121, 182–3
chick-embryo technique 15
 in vitro 19
Children's Hospital of Philadelphia 14, 92
chronic hepatitis 138, 165, 173
Ciba (Swiss drug company) 33–4
clinical experiments: design of 73–5
 culture of 117
 gambling analogy 170
clinical trials
 double-blind 74, 138
 growth industry 174
 historical controls 124, 132
 interferon breast cancer 138, 166
 interferon common cold 66, 81–3
 interferon hairy cell leukemia 166
 interferon leukaemia 116
 interferon MS 171, 177–9
 interferon osteosarcoma 119–20, 125, 131
 interferon shingles 138
 interferon vaccinia 74–6
 historical controls 123–4, 126, 132
Lancet report (1962) 76–80

258   Index

randomized 74, 123, 126, 131, 133, 138, 158
  Workshop on (1978) 138
cloning 143–4, 163
  of interferon gene(s) 144, 146, 150–1, 160–1, 186–7
Cold Spring Harbour Symposium (1952) 17
Coley's toxins 121
Collaboration Agreement (NRDC) 64, 67
Common Cold Unit 45, 46, 66, 76, 81–2, 99, 163
common cold 66, 81, 82, 83, 84, 163, 170
consumer organizations 189–90
cortisone 39, 152, 185
cost of interferon 1, 137, 151, 173
crafting: between bench and bedside 178–9
Crick, F 17
cultural process: healing as 3
cytokine(-s) 146, 168, 170, 171, 172, 181, 188, 191
  and beyond 191
  network 168
  naturalizing interferons as 168

Daily Express (British newspaper) 37
Daily Telegraph (British newspaper) 38
defense system. See immune system
de Maeyer, E. 56, 90
de Maeyer Guignard, J. 56
de Somer, P. (Rega Institute) 138–9
Delbrück, M. 90
depression as side-effect of interferon 177
DeVita, V. (National Cancer Institute) 130, 132, 147
distribution: as unlicensed drug 155
DNA recombinant technology 143–4, 160, 162, 187. See genetic engineering
doctoring the media 183–5
Donald, H. 21
dosage considerations 69, 86, 117, 118, 119, 137–8
drug
  blockbuster 164
  career (-paths) 4–5
  cultural conceptions 3
  life-cycle 4
  looking for a disease 164–8

  recombinant protein 5
  regulatory affairs 58, 166–8, 188–9
drug companies 33, 34–5, 58, 60–1, 80–1, 84–5, 87–88, 152, 164, 165, 174, 183, 187 *see also* pharmaceutical companies
drug industry 5, 7, 128, 171, 172, 183, 190 *see also* pharmaceutical industry
Dulbecco, R. 90

Eastern Equine Encephalitis (EEE) 50
E. coli 146
Edelhart, M. 1, 3
Edney, M. 16
Ehrlich, P. 3, 61, 88
  magic bullet concept 3, 39, 61, 183
electron micrographs (electron-microscopic images) 22, 24
electron microscopy 11, 21, 24
Enders, J.F. 42–5, 49, 51, 52–3, 69
endogenous interferon: by non-viral means 86–7
endogenous research: versus *exogenous* research 99–100, 102
Epstein, L. (Univ. of California) 161
Epstein, S. 190
Erasmus Hospital (Rotterdam, The Netherlands) 138
ethical committees: absence of 59
Evans, D. 105, 106,
Evening Standard 83–4
exogenous interferon: clinical potential 84, 103
Executive Body 63–4, 69, 88
experiment in scientific communication 96–8
experimental therapies 117–8, 162–3;
  mice and malignancy 111–16
experimental system (arrangement) 26, 30, 49
Eye on Research (BBC) 38

Farber, S. 112
FDA. See Food and Drug Administration
Findlay, G.M. 14, 18
Finnish Red Cross Blood Transfusion Service (Helsinki) 102, 103
Finter, N. 81, 85–6, 95, 99, 100, 101–3
  tension with Isaacs 82
first international interferon congress (1964) 54–5
Flash Gordon comic strip 40, 151

flu 13, 34, 36
flu-like side effects 139, 177
Food and Drug Administration (FDA) 164–168, 170, 172, 188
Friend, C. 112
Friend leukaemia virus 112, 114
Fulton, F. 20, 21
future prospects: and retrospective 177–91

Galasso, G. 121
gene splicing groups 146
Genentech 144, 159
Genex 144
genetic engineering (technology) 4, 150–1, 160, 163, 180, 186
genetic engineers 187
genetic medicine 1
  toward 186–9
genital warts 165, 172
gift culture
  Cantell and Paucker 92–3
  informal culture of exchange 92–4, 180
  problems in data exchange 93
Gilbert, W. 150
Glaxo (Laboratories, British drug company) 37, 59, 60, 63, 66, 68, 73, 76, 79, 82, 98, 103
Good, R. 122
Gresser, I. 111, 112, 116, 122, 123
  anti-tumour claim 120
  International Interferon Conference (Lyon, 1969), 115
growth hormone release inhibiting factor (somatostatin) 144
Gustav Stern Symposium on Perspectives in Virology 51, 52
Gutterman, J. 2, 135, 136, 137, 138, 148, 149
  Hairy cell leukaemia trial 165–6, 172

Habif, D.V. (Columbia University) 138
haemagglutination titration 22–3
Hairy cell leukaemia: successful trial 165–6, 172
Harington, Sir C. 9, 59, 61, 70, 77
healing: as a cultural process 3
Henle, G. 14–15
Henle, W. 14–15, 42–3, 45
Hepatitis 138, 165, 173
  risk of infecting blood products 103–4
Hershey, A.: and Chase, M. 17

high-performance liquid chromatography (HPLC) 145
Hilleman, M. (MSD) 59, 85, 86, 87, 99, 102
Himsworth, Sir H. (MRC) 35, 59, 70, 88
Hixson, J. 142
Ho, M. 42, 44, 45, 49, 51, 95
Hoffmann-La Roche (Roche) 144, 163, 165, 166, 170
Hôpital St. Vincent-de-Paul (Paris) 98
hormones 60, 146
House Select Committee on Aging 147
Huebner R.J. 90
human leukaemia virus 111
Hygiene Institute, Zürich 11, 33, 36

ICI *see* Imperial Chemical Industries (British drug company)
ICRF *see* Imperial Cancer Research Fund
IEG 6 *see* Information Exchange Group no. 6
immune defence modifier 127
immune system 123, 127, 135, 161
immunotherapy 2, 122, 123, 127, 135, 161, 164, 171, 188, 191
Imperial Chemical Industries (ICI) 63, 68, 81, 82, 87, 98, 99
Imperial Cancer Research Fund (ICRF) 154, 155
influenza 15
  nature and epidemiology 13
  pandemic 14
  see also flu
influenza virus 14, 15, 17, 18, 33, 38, 47
  electron micrographs 21–2
  heat inactivated virus 16, 22, 27, 28, 31, 48
  growing and harvesting 29
Information Exchange Group no. 6 (IEG 6, NIH) 96–8
innovation: biomedical 4
Institute of Tumor Biology 120
interference: viral 15
Interferon: The New Hope for Cancer (Edelhart) 1
interferon
  Antiviral penicillin 1, 36–9, 61, 154, 173, 182
  definition 52
  derivation of name 55

deviant (abnormal) viral entity 30, 56, 180
discovery of 18–28
family of proteins 145, 109, 160, 166
fibroblast 145
first trial 73–6
inflation of expectations 150–4
invention of term 25
going public 33–6
leukocyte 104, 105, 108, 137, 139, 145, 146
as magic bullet 39, 61
'Misinterpreton' 41, 53
media frenzy 1
mode of action 45–9, 126–7, 146, 161, 168–170
mouse interferon: production of 112
nasal spray 163
new kind of viral interfering agent 26–7
non-toxicity hypothesis ('dogma') 37, 62, 67, 68, 139
phenomenological aspects of research 30
practical medical application 32, 172–3
preparation for clinical trials with volunteers 68–70
product of the cell 49, 56, 181
public football 1
purification of 66, 85, 145, 159, 162
safeguarding British interests 36, 60, 98
side-effects 118, 119, 139, 148–9, 155, 162, 177
specificity effect 45–9
specificity issue 65
interferon B 50–1
Interferon Crusade (Panem) 5
Interferon Scientific Memorandum 97–8
'Interferonologists' 90, 135
International Congress for Interferon Research (Washington DC, 1980) 159
International Congress for Microbiology (Montreal, 1962) 54
International Society for Interferon Research 168, 172
International Society for Interferon and Cytokine Research 172
International Workshop on Interferon
first (NY, 1975) 125–8
second (NY, 1978) 145–47

Internet technology 190
Isaacs, A. 6, 13–15, 34–39, 43–53, 55–56, 59, 62, 64–65, 67–73, 76–77, 79, 81, 86, 98–99, 181–182
death of 100
decision to collaborate with Lindenmann 17–18
discovery of interferon 18–28
dissatisfaction with Interferon Collaboration 87–88
illness of 2, 41, 84–5, 89
interferon 'kitchen' 29–32
quest for scientific credibility 41–2
tension with Finter 82
ITV (British television network) 152–3

Japan 50, 150, 152
Jephcott, Sir H. 63
Johns Hopkins Medical School 50
Johnson & Johnson (American drug company) 59
Journal of Interferon and Cytokine Research 172
Journal of Interferon Research 159, 168

Karolinska Hospital (Radiumhemmet, Stockholm) 117, 123, 132, 165
Karpf, A. 184
key-enzyme hypothesis: viral interference 15–16
Kennedy, T. (US Senator) 149
kidney cancer 172
Kitasato Institute (Tokyo) 50
Knight, P. 160
Kojima, Y.: and Nagano, Y. 50
Koprowski, H. 55, 56
Kradolfer, Professor 33, 34
Krim, A. 110
Krim, M. 7, 110, 130, 133, 135, 136, 141, 144, 148, 149, 184, 185
criticism of 131
impact of interferon lobby 132
interferon lobby 120–4
letter to *New York Times* 158–9
workshops on interferon 124–8, 145–7

Laboratory for Research on Interferon 89
Lancet 37, 76, 77, 79
report of first clinical trial: (1962) 76–79
laryngeal papilloma 138, 165

Lasagna, L. 188–9
Lasker, M. 110, 132, 133, 136, 137, 140, 147
  Foundation 137
'Laskerites' 110, 147
leukocyte interferon 104, 105, 108, 137, 139, 145, 146
Levine, A. 132
Levy, H. 115
Life Magazine 147
Lindenmann, J. 6, 9, 13–17, 30, 32, 34, 36, 39, 41, 43, 52, 55
  discovery of interferon 18–28
  Fellowship Swiss Academy of Medical Sciences 12
  Hygiene Institut (Zürich) 11, 33, 36
  introduction of the name 'interferon' 28–9
  Swiss Federal Health Department 36
Löwy, I. 123, 124
loyalty: patient versus research 123–4
lymphoma 117–8, 120, 172

M. D. Anderson Hospital (Houston) 135, 148
MacCallum, G. 14
MacFarlane-Burnet, Sir F. 17
magic bullet 1, 39, 181, 188
  drugs as 3
  Paul Ehrlich's concept 3, 61, 88, 183
  interferon as no 85–90, 163
  in search of 182–3
  'Zauberkugeln' 3
mangle of practice and culture 5, 6, 7–8, 68, 191
marketing medicines 159, 159, 171–2, 174
marketplace medicine 191
Mayo Clinic 156
measles 66
Medawar, P. Sir 87, 88
media 37–8, 128, 134, 140, 145, 147, 149, 150–154, 158, 161, 163, 170, 173, 190
  coverage of interferon news 157, 184–5
  doctoring of 183–4
  frenzy 1, 156
  information metaphors 142–3
  medical approach 184
  preference for human interest stories 185
  unparalleled media attention 157

medical counter-culture: USA 136, 183–4
medical 'cyber mall' 190
medical market 3, 4
Medical Research Council (MRC) 35–36, 41, 59, 60–66, 68, 70–71, 74, 76–77, 81, 155, 182–183
  Annual Report 37
  patent adviser 37
Medical Tribune 142, 143
Medicines Act (1968) 59
Memorial Sloan Kettering Cancer Center 122, 125, 141
Merck Sharpe & Dohme (MSD, American drug comapny) 59, 80, 81, 84–88, 144
Merigan, T. 94, 95, 124, 126, 130, 138, 140
mice and malignancy
  experimental therapies 110–33
  experiments with RC19 tumour cells 114–15
  hope for similar anti-tumour activity with humans 116–20
  virus inhibitory effect of interferon 113
microbiology 11, 13
miracle drug 39, 151, 152, 153, 154, 158
  devine quality 4
  hailing of a 147–50, 186
  requisite characteristics 185–6
miracle cure 2, 141
'Misinterpreton' 41, 53
Moertel, C. 155
molecular biology 49, 144, 145, 151, 159, 160, 161, 187
molecular medicine 169
molecularization: of medicine 5, 188
Mooser, H. 11, 15, 16, 36
moulding: notion of 178
MS *see* multiple sclerosis
MRC *see* Medical Research Council
MSD *see* Merck Sharpe & Dohme
multi modality approach 160–1, 164, 171–2, 188
Multiple Sclerosis (MS) 171, 172, 177, 189
myxomatosis: virus 12

Nagano, Y.: and Kojima, Y. 50, 55–6
Namalwa cells 146
Nat King Cole 147

Index

National Cancer Act (1971) 110
National Cancer Institute (NCI) 128, 130, 133, 136, 138, 140, 148, 163
  Board of Scientific Counselors 131, 162
  BRM Program 149, 162
  Division of Cancer Biology and Diagnosis 126
  Division of Cancer Treatment 132, 147
  interferon funding 120–1, 124–5, 130–2, 149
  National Cancer Advisory Board (NCAB) 131–2
National Institute of Allergy and Infectious Diseases (NIAID) 96, 97, 120, 121, 122, 135, 138,
National Institutes of Health (NIH) 95, 96, 97, 98, 107, 108, 121, 124, 125, 126, 133, 138, 139
  Reference Reagents Branch 108
National Institute for Medical Research (NIMR, London) 6, 11, 39, 42, 46, 59, 61, 62, 63, 68, 73, 76, 87
  Animals Division 10
  Division of Bacteriology and Virus Research 70, 89
  Division of Biological Standards 69, 98, 100, 101, 105, 107, 108
  Division of Biophysics and Optics 18
  Immunological Products Control Laboratory 74
  research agenda 9
National Program for the Conquest of Cancer 110
National Research Development Corporation (NRDC) 35, 37, 60, 62, 63, 64
  NRDC's Patent Holdings Company 63–4, 88
naturalizing interferons
  as biological response modifiers (BRM) 168–75
  as cytokines 168–75
Nature 130
NBC (American television network) 140
NCI *see* National Cancer Institute
Nevanlinna, H. 102
New England Journal of Medicine 131
New Scientist 70, 161
New York Magazine 142
New York Times 163

NIAID *see* National Institute of Allergy and Infectious Diseases
Nickles, T. 191
NIH *see* National Institutes of Health
NIMR *see* National Institute for Medical Research
Nixon, President R. 110
non-Hodgkin's lymphoma 172
NRDC *see* National Research Development Corporation
nucleic acid hypothesis 24

The Observer 150
O'Connor, T. 124, 125
Oldham, R. 162
oncology 117, 123, 124, 137, 156, 161, 171, 188
Osteogenic sarcoma 118–20, 125–6, 132–3
Osteosarcoma *see* Osteogenic sarcoma

Panem, S. 5, 133, 151–2
pandemic: influenza (1918) 14
papilloma. See laryngeal papilloma
patents
  drugs (expiration) 4, 164
  interferon (position) 62, 64, 67, 69
patenting scientific findings (Britain) 35
Patents Holdings Company 63–4, 88
patients: at risk 189–91
patient activists 189
Paucker, K.: relationship with Cantell 92–3
penicillin
  associated imagery 182
  British failure 34–5, 59–60
  comparison with 1, 2, 34, 37, 38, 59, 153, 173
  as cultural symbol 7, 39, 182
  licensing fees 35
  'syndrome' (British trauma) 39, 60, 80, 154
Pepper, C. 136
Perkins, F. 106
pharmaceutical companies 8, 37, 58, 59, 60, 62–3, 87–88, 125, 144, 163, 167, 173, 182, 183 *see also* drug companies
pharmaceutical industry 3–4, 60, 71, 91, 144, 169, 174 *see also* drug industry
pipeline 187
Pickering, A. 6
Pieters T. 196, 207, 213, 219, 227, 234
polio virus 12, 13

Porter, R. 186
pre-AIDS therapeutic framework 189
Press coverage *see* media coverage
Proceedings of the Japanese Academy 150
Proceedings of the National Academy of Science 44
Proceedings of the Royal Society 28, 36
production of interferon
   human leucocyte interferon 101–104, 137, 144
   monkey interferon 70–6, 79
   r-DNA interferons 150, 159, 160, 162
protein 'soup' 53
public demand: as threat to drug evaluation procedures 155
public sphere 4, 6
purification process: improvement 85, 119, 145, 162
purified interferon 66, 85, 162
purity question 166

Radiumhemmet *see* Karolinska Hospital
randomized clinical trials *see* clinical trials randomized
randomized controlled trial lobby 133
Rauscher, F. 124, 130, 156
Rauscher leukaemia virus 114
receptor hypothesis: viral interference 15–16
recombinant DNA 143, 162
   commercial applications 143–4
recombinant protein drugs 5, 163, 187
red cell ghosts: use in original experiment 21, 23
Regelson, W. (Medical College of Virginia) 99
retrospective and future prospects 177–91
Rettig, R. 131–2
Roche *see* Hoffmann-La Roche
Rockefeller Foundation 125
Rockefeller Institute for Medical Research (New York) 47
roller-tube method 20
Royal Society 37, 38
   Annual Conversazione 37
   Fellow of 28
   Proceedings of 28, 29, 36
Rubin, H. 41, 51, 90
Russian Academy of Medical Sciences (Moscow) 102

safety issues 70–3, 102, 103, 118, 139, 162, 166–7
sale figures 172–3
SARS 2, 173
Schellekens, H. 138–9, 170
Schering-Plough (American drug company) 144, 150, 151, 163, 165, 166, 170
Science 90, 158
Scientific American 53, 54, 69, 70
scientists and media 140–7
SDS polyacrylamide gel electrophoresis 145
Second International Workshop on Interferons 144–5
Sendai virus 46, 103
Shah of Iran 149
Sherwin, S. 161
shingles 138
shopping for health 190
side-effects 118, 119, 139, 148–9, 155, 162, 177
Sloan Kettering Institute for Cancer Research (N.Y.) 122, 125, 141
smallpox: and interferon 66
Society for Microbiology 12, 13
Soloviev, V.(Russian Academy of Med. Sciences) 102
standardization
   adoption of research standard for human interferon 106, 108
   benefits 180
   benefits for research 109
   creating a breeding ground for 94–5
   establishing new differences 109
   experiment in scientific communication 96–8
   Expert Committee on Biological Standardization (WHO) 108
   formal 180
   informal 180
   international workshop 108
   national interests 98, 100, 105, 106–8
   need for 91–2
   problems of co-ordination 105–6
   research standards and Cantell 101–4, 106–7
   research standards and Chany 98–100, 106–7
   timescale 109
standardized production: need for 68–70

Stanford University School of Medicine (CA) 94
Statolon 98
sterile conditions: production area 103
Strander, H. 102, 117, 123, 124, 125, 128, 148, 153–4
  criticism of trial data 126, 131, 132, 133
  human cancer trials with Cantell 118–20
Sunday Mirror 153
Symposium of Applied Virology and Medicine (Second, Florida 1966) 99
synthetic interferons 163–4, 187

Taubes, G. 159
Temin, H. 41
Thalidomide tragedy 4, 59
therapeutic life-cycle 4
therapeutic profile: current diversity 172
Therapeutic Research Corporation (TRC) 63
Tishler, M. (Merck Sharpe & Dohme) 81, 85
Tizard, Sir H. 59
trachoma virus: interferon's effect against 66, 68
treatments: social and cultural dimension 175
TV Eye: Cancer: The new weapon (ITV) 154
Tyrrell, D.(Common Cold Unit) 45, 46–7, 48, 49, 73, 78, 81, 82, 83, 85

United States Congress: additional interferon funding 149
University of Zurich 6 *see also Hygiene Institut*
unorthodox cancer remedies: capitalising on demand 135–40
Upton, A. 147

vaccination 61
vaccinia virus 30, 48, 50, 66, 67, 68
Valentine, R.C. 24
venture capitalists 144
Vietnam war: influence 135
Vilcek, J. 130; flight to West 55
viral chemotherapy: chemical means 87
viral infection: cellular/host resistance 53

viral inhibitory factor (VIF) 42, 45, 52–3
primary profile 43
viral inhibitory fluid 44; effect of 48
viral interference 14
  competition hypothesis 33
  key-enzyme hypothesis 15–16
  receptor hypothesis 15–16
  unknown mechanism 17
viral interfering factor 56, 180
Virology 45, 90
virus: growth inhibition 25
  and interferon differences 30
  myxomatosis 12
  perceptions 11
  polio 12
virus-coated ghosts 22, 23
visualization: electron microscopy 24
von Hoffman, N. 149

Wade, N. 151
Wagner, R.(Johns Hopkins Medical School) 50–1
Walker, J. 29
Wallstreet 144
Wallstreet Journal 150
Walter and Eliza Hall Institute (Melbourne) 17
warts 165
Washington Post 149, 161
Watson, J. 17
Wayne, J. (movie star) 149
Weissmann, C. 159–60
Wellcome (laboratories, British drug company) 60, 63, 66, 68, 76, 77, 81, 82, 87, 98, 146, 163, 168
Wheelock, E.F. 109
WHO *see* World Health Organization
wonder cure 156
wonder drug 7, 149, 153, 156, 186
  see also miracle drug
wonder medicine 2
World Influenza Centre 13–14, 18, 36, 62
World Health Organization 13, 18, 100, 105, 108
world market interferons 172–3

Young, F. 167, 168

Zoon, K. 167